AF834705

Flow and Transport in Porous Media

Summer School on

Flow and Transport in Porous Media

Beijing, China 8-26 August 1988

Editor
Xiao Shutie
Department of Applied Mathematics
Tsinghua University
Beijing

World Scientific
Singapore • New Jersey • London • Hong Kong

Published by

World Scientific Publishing Co. Pte. Ltd.

P O Box 128, Farrer Road, Singapore 9128

USA office: Suite 1B, 1060 Main Street, River Edge, NJ 07661

UK office: 73 Lynton Mead, Totteridge, London N20 8DH

Library of Congress Cataloging-in-Publication Data is available.

FLOW AND TRANSPORT IN POROUS MEDIA

Copyright © 1992 by World Scientific Publishing Co. Pte. Ltd.

All rights reserved. This book, or parts thereof, may not be reproduced in any form or by any means, electronic or mechanical, including photocopying, recording or any information storage and retrieval system now known or to be invented, without written permission from the Publisher.

ISBN 9971-50-934-2

Printed in Singapore by JBW Printers & Binders Pte. Ltd.

PREFACE

This book stems from a Summer School on "Flow and Transport in Porous Media" organized by C. J. van Duijn (Delft), U. Hornung (Munchen), and S. Xiao (Beijing) which was held in the Summer of 1988 at Tsinghua University.

The 40 hours lectures were contributed by four speakers, Prof. U. Hornung, Prof. C. J. van Duijn, Dr. P. Knabner, and Dr. B. H. Gilding. These lectures cover the most important mathematical aspects of the flow and transport in porous media. They had been well received by the attentive audience who came from universities and research institutes. The publication of this book will be beneficial to the researchers in this field.

I wish to thank the publisher for putting up with the delay in publication.

March 1992

Xiao Shutie
Tsinghua University
Beijing

CONTENTS

CONTENTS

MATHEMATICAL MODELLING OF SATURATED AND UNSATURATED GROUNDWATER FLOW

B.H. Gilding

Faculty of Applied Mathematics
University of Twente
Enschede
The Netherlands

Preface

These notes follow the lectures given by myself at the Summer School "Flow and Transport in Porous Media" held at Tsinghua University, Beijing, from 8 to 26 August 1988. The contents reflect the work on the mathematical modelling of groundwater flow which I have carried out during the past sixteen years. There are two sides to this work. One side is the practical application of mathematical models for water resources evaluation in hydrology. The other side is the theoretical mathematical analysis of such models. The element common to both sides is the study of a certain class of partial differential equations. The lectures were intended as an introduction to the current way of thinking about applying mathematical models in groundwater resources evaluation and about analysing such models in general. They were aimed at engineers and starting researchers with little or no experience in this kind of work. These notes have been compiled in accordance with these aims.

I am aware that many of the statements made about applying groundwater models in practice are in a sense no more than 'folklore' among those working in the field, and if I was asked to substantiate them with documented literature I would find it hard to do so. Likewise, here and there I have been somewhat free in the interpretation of certain mathematical concepts and results. Also, no attempt has been made to provide an exhaustive list of references. Intentionally the emphasis has been placed on introducing ideas. The references are intended to be a gateway to the wealth of relevant literature available. By focusing these notes in this way I hope that I have succeeded in best conveying the spirit of the work involved. Certainly in this way I am able to provide the most faithful reproduction of the lectures as they were presented in Beijing.

I would very much like to thank the Chinese State Commission for Education and the Dutch Ministry of Education and Science who together financially supported my participation in the Summer School in Beijing in 1988. The typing of these notes was carried out within the Faculty of Applied Mathematics of the University of Twente. The figures were prepared with the support of the Commission of the European Communities Stimulation Action Twinning Programme "Filtration and Nonlinear Diffusion Processes" by KGC Drawings.

1 Saturated groundwater flow modelling - general

1.1 The hydrological cycle

In these lecture notes we will be concerned with the flow and transport of one particular fluid - water - in one particular porous medium - the ground. In the first four sections we will be specifically concerned with the mathematical modelling of this phenomenon from the view-point of the hydrologist. That is to say, from the point of view of trying to quantify the transport and storage of water in the ground for the purposes of water resources management.

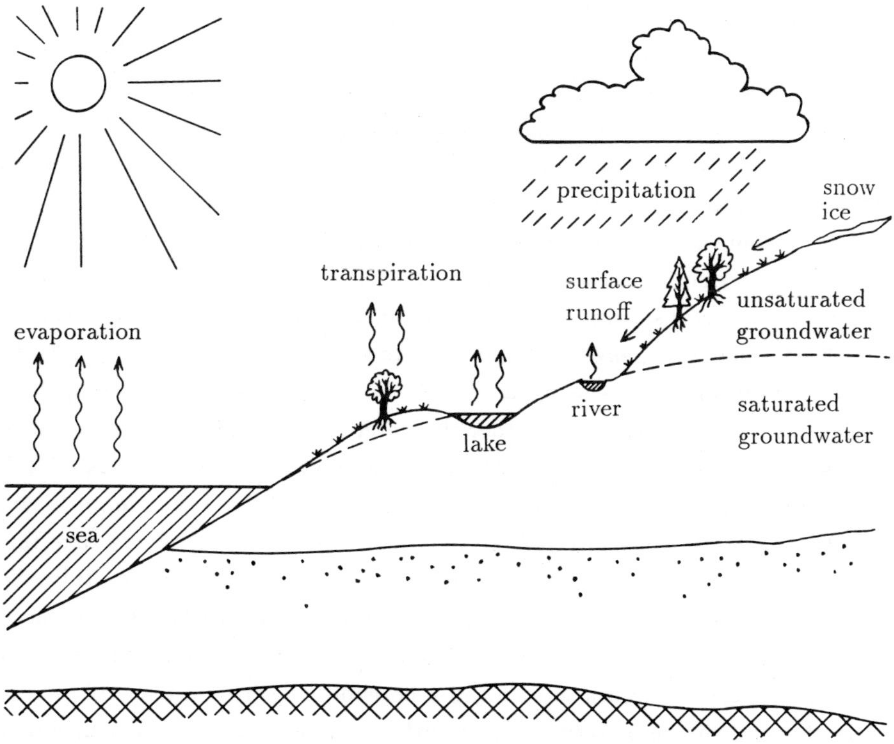

Figure 1.1: The hydrological cycle

Figure 1.1 displays a classical picture of the hydrological cycle. In this the occurrence of water on earth is seen as part of a circulation process. This begins, for instance, with the fall of precipitation in the form of rainfall, hail or snow. Not all the precipitation reaches the ground though. Some is intercepted by the leaves and branches of vegetation. Of the rainfall which does reach the ground, some will infiltrate into the subsurface as ***groundwater***. The remainder will flow overland, maybe to become groundwater at some other location, or maybe to feed rivers, streams and lakes. Snow which falls may rest on plants or on the land surface for some length of time before it eventually follows the same course as the rainfall. Groundwater may later be abstracted by plant roots. It may also flow underground to seep into rivers or lakes, or, more spectacularly, give rise to springs which themselves generate rivers or lakes. Equally as well, water in rivers and lakes may seep into the subsurface. From rivers and lakes, water will flow to the seas and oceans. The water intercepted by the plants will evaporate in time. Most of the groundwater taken up by the plant roots will be utilized for the transpiration process of the plants. Finally, evaporation of water on the land surface, in rivers and lakes and in the sea will take place. In these ways the water is returned to the atmosphere. From here it may again fall as precipitation, and the cycle is complete [19,31,109,112].

For convenience in modelling [2,10,19,58,69,76,93], hydrologists generally make a distinction between ***saturated groundwater*** flow and ***unsaturated groundwater flow***. Strictly speaking, saturated groundwater flow is the flow of water in the ground where the pores in the soil matrix are fully occupied by water. If the pores contain air the flow is unsaturated. For all intents and purposes, this is equivalent to the criterion of whether the water pressure-head is above or below atmospheric pressure. For details of the distinction we refer to [18,19,24,47,111].

In a given ground profile, unsaturated flow is viewed as occurring in a zone directly under the land surface, and it is from this zone that plants take their water supply. Saturated groundwater flow occurs in a zone below this. See Figure 1.2. The two zones are separated by an imaginary surface known as the phreatic surface or the water-table. The latter can be equated with the level at which water first comes to stand in a borehole dug from the surface.

In this and the next section our attention will be focussed on mathematical modelling of the saturated groundwater zone.

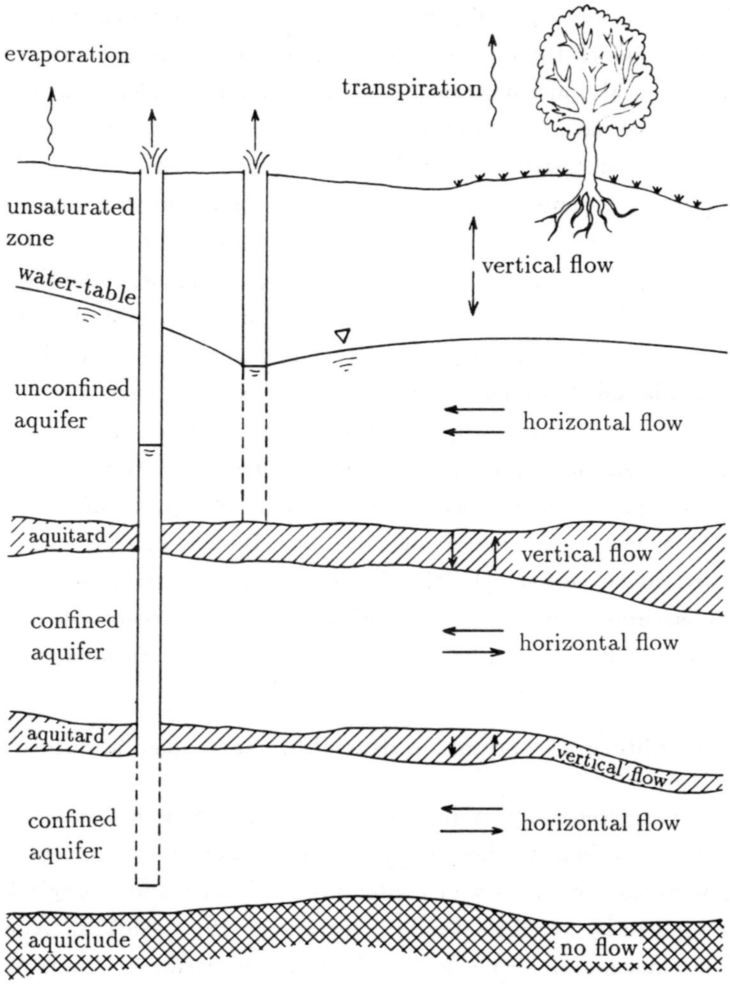

Figure 1.2: Schematization of groundwater flow

The saturated groundwater zone is important to hydrologists as a source of water for drinking, industrial and agricultural purposes, and also as one of the components in the hydrological cycle which is the easiest to control.

In this respect one is generally concerned with groundwater resources in a given area in which the horizontal dimensions are much larger than the corresponding vertical dimensions [69]. Out of practical consideration it is therefore helpful to view saturated groundwater flow as a two-dimensional areal phenomenon in a schematized ground profile such as that shown in Figure 1.2. In this picture, groundwater flow is seen as occurring predominantly in water-bearing layers, for instance sand formations, known as *aquifers*. These layers may be separated by poorly-permeable layers, for instance clay layers, which are termed *aquitards*. Layers which may to all intents and purposes be viewed as impermeable are called *aquicludes*. If an aquifer is covered by an impermeable formation it is referred to as being a *confined* (or an *artesian*) *aquifer*. An aquifer covered by a poorly-permeable aquitard is called a *semi-confined* or *leaky aquifer*. An aquifer with no overlying restriction is termed an *unconfined* or *phreatic aquifer*. In this scheme, groundwater flow in aquifers is considered to be an essentially horizontal phenomenon. Whilst, in aquitards the flow is considered to be vertical [19,26,118]. It is also convenient to view the flow in the unsaturated zone as an essentially vertical phenomenon on this scale [2,10,58,69,76,93]. This is a point to which we shall return later.

1.2 The differential equation describing saturated groundwater flow

When considered as a two-dimensional phenomenon in a horizontal direction, differential equations to describe groundwater flow may be derived directly by consideration of the law of conservation of mass and Darcy's law. Such derivations can be found in standard textbooks [19,24,109,112]. Alternatively, the same equations may be derived from consideration of the corresponding laws in three dimensions. We shall follow this latter approach, for in this way we obtain a little more insight into the assumptions involved, and also set the scene for later considerations.

We consider here only the flow of water with constant density which we assume to take the value 1 (unity). Furthermore, we neglect the effects of any thermal or chemical processes.

In this case, if we consider an elementary representative volume of porous media (see Figure 1.3) we may compose a mass balance equation. This basically says that in any time interval, the change in water content in the volume is equal to the difference between the amount of water which has flowed in and out of each side of the volume minus the amount of water which has been abstracted. Referring to Figure 1.3 this gives us the relation

$$\frac{\Delta\theta}{\Delta t} + \frac{\Delta q_x}{\Delta x} + \frac{\Delta q_y}{\Delta y} + \frac{\Delta q_z}{\Delta z} + w = 0,$$

where θ denotes water content (dim. L^3/L^3), t time (dim. T), q_x, q_y and q_z flow components (dim. $(L^3/T)/L^2$), w abstraction (dim. $(L^3/T)/L^3$), and the prefix Δ denotes the change in these quantities. In the limit of the representative volume becoming infinitesimally small, this yields the relation

$$\frac{\partial\theta}{\partial t} + \operatorname{div} \mathbf{q} + w = 0 \tag{1.1}$$

where $\mathbf{q}$ denotes the soil-moisture flux vector (dim. L/T), $\frac{\partial}{\partial t}$ the partial derivative with respect to t, and 'div' the divergence operator in three spatial dimensions (dim. 1/L). The relation (1.1) is the *law of conservation of mass.* Often this is referred to as the *continuity equation.*

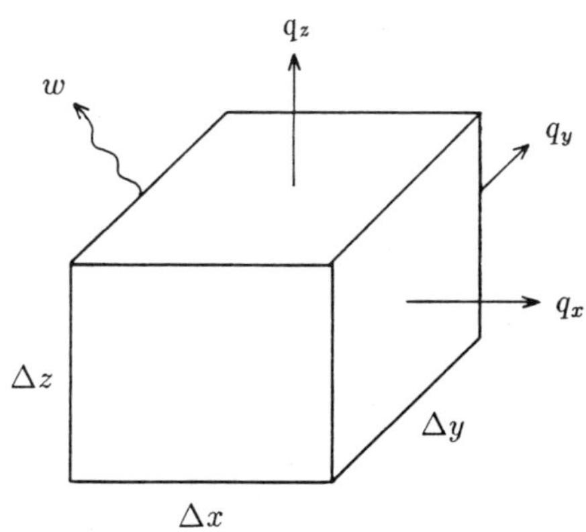

Figure 1.3: Representative elementary volume for mass balance

The second law is that known as Darcy's law. This is named after a French engineer who in the last century discovered a relation for the losses of water from the fountains of Dijon [33]. It was given a firm footing in 1940 by Hubbert [73]. The latter described the following experiment. Consider two reservoirs with different water-levels connected by a pipe containing some porous material. The porous material is held in place by filters. See Figure 1.4. Then under steady conditions the flow through the pipe satisfies

$$\text{flow} \ = \frac{\Phi_1 - \Phi_2}{L} \mathcal{A}\, K$$

where $\Phi_1 - \Phi_2$ denotes the difference in water-levels in the two reservoirs, L the length of the porous material in the pipe, and $\mathcal{A}$ the cross-sectional area of the porous material. The constant K is an empirical constant, nowadays called the *hydraulic conductivity*, which relates only to the material in the pipe. In three-dimensional differential form this empirical law becomes

$$\mathbf{q} = -K \text{ grad } \Phi \tag{1.2}$$

where K becomes the hydraulic conductivity tensor which is symmetric (dim. L/T), Φ denotes the potential (dim. L) and 'grad' the gradient operator in three spatial dimensions (dim. 1/L). We shall refer to (1.2) as *Darcy's law.*

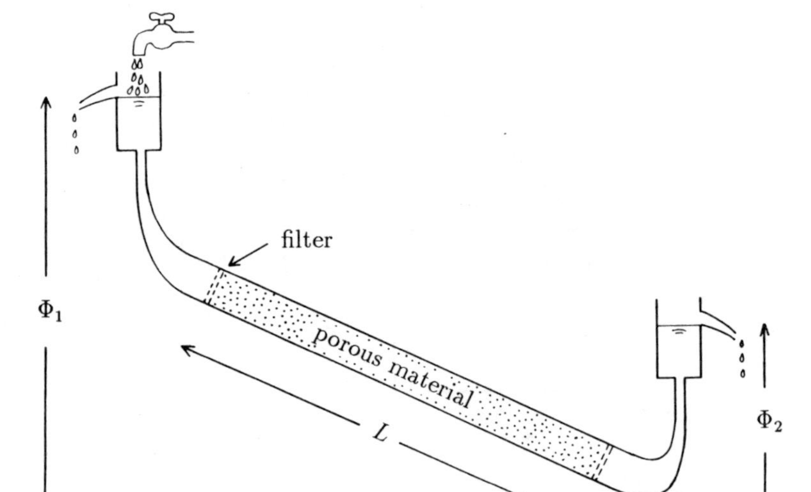

Figure 1.4: Simplified set-up for verification of Darcy's law

For a fuller discussion of the law of conservation of mass and Darcy's law, we refer to the work of U. Hornung in this book.

Combining (1.1) and (1.2) yields the equation of motion

$$\frac{\partial \theta}{\partial t} = \text{div } (K \text{ grad } \Phi) - w. \tag{1.3}$$

Subsequently, considering only saturated flow, and defining *specific storativity* s (dim. 1/L) as the change in water content per unit change in potential, (1.3) becomes

$$s\frac{\partial \Phi}{\partial t} = \text{div } (K \text{ grad } \Phi) - w. \tag{1.4}$$

Consider now an aquifer in a two-dimensional (x, y) Cartesian coordinate system. Suppose that the aquifer is bounded below and above by the elevations $\eta(x, y, t)$ and $\zeta(x, y, t)$ respectively. Define

$$h = \int_\eta^\zeta \Phi dz/(\zeta - \eta),$$

i.e. h is the average potential or the *head* in the aquifer (dim. L). The head is the water-level which would be measured at rest in a borehole sunk through the entire thickness of the aquifer. The idea is now to integrate (1.4) with respect to z from η to ζ [19].

Assuming that specific storativity is independent of depth

$$\begin{aligned}
\int_\eta^\zeta s\frac{\partial \Phi}{\partial t}dz &= s\int_\eta^\zeta \frac{\partial \Phi}{\partial t}dz \\
&= s\left\{\frac{\partial}{\partial t}\int_\eta^\zeta \Phi dz - \Phi|_{z=\zeta}\frac{\partial \zeta}{\partial t} + \Phi|_{z=\eta}\frac{\partial \eta}{\partial t}\right\} \\
&= s\left\{\frac{\partial}{\partial t}h(\zeta - \eta) - \Phi|_{z=\zeta}\frac{\partial \zeta}{\partial t} + \Phi|_{z=\eta}\frac{\partial \eta}{\partial t}\right\} \\
&= s\left\{(\zeta - \eta)\frac{\partial h}{\partial t} + (h - \Phi|_{z=\zeta})\frac{\partial \zeta}{\partial t} - (h - \Phi|_{z=\eta})\frac{\partial \eta}{\partial t}\right\}. \qquad (1.5)
\end{aligned}$$

Whilst

$$\begin{aligned}
&\int_\eta^\zeta \{\text{div } (K \text{ grad } \Phi) - w\}dz \\
&= \int_\eta^\zeta \{- \text{ div } \mathbf{q}\}dz - \int_\eta^\zeta w dz \\
&= -\int_\eta^\zeta \left\{\frac{\partial}{\partial x}q_x + \frac{\partial}{\partial y}q_y\right\} dz - q_z|_{z=\zeta} + q_z|_{z=\eta} - \int_\eta^\zeta w dz
\end{aligned}$$

$$
\begin{aligned}
&= -\frac{\partial}{\partial x}\int_\eta^\zeta q_x dz + q_x|_{z=\zeta}\frac{\partial \zeta}{\partial x} - q_x|_{z=\eta}\frac{\partial \eta}{\partial x} \\
&\quad -\frac{\partial}{\partial y}\int_\eta^\zeta q_y dz + q_y|_{z=\zeta}\frac{\partial \zeta}{\partial y} - q_y|_{z=\eta}\frac{\partial \eta}{\partial y} \\
&\quad -q_z|_{z=\zeta} + q_z|_{z=\eta} - \int_\eta^\zeta w dz \\
&= -\frac{\partial}{\partial x}\int_\eta^\zeta q_x dz - \frac{\partial}{\partial y}\int_\eta^\zeta q_y dz - \int_\eta^\zeta w dz \\
&\quad -\mathbf{q}|_{z=\zeta}\cdot \text{ grad } (z-\zeta) + \mathbf{q}|_{z=\eta}\cdot \text{ grad } (z-\eta). \qquad (1.6)
\end{aligned}
$$

Next, assuming horizontal components of hydraulic conductivity are independent of depth

$$
\begin{aligned}
\int_\eta^\zeta q_x dz &= -\int_\eta^\zeta \left\{K_{xx}\frac{\partial \Phi}{\partial x} + K_{xy}\frac{\partial \Phi}{\partial y} + K_{xz}\frac{\partial \Phi}{\partial z}\right\} dz \\
&= -K_{xx}\int_\eta^\zeta \frac{\partial \Phi}{\partial x}dz - K_{xy}\int_\eta^\zeta \frac{\partial \Phi}{\partial y}dz - K_{xz}\int_\eta^\zeta \frac{\partial \Phi}{\partial z}dz \\
&= -K_{xx}\left\{\frac{\partial}{\partial x}\int_\eta^\zeta \Phi dz - \Phi|_{z=\zeta}\frac{\partial \zeta}{\partial x} + \Phi|_{z=\eta}\frac{\partial \eta}{\partial x}\right\} \\
&\quad -K_{xy}\left\{\frac{\partial}{\partial y}\int_\eta^\zeta \Phi dz - \Phi|_{z=\zeta}\frac{\partial \zeta}{\partial y} + \Phi|_{z=\eta}\frac{\partial \eta}{\partial y}\right\} \\
&\quad -K_{xz}\left\{\Phi|_{z=\zeta} - \Phi|_{z=\eta}\right\} \\
&= -K_{xx}\left\{\frac{\partial}{\partial x}(\zeta-\eta)h - \Phi|_{z=\zeta}\frac{\partial \zeta}{\partial x} + \Phi|_{z=\eta}\frac{\partial \eta}{\partial x}\right\} \\
&\quad -K_{xy}\left\{\frac{\partial}{\partial y}(\zeta-\eta)h - \Phi|_{z=\zeta}\frac{\partial \zeta}{\partial x} + \Phi|_{z=\eta}\frac{\partial \eta}{\partial x}\right\} \\
&\quad -K_{xz}\left\{\Phi|_{z=\zeta} - \Phi|_{z=\eta}\right\} \\
&= -K_{xx}\left\{(\zeta-\eta)\frac{\partial h}{\partial x} + (h-\Phi|_{z=\zeta})\frac{\partial \zeta}{\partial x} + (h-\Phi|_{z=\eta})\frac{\partial \eta}{\partial x}\right\} \\
&\quad -K_{xy}\left\{(\zeta-\eta)\frac{\partial h}{\partial y} + (h-\Phi|_{z=\zeta})\frac{\partial \zeta}{\partial y} + (h-\Phi|_{z=\eta})\frac{\partial \eta}{\partial y}\right\} \\
&\quad -K_{xz}\left\{\Phi|_{z=\zeta} - \Phi|_{z=\eta}\right\}. \qquad (1.7)
\end{aligned}
$$

Similarly,

$$
\begin{aligned}
\int_\eta^\zeta q_y dz &= -K_{yx}\left\{(\zeta-\eta)\frac{\partial h}{\partial x} + (h-\Phi|_{z=\zeta})\frac{\partial \zeta}{\partial x} + (h-\Phi|_{z=\eta})\frac{\partial \eta}{\partial x}\right\} \\
&\quad -K_{yy}\left\{(\zeta-\eta)\frac{\partial h}{\partial y} + (h-\Phi|_{z=\zeta})\frac{\partial \zeta}{\partial y} + (h-\Phi|_{z=\eta})\frac{\partial \eta}{\partial y}\right\} \\
&\quad -K_{yz}\left\{\Phi|_{z=\zeta} - \Phi|_{z=\eta}\right\}. \qquad (1.8)
\end{aligned}
$$

Subsequently, assuming that

$$\Phi|_{z=\zeta} = \Phi|_{z=\eta} = h, \tag{1.9}$$

combination of (1.4) - (1.8) yields

$$\begin{aligned} s(\zeta-\eta)\frac{\partial h}{\partial t} &= \frac{\partial}{\partial x}\left\{K_{xx}(\zeta-\eta)\frac{\partial h}{\partial x} + K_{xy}(\zeta-\eta)\frac{\partial h}{\partial y}\right\} \\ &\quad + \frac{\partial}{\partial y}\left\{K_{yx}(\zeta-\eta)\frac{\partial h}{\partial x} + K_{yy}(\zeta-\eta)\frac{\partial h}{\partial y}\right\} - W \\ &\quad - \mathbf{q}|_{z=\zeta}\cdot \text{ grad } (z-\zeta) + \mathbf{q}|_{z=\eta}\cdot \text{ grad } (z-\eta) \end{aligned}$$

where

$$W = \int_{\eta}^{\zeta} w dz$$

denotes the total water abstraction over the thickness of the aquifer (dim. $(L^3/T)/L^2$). To complete the deduction we now use the fact that the surfaces $z = \zeta(x,y,t)$ and $z = \eta(x,y,t)$ must be material surfaces for the flow. Thus [18,19]

$$\mathbf{q}|_{z=\zeta}\cdot \text{ grad } (z-\zeta) = n|_{z=\zeta}\frac{\partial \zeta}{\partial t} - Q_\zeta$$

and

$$\mathbf{q}|_{z=\eta}\cdot \text{ grad } (z-\eta) = n|_{z=\eta}\frac{\partial \eta}{\partial t} - Q_\eta$$

where $n(x,y,z,t)$ denotes the *porosity* of the ground, that is to say the fraction of the porous material in volume which may be occupied by water at the point (x,y,z) and time t (dim. L^3/L^3), and where Q_ζ and Q_η denote the *accretion* through the surfaces ζ and η respectively, that is to say, the supply of additional water through these surfaces per unit horizontal area (dim. $(L^3/T)/L^2$). Assuming that the porosity n is independent of depth gives finally

$$\begin{aligned} s(\zeta-\eta)\frac{\partial h}{\partial t} + n\frac{\partial}{\partial t}(\zeta-\eta) &= \frac{\partial}{\partial y}\left\{K_{xx}(\zeta-\eta)\frac{\partial h}{\partial x} + K_{xy}(\zeta-\eta)\frac{\partial h}{\partial y}\right\} \\ &\quad + \frac{\partial}{\partial y}\left\{K_{yx}(\zeta-\eta)\frac{\partial h}{\partial x} + K_{yy}(\zeta-\eta)\frac{\partial h}{\partial y}\right\} \\ &\quad - W + Q_\zeta + Q_\eta . \end{aligned} \tag{1.10}$$

This is the equation of motion in two spatial dimensions. Note the sign convention with regard to the last three terms. They are positive if water is added to the aquifer (source) and negative if it is withdrawn (sink).

Consider now a confined or semi-confined aquifer. In this case the boundaries ζ and η are fixed. Equation (1.10) is then

$$\begin{aligned} sD\frac{\partial h}{\partial t} &= \frac{\partial}{\partial x}\left\{K_{xx}D\frac{\partial h}{\partial x} + K_{xy}D\frac{\partial h}{\partial y}\right\} \\ &\quad + \frac{\partial}{\partial y}\left\{K_{yx}D\frac{\partial h}{\partial x} + K_{yy}D\frac{\partial h}{\partial y}\right\} \\ &\quad - W + Q_\zeta + Q_\eta \end{aligned} \tag{1.11}$$

where D is the thickness of the aquifer (dim. L). The term

$$S = sD$$

is referred to as the *storage coefficient* of the aquifer (dimensionless), and

$$T_{xx} = K_{xx}D, \quad T_{xy} = K_{xy}D, \quad T_{yx} = K_{yx}D \quad \text{and} \quad T_{yy} = K_{yy}D$$

are the components of the coefficients of *transmissivity* or *transmissibility* of the aquifer (dim. L^2/T). By way of the symmetry of the hydraulic conductivity tensor we have

$$K_{xy} = K_{yx}$$

and therefore also

$$T_{xy} = T_{yx}.$$

If the flow is isotropic

$$T_{xx} = T_{yy} \quad \text{and} \quad T_{xy} = T_{yx} = 0. \tag{1.12}$$

For an unconfined aquifer only the lower boundary η of the aquifer is fixed. However noting that under our assumption of constant unit density the potential in the aquifer

$$\Phi = p + z \tag{1.13}$$

where p denotes the water pressure head (dim. L) and z elevation (dim. L), if the upper boundary of the phreatic aquifer is defined as the surface where the pressure head is atmospheric (i.e. zero), (1.13) says

$$\Phi|_{z=\zeta} = \zeta.$$

Recalling assumption (1.9), combining this with (1.10) yields

$$\begin{aligned}\{s(h-\eta)+n\}\frac{\partial h}{\partial t} &= \frac{\partial}{\partial x}\left\{K_{xx}(h-\eta)\frac{\partial h}{\partial x}+K_{xy}(h-\eta)\frac{\partial h}{\partial y}\right\}\\ &\quad+\frac{\partial}{\partial y}\left\{K_{yx}(h-\eta)\frac{\partial h}{\partial x}+K_{yy}(h-\eta)\frac{\partial h}{\partial y}\right\}\\ &\quad-W+Q_\zeta+Q_\eta.\end{aligned} \tag{1.14}$$

This is the equation for flow in an unconfined aquifer (sometimes called the *Boussinesq equation*). In practice the term

$$\mu = s(h-\eta)+n$$

is often assumed constant and termed the *specific yield* of the aquifer (dimensionless).

Equations (1.11) and (1.14) may be used to model a system of multiple aquifers such as that shown in Figure 1.2. The only detail remaining to be resolved is that of the modelling of the coupling between the different aquifers through the terms Q_ζ and Q_η. The most common approach [19,26,96,102,112, 118] to this is simply to apply (1.4) to an aquitard separating two aquifers under the assumption that storage and abstraction can be neglected and that as mentioned earlier the flow in the aquitard is vertical. In this case, (1.4) becomes

$$0=\frac{\partial}{\partial z}(K_{zz}\frac{\partial \Phi}{\partial z}).$$

Whence integrating

$$0=K_{zz}\frac{\partial \Phi}{\partial z}+q_z$$

where q_z is the flux through the aquitard and is independent of z. Assuming K_{zz} is constant in depth subsequently yields

$$q_z=-K_{zz}\frac{\Phi_2-\Phi_1}{z_2-z_1}$$

where Φ_1 and Φ_2 denote the potentials at elevations z_1 and z_2 respectively. Taking z_1 to be the elevation of the lower boundary of the aquitard which is subsequently hypothesized to be the upper boundary ζ^- of an underlying aquifer and z_2 to be the upper boundary of the aquitard which is then likewise hypothesized to be the lower boundary η^+ of an overlying aquifer and recalling (1.9), this gives

$$q_z = -K_{zz}\frac{h^+ - h^-}{\eta^+ - \zeta^-}$$

where h^+ denotes the head in the overlying aquifer and h^- the head in the underlying aquifer. Thus one can relate the term Q_η, Q_η^+ say, in the overlying aquifer, and Q_ζ, Q_ζ^- say, in the adjacent underlying aquifer by the expression

$$Q_\eta^+ = -Q_\zeta^- = -(h^+ - h^-)/C. \tag{1.15}$$

Here

$$C = (\eta^+ - \zeta^-)/K_{zz}$$

is termed the *resistivity* of the aquitard (dim. T).

For a more refined approach we refer to [26,95,110]. However, since the above-mentioned approach is that most commonly used, we shall restrict our attention to it here.

EXERCISE. It has been suggested that from (1.4), equations (1.11) and (1.14) may also be obtained with the hydraulic parameters s, K and n replaced by

$$\overline{s} = \int_\eta^\zeta dz/(\zeta - \eta),\ \overline{K} = \int_\eta^\zeta K dz/(\zeta - \eta) \text{ and } \overline{n} = \int_\eta^\zeta n dz/(\zeta - \eta)$$

if one assumes that the potential Φ is independent of z. Check the verity of this suggestion.

EXERCISE. Derive the equations corresponding to (1.11) and (1.14) when in place of averaging with respect to z one averages with respect to the orthogonal to an arbitrary surface in three dimensions. See Figure 1.5.

One may care to refer to [1] for the result of this exercise. It has been applied for modelling groundwater flow over steep surfaces such as an aquifer formed by geological debris on a mountain side.

1.3 The finite element method

Having derived equations (1.11), (1.14) and (1.15) to describe the transport of water in a multiple aquifer system, we are faced with the challenge of solving these equations in practice.

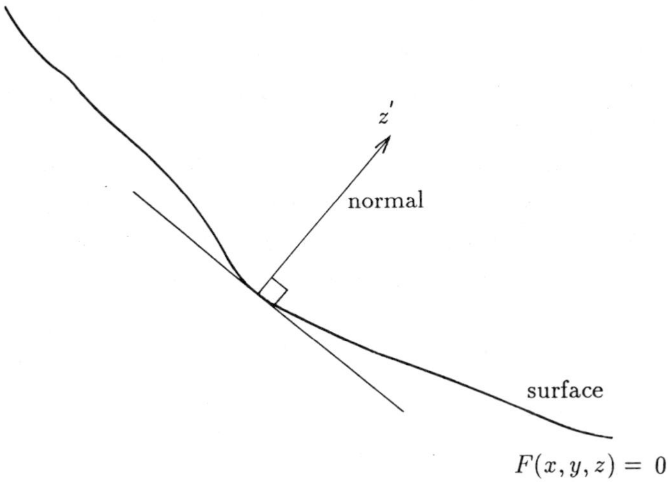

Figure 1.5: Normal to given surface

Under simplified assumptions there are of course a number of analytical solutions which may be obtained for these equations [19,94,96,102,109,112]. These also find their use in hydrology in pumping test analysis. In general though for a given system of aquifers with complicated boundary conditions we are forced to seek a numerical solution of these equations.

A number of numerical methods are available for tackling (1.11) and (1.14) - for instance, finite difference, finite element and boundary element techniques [19,26,93,96,99,102,110,118]. All may be used for equations (1.11) and (1.14) and have their own advantages and disadvantages. Here, we shall describe the finite element method. This is particularly attractive for use in groundwater resources management because of its ability to handle complex geometries, and is therefore widely used [19,47,94,95,118]. According to Zienkiewicz [117] "The finite element method is essentially a process through which a continuum with infinite degrees of freedom can be approximated to by an assemblage of subregions (or elements) each with a specified but now finite number of unknowns". In reality a groundwater system is highly heterogeneous in nature and as such can only be faithfully described if the variables which characterize the system are quantified at each and every point. In a finite element model one essentially seeks to divide an aquifer into a finite number of subregions in each of which the hydraulic variables are described in a simple mathematical form, but in such a way that the essential hydraulic behaviour of the system is preserved by imposing continuity conditions across the boundaries of the subregions and the governing flow equations.

There are several approaches to finite element modelling in subsurface hydrology. The most well established of these are the *Galerkin* and *Rayleigh-Ritz* procedures [95,99]. Generally, the Galerkin method (the more recent development) is the superior procedure, for it is not only mathematically simpler but has a wider area of application. However, the two procedures are equivalent as applied to groundwater flow problems and both will be considered below.

To illustrate the basic principles of the finite element method, for simplicity, the discussion is restricted to steady-state flow in a confined isotropic aquifer. The extension of the theory to transient or more complex situations may take a number of different directions. The presentation of the theory, although based on sound mathematical principles, will be largely heuristic. For a more detailed treatment reference is made to the work of P. Knabner in this book and for a rigorous mathematical treatment to the monograph by Strang and Fix [104].

Let h denote the piezometric head in a rigid isotropic confined aquifer of area Ω. Then, cf. (1.11) and (1.12), if the flow in the aquifer is assumed to be steady and the aquifer isotropic the flow is governed by the equation

$$\frac{\partial}{\partial x}\left(T\frac{\partial h}{\partial x}\right)+\frac{\partial}{\partial y}\left(T\frac{\partial h}{\partial y}\right)=W. \tag{1.16}$$

Here, for convenience, we drop the subscripts from the components of the transmissivity tensor. The problem is, assuming that the transmissivity, recharge or discharge and boundary flow of the aquifer are known and the head is defined at certain points within or on the boundaries of the aquifer, to solve equation (1.16) to determine the piezometry, h.

To apply the finite element method the first requirement is to choose a complete set of ***basis functions*** on Ω. That is to say, a set of continuous and piecewise continuously differentiable functions ϕ_i from which any continuously differentiable function in Ω may be approximated arbitrarily closely by a suitable linear combination.

Consider a simple example in one space dimension as an illustration. It is possible to construct a set of basis functions on the interval $0 \leq x \leq 1$ in the following manner. Dividing the interval in half, take the functions ϕ_1, ϕ_2 and

ϕ_3 shown in Figure 1.6(b). The function ϕ_1 is piecewise-linear and defined to be one at $x = 0$ and zero at $x = \frac{1}{2}, 1$; similarly ϕ_2 is piecewise-linear, one at $x = \frac{1}{2}$ and zero at $x = 0, 1$; and ϕ_3 is one at $x = 1$ and zero at $x = 0, \frac{1}{2}$. Next, dividing the subintervals in half, take the functions ϕ_4 to ϕ_8 shown in Figure 1.6(d). The function ϕ_4 is piecewise-linear and defined to be one at $x = 0$ and zero at $x = \frac{1}{4}, \frac{1}{2}, \frac{3}{4}, 1$; ϕ_5 is defined to be one at $x = \frac{1}{4}$ and zero at $x = 0, \frac{1}{2}, \frac{3}{4}, 1$; and so on. Upon further halving the subintervals, the functions shown in Figure 1.6(f) are incorporated in the series. The subsequent inclusion of further functions defined analogously on finer and finer subintervals leads to an infinite series of basis functions. To illustrate that the series of functions is indeed a complete set of basis functions on the interval $0 \leq x \leq 1$ reference is made to the function $F(x)$ shown in Figure 1.6(a). The linear combination $F(0)\,\phi_1 + F(\frac{1}{2})\,\phi_2 + F(1)\,\phi_3$ is the approximation to F shown in Figure 1.6(c). Whilst a linear combination of ϕ_4 through ϕ_8 produces the approximation to F illustrated in Figure 1.6(e). If the next group of basis functions is considered the approximation to F shown in Figure 1.6(g) is obtained. It is clear that by considering further basis functions it is possible to approximate $F(x)$ arbitrarily closely.

Other examples of basis functions include the well-known Fourier series. The basis functions generally used in two dimensions are 'hat shaped' functions constructed from polynomials restricted to a confined area. They are two-dimensional analogues of the functions shown in Figure 1.6. A few examples are illustrated in Figure 1.7.

Returning to the central discussion, having chosen a complete set of basis functions on Ω, the objective of the finite element method, is to construct an approximation to the head h in the aquifer Ω in terms of these functions. (Recall, that by definition, any continuous function in Ω may be approximated arbitrarily closely by a linear combination of the basis functions.) Consider therefore a combination of the form

$$\hat{h}(x, y) = \sum_{i=1}^{I} h_i \phi_i(x, y) \tag{1.17}$$

composed from I linearly independent basis functions $\{\phi_i\}_{i=1}^{I}$. The problem is now to find the values $h_1,\ h_2,\ \ldots,\ h_I$ such that $\hat{h}$ most closely resembles the true solution of the problem h. This is the crux of the various finite element methods. It is resolved by imposing on $\hat{h}$ a characteristic of the true solution.

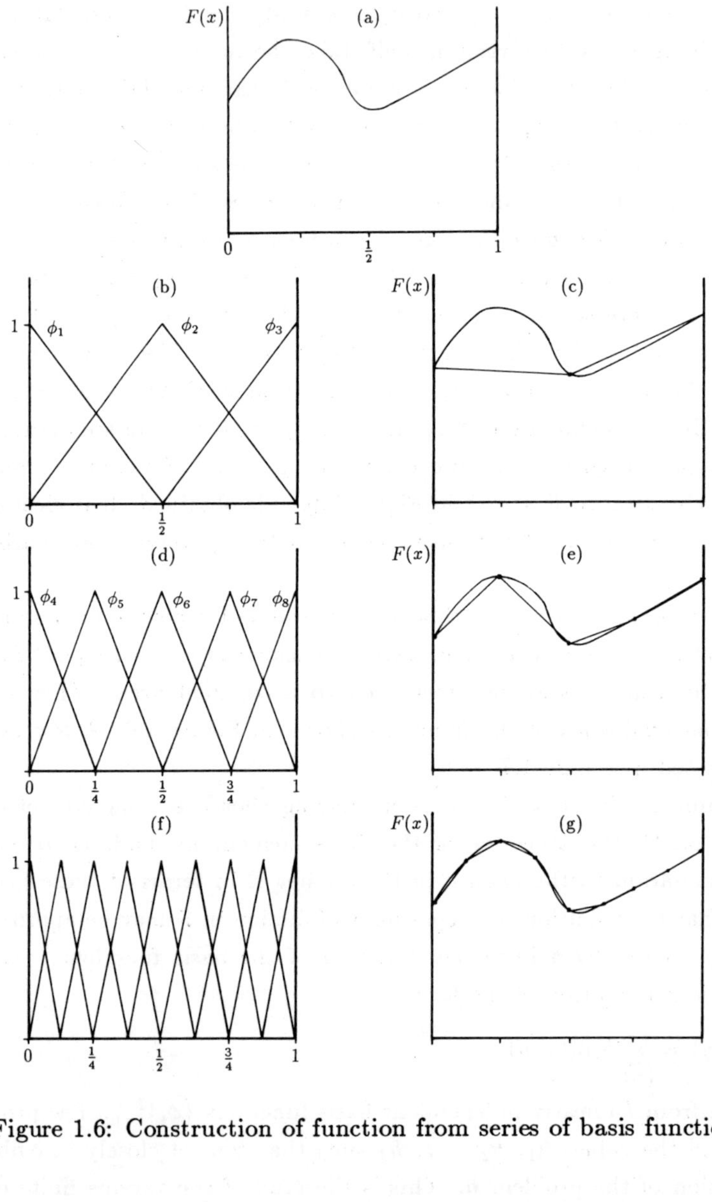

Figure 1.6: Construction of function from series of basis functions

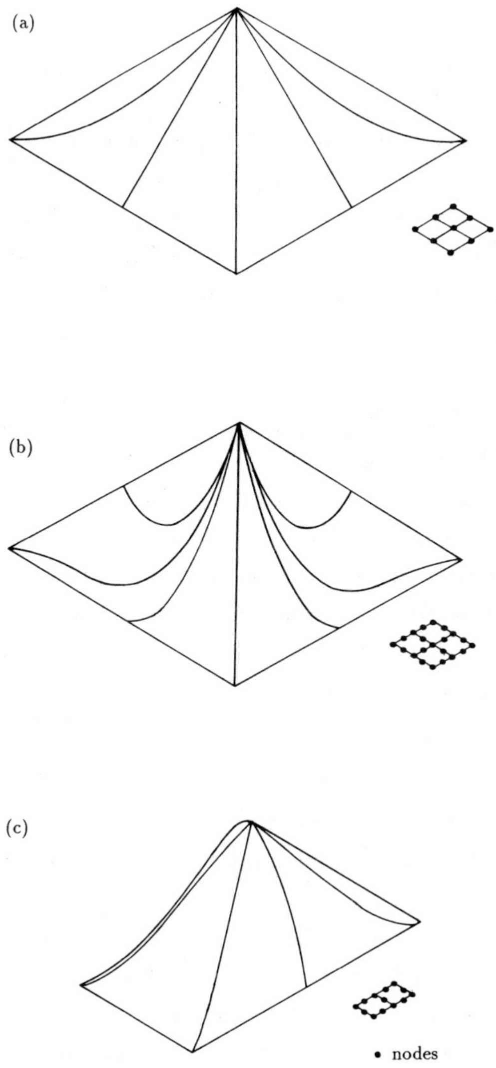

Figure 1.7: Examples of two-dimensional basis functions

Observe that if h is a solution of equation (1.16), given any differentiable function ψ defined in Ω, multiplying equation (1.16) by ψ and integrating by parts gives

$$\int_{\Omega} T\left\{\frac{\partial\psi}{\partial x}\frac{\partial h}{\partial x}+\frac{\partial\psi}{\partial y}\frac{\partial h}{\partial y}\right\}dx\,dy=\int_{\Gamma}\psi W_n ds-\int_{\Omega}\psi W dx\,dy, \tag{1.18}$$

where Γ is the boundary of the aquifer Ω, W_n denotes the boundary flow and ds denotes the differential arc length along Γ. Conversely if a sufficiently smooth function h satisfies (1.18) for any test function ψ then h is a solution of equation (1.16) with the given boundary flow [99]. In the Galerkin finite element method one prescribes that the approximate solution holds this inherent property of the true solution in satisfying the identity (1.18) for the test functions $\{\phi_j\}_{j=1}^{I}$.

Substituting $\hat{h}$ into (1.18) yields

$$\sum_{i=1}^{I} h_i \int_{\Omega} T\left\{\frac{\partial\psi}{\partial x}\frac{\partial\phi_i}{\partial x}+\frac{\partial\psi}{\partial y}\frac{\partial\phi_i}{\partial y}\right\}dx\,dy=\int_{\Gamma}\psi W_n ds-\int_{\Omega}\psi W dx\,dy. \tag{1.19}$$

(Here, strictly mathematically speaking the partial derivatives are generalized.) Note that the integrals in equation (1.19) depend only on ψ, the known functions $\{\phi_i\}_{i=1}^{I}$, the known aquifer transmissivity, the known boundary flow and the known recharge or discharge. Thus, if the basis functions $\{\phi_j\}_{j=1}^{I}$ are used as test functions in (1.19), one obtains

$$\sum_{i=1}^{N} h_i \int_{\Omega} T\left\{\frac{\partial\phi_j}{\partial x}\frac{\partial\phi_i}{\partial x}+\frac{\partial\phi_j}{\partial y}\frac{\partial\phi_i}{\partial y}\right\}dx\,dy \;=\; \int_{\Gamma}\phi_j W_n ds-\int_{\Omega}\phi_j W dx\,dy$$
$$\text{for } j=1,\,2,\,\ldots,\,I, \tag{1.20}$$

and hence, n equations in the n unknowns - $h_1,\ h_2,\ \ldots,\ h_I$. Therefore, by solving for $h_1,\ h_2,\ \ldots,\ h_I$ an approximate solution to the problem - $\hat{h}=\sum_{i=1}^{I} h_i\phi_i$, may be obtained.

The more terms that are included in the finite series approximation the closer $\hat{h}$ approximates the true solution of the problem h. Thus by considering a sufficiently large number of linearly independent basis functions in the expansion (1.17) and solving the resulting equations (1.20) h may be determined arbitrarily closely. This is the foundation of the finite element method.

The proof that as further basis functions are included in the expansion (1.17) the calculated solution $\hat{h}$ converges, in some sense, to the true solution h involves somewhat abstract mathematical concepts and reference is made to [95,104] for verification. Moreover, it should be mentioned that it is only true under certain conditions on the basis functions [95,104]. However these are not a serious restriction.

The Rayleigh-Ritz method provides an alternative approach for deriving the equations (1.20). This is founded on the correspondence between problems of variational calculus and differential equations. In the present field the correspondence takes the following form. If a function h minimizes the quantity

$$J(h) = \int_\Omega \left[\tfrac{1}{2} T \left\{ \left(\frac{\partial h}{\partial x} \right)^2 + \left(\frac{\partial h}{\partial y} \right)^2 \right\} + hW \right] dx\, dy - \int_\Gamma h W_n \, ds \qquad (1.21)$$

then h is a solution of equation (1.16) in Ω with the boundary flux W_n [99]. (The notation is as previously used.) Thus instead of looking for a solution of equation (1.16) with the given boundary flow the problem may be converted to that of finding a function h which minimizes J. The strategy of the Rayleigh-Ritz finite element method is of looking for an approximate solution of equation (1.16) of the form (1.17), by choosing the coefficients h_1, h_2, ..., h_I such that $J(\hat{h})$ is minimized. Substituting $\hat{h}$ into (1.21) gives

$$\begin{aligned} J(\hat{h}) &= \int_\Omega \left[\tfrac{1}{2} T \left\{ \left(\sum_{i=1}^{I} h_i \frac{\partial \phi_i}{\partial x} \right)^2 + \left(\sum_{i=1}^{I} h_i \frac{\partial \phi_i}{\partial y} \right)^2 \right\} + \sum_{i=1}^{I} h_i \phi_i W \right] dx\, dy \\ &\quad - \int_\Gamma \sum_{i=1}^{I} h_i \phi_i W_n \, ds. \end{aligned}$$

Now by elementary calculus, the coefficients h_1, h_2, ..., h_I which cause $J(\hat{h})$ to be minimized are determined by the equations

$$\frac{\partial J}{\partial h_j}(\hat{h}) = 0 \quad \text{for } j = 1,\ 2,\ \ldots,\ I,$$

or, in other words, rewriting in full, by the equations

$$\int_\Omega \left[\sum_{i=1}^{I} T h_i \left\{ \frac{\partial \phi_j}{\partial x} \frac{\partial \phi_i}{\partial x} + \frac{\partial \phi_j}{\partial y} \frac{\partial \phi_i}{\partial y} \right\} + \phi_j W \right] dx\, dy \;-\; \int_\Gamma \phi_j W_n \, ds = 0$$
$$\text{for } j = 1, 2, \ldots, I.$$

It is seen that these equations are identical to (1.20). Thus by an alternative approach the previous approximate basis function formed solution is again derived.

So far little mention has been made of any specific choice of basis functions. However, an attractive method of generating basis functions has been developed for use in engineering practice. The established procedure is as follows. First the aquifer is divided into elements and nodes are allotted to corner points and other strategic points in each element. See, for example, Figure 1.8 where an aquifer is divided into quadrilateral elements whose sides are linear. This example comes from a real study of one of the islands in the Netherlands. Then for each node a basis function is defined with the following properties:

- The function takes the value 1 at the specified node.
- The function takes the value 0 at all other nodes.
- The function vanishes in every element not containing the specified node.

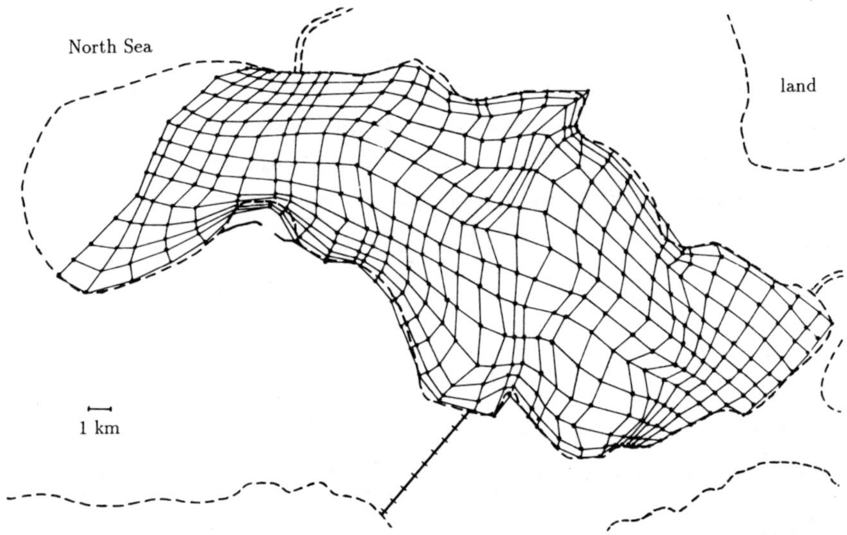

Figure 1.8: Example of finite element network

Traditionally, for simplicity, but this is not necessary, the basis functions are polynomial within each element. The basis functions displayed in Figure 1.7 are examples. Figure 1.7(a) represents a basis function defined at the intersection of four square elements, each with only corner nodes. Figure 1.7(b) shows a basis function defined at the confluence of four square elements with additional mid-side nodes. Whereas Figure 1.7(c) represents a basis function associated with one of these mid-side nodes. The benefits of this selection procedure are evident, viz.:

- The coefficient of a basis function as calculated in the expansion (1.17) is precisely the head computed at the associated nodal point.
- Cross terms in equation (1.20) involving basis functions defined on different elements disappear.

Plainly, more basis functions are introduced by dividing the network more finely. (See, for example, Figure 1.6.) Thus, the finer the network, the more accurately the computed solution approximates the true solution. For instance, if the basis functions are continuous piecewise-linear functions then the error between the two solutions is at most proportional to the square of the maximum dimension of the elements [104]. In application accuracy can be increased by a judicious design of the network.

In the preceding discussion it has been tacitly assumed that the transmissivity of the aquifer is explicitly known throughout the formation. In practice means must be found of extemporizing this information. There are two standard techniques [95]. One is to specify a uniform transmissivity within each element. The other is to express the transmissivity T of the aquifer in terms of the basis functions, i.e. by setting

$$T(x, y) = \sum_{k=1}^{I} T^{(k)} \phi_k(x, y).$$

In this latter case, due to the convenient choice of the basis functions this is equivalent to specifying the transmissivity tensor components $T^{(k)}$ at each nodal point. Likewise, it has been implicitly assumed that the recharge or discharge of the aquifer is known at every point. In practice means must also be found of extemporizing these data. An acceptable solution is to consider recharge or discharge respectively as a point source or sink located at a nodal point, or, in the case of inflow from an overlying leaky aquitard to approximate analytically and average over an element [95].

EXERCISE. Derive the equations of the finite element method for an anisotropic steady-state aquifer.

EXERCISE. Work out the finite element equations when the recharge term

$$Q = (h^+ - h)/C$$

is included in (1.16) with h^+ and C known distributions. This is, of course, the component of leakage from one aquifer to another in a multi-layered system. Hint: develop h^+ analogous to h in terms of the basis functions.

2 Saturated groundwater flow modelling - case study

2.1 Steps in mathematical modelling

In any mathematical modelling study, be it for groundwater flow or any other problem area, there are a number of actions which are always involved.

Step 1. The making of a first analysis.
This is the identification of the crucial question which has to be answered by the modelling.

Step 2. Summing up and reducing the factors involved.
An overview of all the factors which are important to the central question should be made and the interrelation between these factors established. Thereafter, factors which are important can be distinguished from those which are not.

Step 3. Simplification.
This is the making of assumptions for the purpose of deriving a simpler model, in so far as this is useful and admissible.

Step 4. Setting up of the model.
Variables have to be named. The relations or the equations between the variables have to be defined, and where necessary the boundary conditions have to be specified.

Step 5. Testing of the model.
This is in fact the step of solving the equations set up in the previous step, and testing the results in simple known situations. One should acquire the confidence that the model does what is expected in ideal situations related to that under study.

Step 6. Calibration and verification.
This, in contrast to the previous step in which the model was tested against simple idealized situations, is testing the model for the situation one actually wishes to model. In theory, one puts relevant data into the model and confirms that the ensuing results agree with reality. In practice, there

are often discrepancies related to uncertainties in the modelling or in the data, and one may have to adjust certain assumptions or values to obtain acceptable results. This process is often referred to as model calibration. If calibration is not necessary, one may speak of *verification* of the model.

Step 7. Application.

This is of course the point of the whole exercise. The model is applied to provide an answer to the original question. In groundwater modelling this step often becomes the step of prediction, i.e. establishing what will happen in some proposed future situation.

Step 8. Reporting.

This last step is often not seen as part of the modelling process. However, it is an integral part of a modelling exercise. After all, if one is not going to report to anyone about the results of the modelling, what was the point of doing it? It is at this stage that a model can be assessed in its totality, one can critically evaluate its shortcomings, and can indicate the trustworthiness of the results.

Note that this division in steps is not really clear-cut. There is often an interaction between the various steps. Very often in the course of a modelling exercise one is forced to reconsider various aspects and reassess earlier steps. A simple illustration could be that in the course of the testing of the model (Step 6) one finds that it does not perform as expected. One then has to return to examine whether the assumptions made (in Step 3) have not been too restrictive, or even worse, whether important factors have been neglected (in Step 2). Another possibility is that one finds that one has insufficient data for the model calibration (in Step 6) and then has to reassess the whole model. It can even occur that one discovers that under the given circumstances the original question posed is unanswerable and the whole discussion can begin again at Step 1. Perhaps in this case another question could better have been posed. This all is where mathematical modelling enters the realm of an art.

In the next subsection, we shall see an illustration of the above-mentioned steps in mathematical modelling, in a practical groundwater modelling study.

2.2 A silt-depot study

The harbour of Rotterdam is the largest in the world, and the surrounding inland waterways are heavily trafficked. A continual problem for the harbour management and local authorities is the deposition of silt from the River Maas and the River Rhine in the harbour and surrounding waterways. Continual dredging is necessary. Much of the dredged material can be deposited at sea. However, an increasing difficulty is being caused by the quantity of pollutants present in the silt. Wastes from chemical and industrial plants further upstream are discharged into the rivers and heavy metals and organic micro-pollutants which they contain become attached to the silt. These are deposited in considerable quantities along with the silt. Alternative ways of disposing of this contaminated sludge in the future are currently being sought.

One of the possibilities for disposing of the contaminated silt is that of depositing it on land in specially constructed depots. The idea is to surround a given area with artificial dikes and use the so-created basin to store the silt. (See Figure 2.1.) Initially any water released from the silt can be captured and disposed of. Also water exuding from the depot through the dikes forming its side-walls can be intercepted by a system of ditches around the depot and from there transported away. What are the long term effects though?

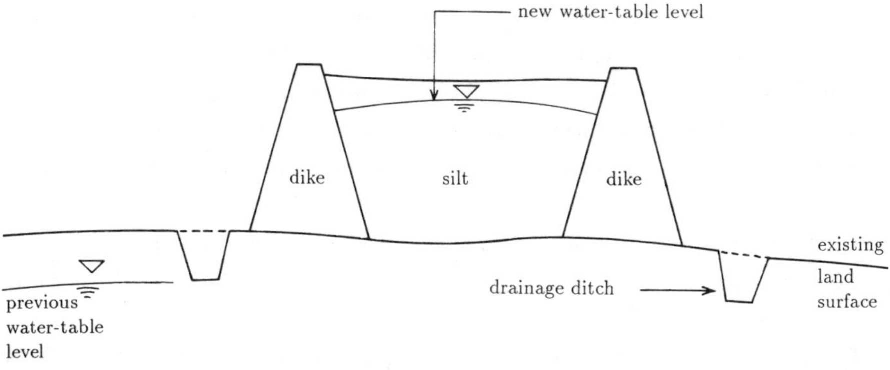

Figure 2.1: Cross-section of silt-depot

One of the effects that will occur is that within the depot the water-table will rise considerably above its present elevation at the site. This will alter the groundwater flow pattern in the neighbourhood. More to the point, it will generate infiltration out of the depot into the underlying subsurface. The questions are then: Where will the contaminated groundwater infiltrating out of the depot eventually end up, and in what concentrations will it contain its contaminants? Is is conceivable that leakage from the depot could eventually lead to the pollution of neighbouring or even distant arable lands and nature conservation areas.

The objective of the study which will now be described was to predict the change in groundwater flow which would result from the proposed construction of one such depot, and simultaneously to identify the possible destination of any contaminants leaching out of the depot. Further study was devoted to the investigation of the chemical reactions between contaminants in the leaking groundwater and the subsoil taking into account the effect of absorption in the subsoil [120]. Here though, we shall concern ourselves exclusively with the quantitative groundwater flow study.

The general area for location of the silt-depot is shown in Figure 2.2. The precise location between the city of Rotterdam and the town of Capelle aan den IJssel is shown in Figure 2.3. In this figure too, one can see the shape of the proposed depot which has a surface area of about 70 hectares. Upon completion and after consolidation the silt in the depot will be 4 m thick. Given the objective of the study described above, this presents us with the question we must answer: Supposing that the proposed silt depot is constructed, how will the groundwater in the area of the site flow in the long term?

Having identified the question that must be answered, the second step described in the previous subsection is to sum up and reduce all the factors involved in this question. Obviously, besides the indicated construction of the depot, the most important factors in this study are the (existing) hydrogeology of the area, and the hydraulic characteristics of the silt which is to be deposited in the depot.

To begin with the hydrogeology, looking at Figures 2.2 and 2.3 the most striking features are the large waterways running through the depicted area.

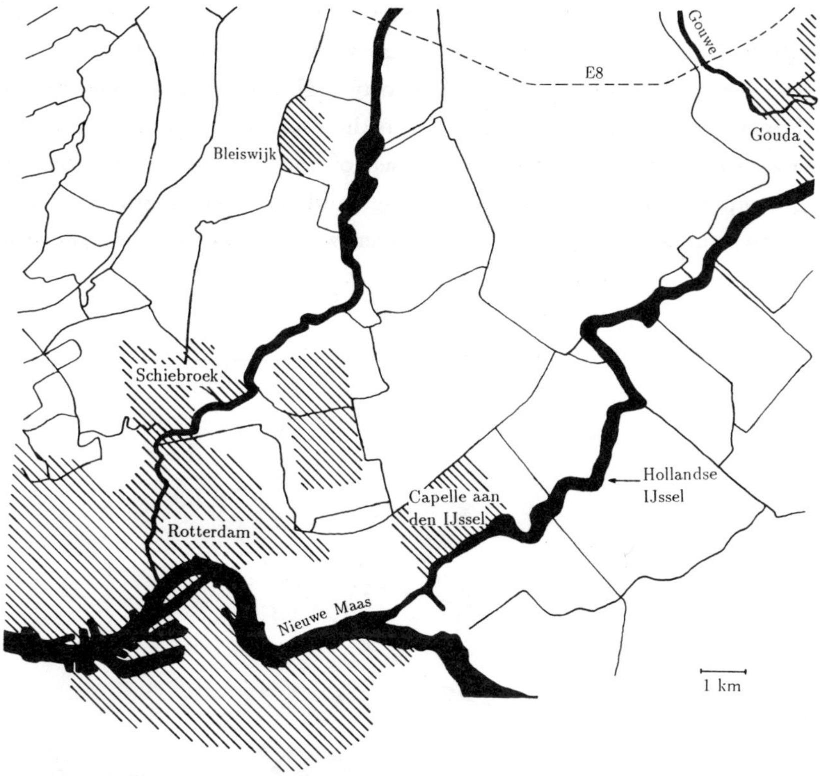

Figure 2.2: Location of silt-depot

Specifically, the Hollandse IJssel and the Nieuwe Maas underneath in these figures. They have an average water level of 0·47 m above mean sea level. The present land surface at the site is 1·75 m below mean sea level and consists of pasture. Inundation is prevented by a drainage system in which the water level is maintained at 2·30 m below mean sea level. Looking further afield the land surface approaches a depth of more than 6 m below mean sea level with a water-table which varies in depth from 2 m up to 7 m below mean sea level. (In fact, this is the most low-lying part of the Netherlands.) Generally speaking, the large rivers and their immediate environs function as sources of water which eventually has to be drained from the ground in the surrounding lower areas. A geological cross-section running roughly from south to north through the study area is shown in Figure 2.4. Here too the topography can be discerned. The ground immediately below the surface is composed of poorly permeable Holocene peat and clay layers with a total thickness of 10 to 14 metres. Beneath this layer, there is a fairly highly permeable sand formation which increases in thickness from 10 m in the south to 45 m in the north. The upper strata of the sand formation consist of fine sands with the lower strata being composed of fairly coarse to coarse sands containing gravel. The sand formation is underlain by a clay-loam layer which demarks the transition to

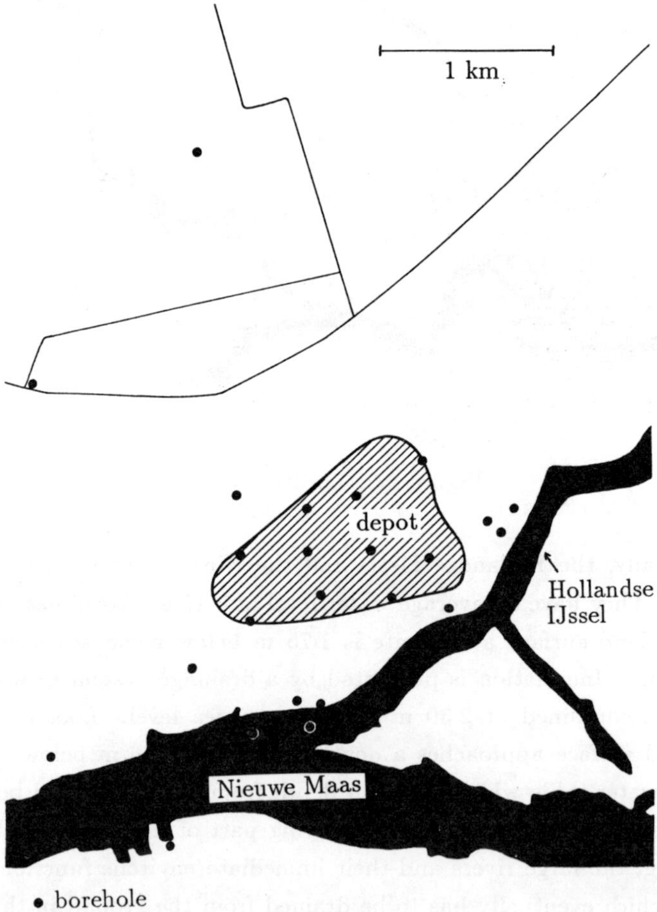

Figure 2.3: Detail of silt-depot site

the deeper formations of Kedichem and Tegelen. From other studies, the hydraulic conductivity of the sand formation has been calculated to lie in the range of 15 to 50 m/day and the resistivity of the overlying Holocene formations has been estimated to lie between 2500 and 5000 days. By constructing a system of drainage channels within the depot the plan is to maintain the phreatic surface within the depot at 1 m above mean sea level. This gives a rise of more than 3 m with respect to the present water-table level. With regard to the silt in the depot, the hydraulic conductivity of this material has been determined in the laboratory to be 0·00025 m/day.

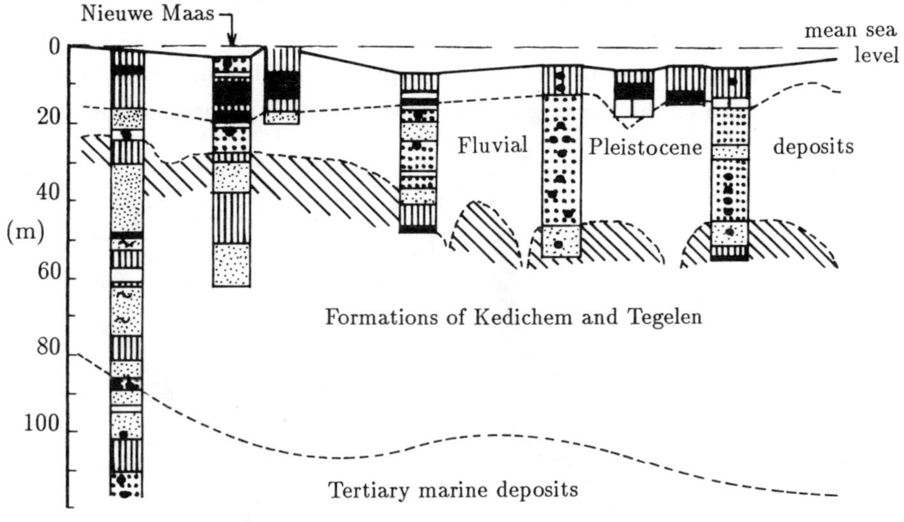

Figure 2.4: Geological cross-section

Further details of the hydrogeology of the area can be found in [98,115,119]. Here, to be honest, we have already practised a certain reduction of all possible factors involved, indicating only the most relevant.

Our next step is to make a number of assumptions which will lead to a tractable mathematical model.

Our first assumption is that the large waterways in Figures 2.2 and 2.3 act as watersheds and therefore can be taken as fixed known boundaries for any modelling study. Secondly, we assume that the influence of the depot will not be felt at a distance of more that 8 km from the depot site. This provides a second criterion with which we can restrict the area to be considered. In a vertical direction we suppose that the groundwater flow occurs predominantly through the Pleistocene sand formation. Hereby, the clay-loam layers at the bottom of this formation function as an impermeable layer. Finally the overlying Holocene formations act as a semi-confining layer. Thus, we can use the equations for a single leaky aquifer described in Subsection 1.2 to model the groundwater flow. In adopting this approach, we view the controlled surface water-levels as belonging to a fictive overlying aquifer. We note that these suppositions are motivated and justified by our consideration of the hydrogeology. For want of any other information we assume the aquifer to be isotropic. Finally, we assume that by considering only the long term effects and the yearly average conditions we may view the situation as being in equilibrium, i.e. steady-state.

We now turn to the working out of these ideas in a concrete model.

Figure 2.5: Finite element network

At a very early stage, the choice was made to apply a computer program [118] based on the finite element method for solving the equations of groundwater flow as outlined in Subsection 1.3. This choice was largely based on the fact that this program had a long history of previous successful applications [69,118]. So it could be used with confidence. (In this respect as it were, Step 5 mentioned in the previous subsection had long ago been carried out.)

The finite element network designed for computing the groundwater flow in this study is shown in Figure 2.5. Here the consequences of the decision to bound the study area by the large waterways and by a suitable distance from the depot can be clearly seen. Note the refinement of the network in the vicinity of the depot site. This is for a better resolution of the results in that area in which the greatest changes are to be expected. A close-up of the network in this vicinity is shown in Figure 2.6. The shape of the depot is clearly embodied in its design.

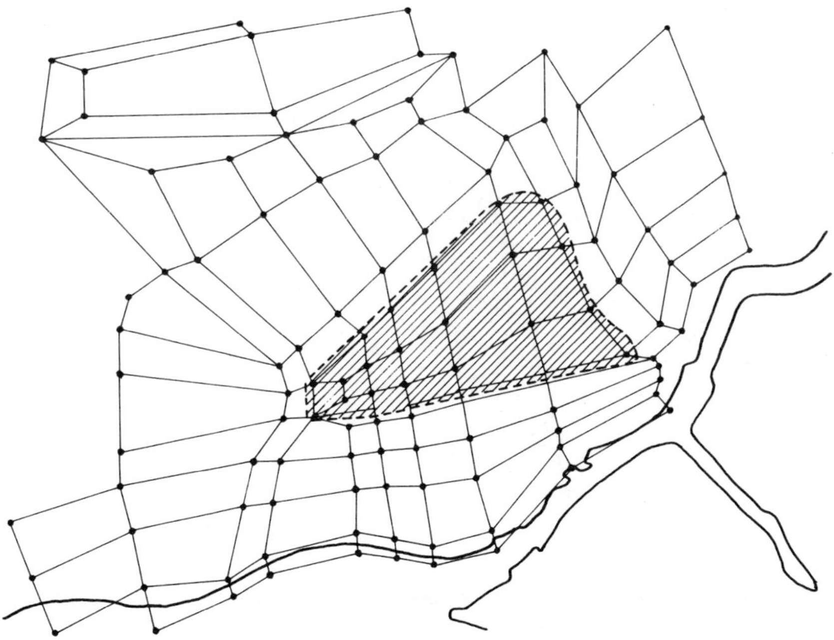

Figure 2.6: Detail of network

As mentioned above the equation for groundwater flow in a single semi-confined aquifer will be solved on this network. The existing controlled water-table levels will be taken as the head in the adjacent overlying aquifer from which leakage will take place. These are known and therefore fixed. For the flow in the main aquifer itself, the boundary condition along the large waterways will be taken as a specified head equal to the known average water-level in the rivers. This is justified by the argument that the rivers are sufficiently deep to be in good hydraulic contact with the main aquifer. Having decided that the remaining boundaries are so far away as not to be influenced by construction of the depot, here also a known potential may be assumed - namely that indicated by the present conditions.

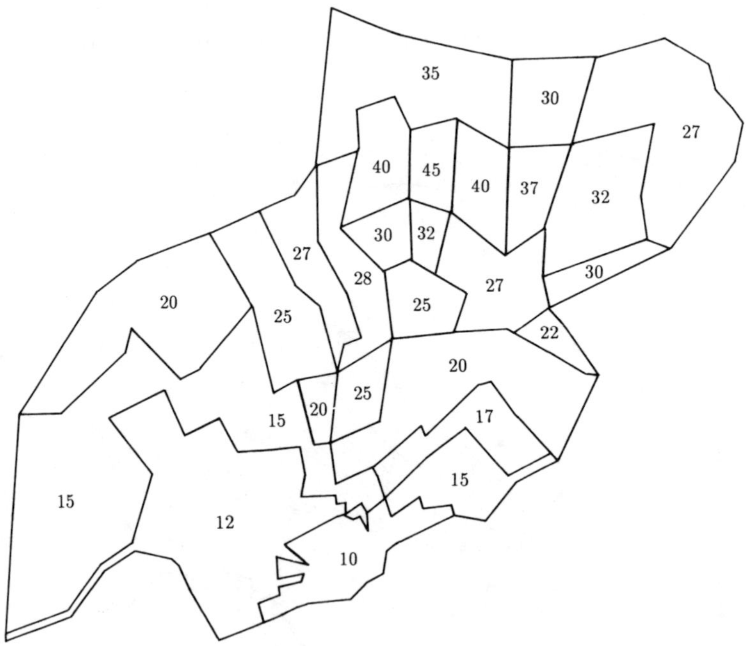

Figure 2.7: Thickness values (in m)

Thus we have almost everything we need to apply the finite element method. The exception is data on the hydraulic parameters occurring in equation (1.11) and (1.15). In other words, we still have to introduce the parameters of transmissivity, T, and resistivity, C. To remove uncertainty in the transmissivity, it was decided to use borehole logs to construct a picture of the thickness of the sand formation in the study area. Using the relation $T = KD$ where K denotes the assumed isotropic hydraulic conductivity of the aquifer and D the thickness, uncertainties in the transmissivity values could be reduced to uncertainties in the hydraulic conductivity values K. The values for the thickness D fed into the model per element are shown in Figure 2.7. In the first instance a uniform K value of 50 m/day and a uniform C value of 5000 days were fed into the model, and the would-be groundwater potential in the area under the present circumstances was computed. The results were far from satisfactory. This indicated that some calibration of the model was necessary.

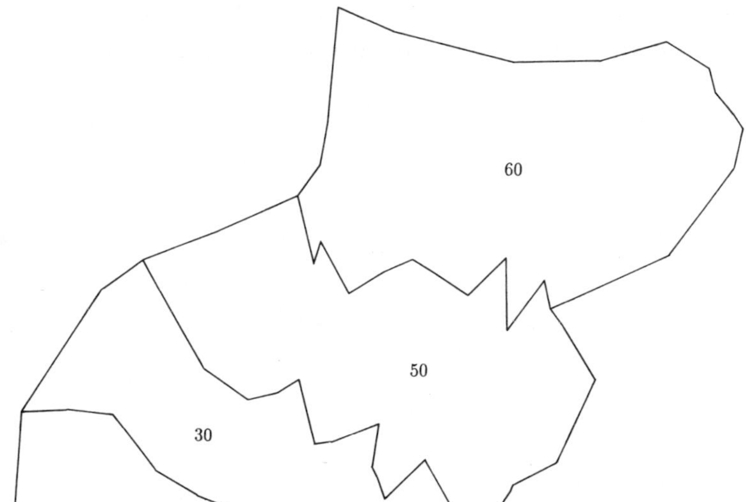

Figure 2.8: Hydraulic conductivity values (in m/day)

The results were partly unsatisfactory in the sense that when compared to the measured groundwater levels the gradient of the computed groundwater head was too shallow in the vicinity of the proposed depot site, whereas it was too steep in the north. This motivated lowering the hydraulic conductivity in the south and raising it in the north so that the computed groundwater level gradients were in closer agreement with the measurements. The final distribution of values so introduced in the network is shown in Figure 2.8.

Although through the adjustment of the hydraulic conductivity values, the groundwater gradients were in better agreement with measurements, on the whole the groundwater potentials were still considered to be too high. One factor contributing to this was identified to be the resistivity values in the vicinity of the large rivers. Because under the rivers the Holocene layer will be thinner than in the environs, the resistivity could be lower here than elsewhere. So, here the values of C were decreased. By further playing with the values of C in the whole domain, the computed groundwater levels were gradually brought into increasingly better agreement with the field data. In Figure 2.9 the finally-used resistivity values are displayed. The corresponding computed and measured groundwater levels are depicted in Figure 2.10.

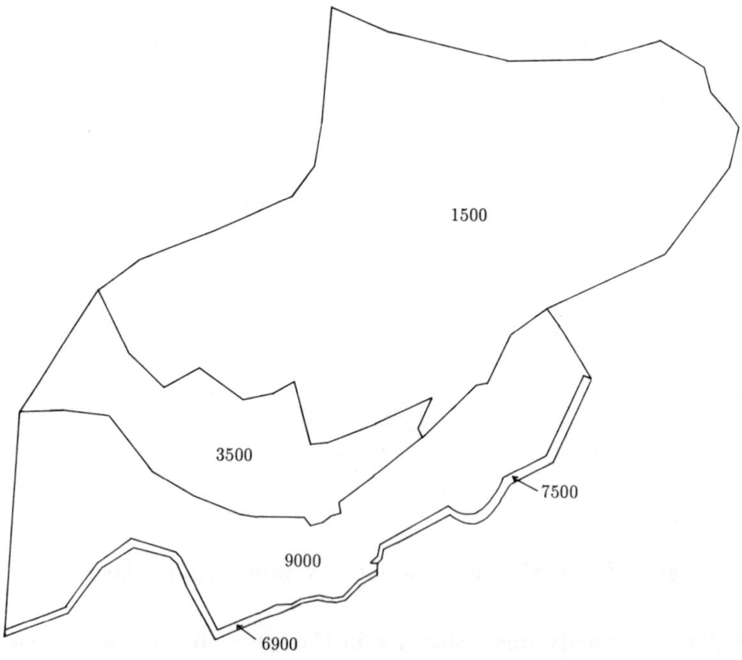

Figure 2.9: Resistivity values (in days)

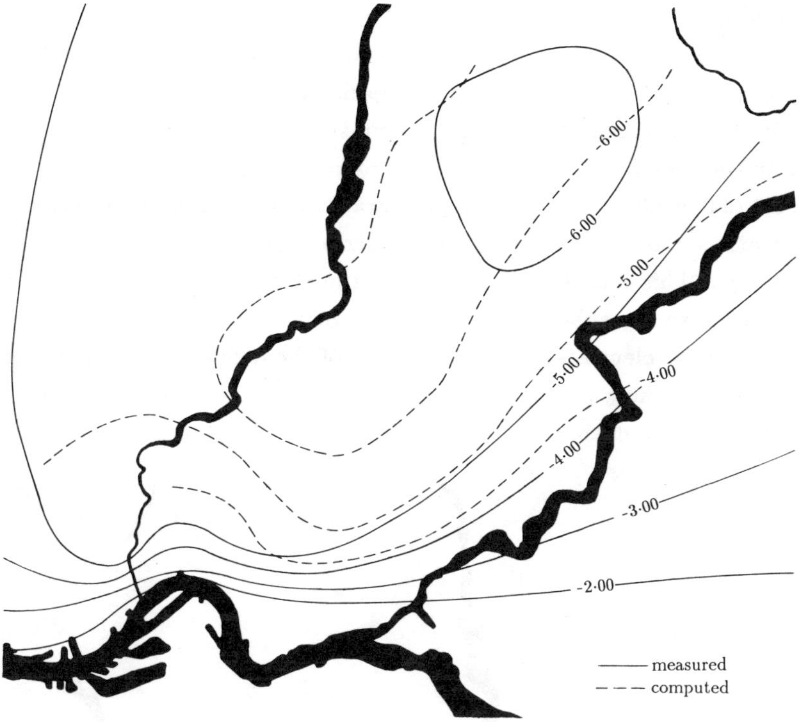

Figure 2.10: Computed and measured groundwater levels in the present situation (in m above mean sea level)

The discerning reader will notice some discrepancy between the computed and measured isohypsen (i.e. - contours of equal groundwater heads) in Figure 2.10. This could have been improved by further adjustment of the K and C values. This was not done though because it was considered irresponsible. The values already in use in Figures 2.8 and 2.9 cover the entire range of values indicated by the field studies. It was felt that taking any lower or higher values would be pushing the limits to values which were physically unrealistic.

We note that in general by playing with models it is often possible to obtain almost perfect results at the expense of plausibility. The art is in recognizing where these limits are. This is where not only mathematical knowledge, but also geohydrological insight is indispensable in practical groundwater modelling.

The main remaining step was now prediction of the change in groundwater flow pattern due to the construction of the depot. Firstly, this involved raising the specified groundwater potential in the overburden above the depot site in accordance with the expected rise in phreatic surface. As the phreatic surface would be raised though, so also would the resistivity of the intervening layer be raised by the presence of the silt. With a representative hydraulic conductivity of the silt of 0·00025 m/day and a representative thickness of the silt layer under the phreatic surface of 2 m, this gave an increase in the resistivity of 2/0·00025 = 8000 days. This was therefore added to the existing resistivity at the depot site. This led to the substitution of 9000 + 8000 = 17000 days for the values of C in the finite elements placed at the site of the depot. With these changes, the groundwater flow was again computed. The predicted groundwater levels in the principal aquifer after construction of the depot are shown in Figure 2.11.

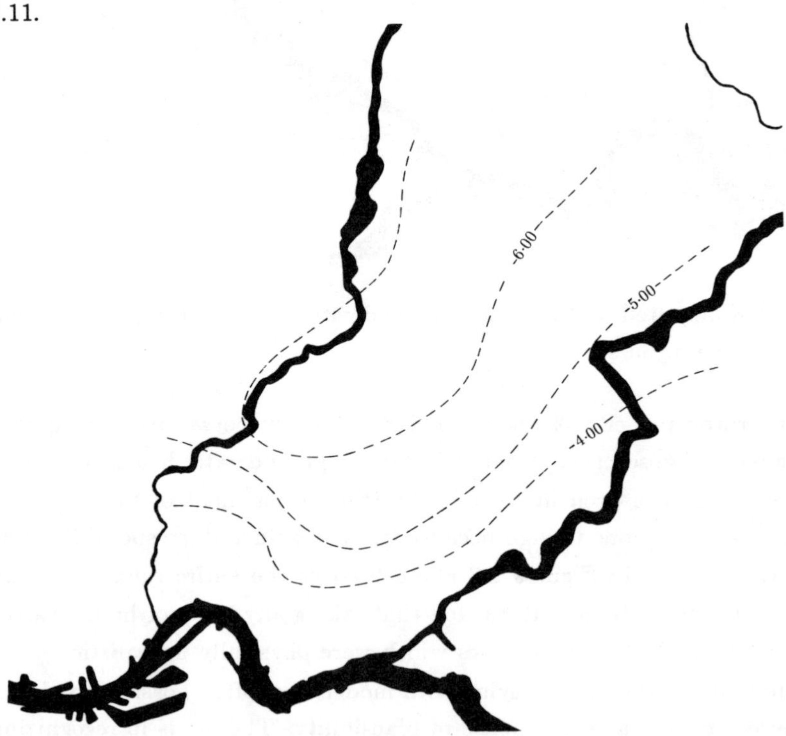

Figure 2.11: Predicted groundwater levels in the proposed situation (in m above mean sea level)

The first thing that we observe is that at a distance of more than 2 km from the proposed depot site, the change in groundwater potential is insignificant. This fortunately confirms that our truncation of the study area at a distance of 8 km from the sight was acceptable. Had this not been the case we should have had to design a new finite element network and start all over again (cf. the comments at the end of the previous subsection).

Besides drawing the conclusion that the influence of the depot will not be noticeable in groundwater levels at moderate distances from the depot, we can also conclude that the change at the depot site itself will not be too great. To be precise, a rise in groundwater level in the main aquifer of a mere 0·25 m is predicted.

To be on the safe side, a further prediction was done with a rather pessimistic estimation of the uncertain resistivity of the overlying Holocene formation at the depot site. The new value taken for the formation was 5000 days, leading to a resistance of $5000 + 8000 = 13000$ days for the total formation above the main aquifer in the proposed new situation. This did not however lead to any remarkable alteration in the results.

Our conclusions are that the groundwater flow in the area would hardly be affected by the construction of the depot. The predicted flow pattern has been handed over to the geochemists for their calculations on the reactions of the contaminants transported from the depot in the subsurface.

Full details of the study can be found in [120].

In summary, we have seen above how a groundwater modelling study has been carried out, and how many of the steps mentioned in the previous subsection have been involved. The modelling process is a mixture of the use of hydrogeological data, schematization, assumption, and the application of mathematical techniques. Here and there practical considerations such as the availability of data and the accessibility to an existing computer program have influenced the choices. The whole process of coming to a satisfactory conclusion is clearly partly an art in which experience and hydrogeological insight are of paramount help.

2.3 General hints

In the previous section we have seen how a groundwater modelling study was carried out in practice. There are a number of lessons which can be drawn from this exercise which apply in general.

The first of these is that one should always keep a critical eye on one's work. A rather trivial example of where one should keep a critical eye on one's work is in model input. It is easy to make slip-ups. For example, a decimal point may occur in the wrong place. In user-friendly programs, simple slip-ups such as negative hydraulic conductivity values are checked automatically. However, it is as good as impossible to anticipate all the errors that an inexperienced user could make. In those programs with extensive data checking algorithms such as [118], the code for checking the data exceeds the code for performing the actual computations by as much as a factor of two. Mistakes such as a hydraulic conductivity being ten times too large are impossible to trace.

A further point where one should be especially critical is in whether or not the supposed schematization or assumptions are really justified. (Looking at the case study described in the previous section, one might rightly ask whether the discrepancy between computed and measured heads could not be due to an incorrect assumption that the clay and loam layers in the Pleistocene formation constitute an impermeable base for the flow. This is, perhaps, a feasible explanation.)

One should watch out to see that the results are plausible; for instance, that water levels do not rise where they can be expected to fall, that phreatic water levels do not rise above the ground surface, and that water balances are reasonable. In this respect, we cannot stress that geohydrological insight is important too much.

Mathematical models are no panacea for geohydrological problems. They are no more than a tool in groundwater resources evaluation, albeit that they are an extremely useful tool. One must not expect too much from groundwater modelling. It is important not to try to use models to answer questions for which they were not designed. A common beginners mistake is to build a model for computing groundwater heads, such as that described in the previous subsection and then try and use it to compute groundwater fluxes. This is especially difficult with the finite element method described in Subsection 1.3

because groundwater flux is not continuously defined over element boundaries. If one really wishes to compute water balances one should take the required accuracy into account from the beginning. (Mathematically, it can be shown that in deriving fluxes one loses an order of accuracy in the computations.)

The counterpart to the above remarks is that it is a good idea to tailor a model to suit the question in hand. One should make a conscious decision about the amount of detail and schematization required. The art is not to be bound to one's original decisions if these turn out to be unwarranted. (Unfortunately, this last aspect is often neglected in engineering practice. Under pressures of deadlines and budgets, it is often difficult to disregard days, weeks or even months of work and start again. The beginner is often unaware that building a groundwater model is labour intensive. Fortunately in engineering practice, experience can limit the number of cases in which faulty early decisions are made.)

2.4 A few tricks of the trade

In this subsection, we shall discuss a few tricks of the trade which can be used in groundwater modelling.

Simulate an average situation as a steady state

Instead of, for example, trying to simulate a hydrogeological situation which is subject to seasonal fluctuations, one could consider the yearly average situation as a steady-state situation. In this, in place of the fluctuating seasonal recharge the yearly average recharge can be fed in to the model. Instead of a fluctuating groundwater level, one then finds the yearly average head. The great advantage of this trick is that it removes all time-dependence in the problem, so that one no longer needs information on storage parameters.

EXERCISE. Take the equation (1.11) for flow in a confined or semi-confined aquifer in a domain $(x, y) \in \Omega$ over a time period $0 \leq t \leq \tau$ with $\tau > 0$. Define the average head $\bar{h}$ at a point (x, y) by

$$\bar{h}(x, y) = \tfrac{1}{\tau} \int_0^\tau h(x, y, t) dt$$

and the average recharges

$$\bar{W} = \frac{1}{\tau}\int_0^\tau W(x,y,t)dt, \quad \bar{Q}_\varsigma = \frac{1}{\tau}\int_0^\tau Q_\varsigma(x,y,t)dt$$

and

$$\bar{Q}_\eta = \frac{1}{\tau}\int_0^\tau Q_\eta(x,y,t)dt.$$

Show that if

$$h(x,y,0) = h(x,y,\tau) \text{ for all } (x,y) \in \Omega$$

then $\bar{h}$ satisfies equation (1.10) with W, Q_ς and Q_η replaced by $\bar{W}$, $\bar{Q}_\varsigma$ and $\bar{Q}_\eta$ respectively, and with all time-dependence removed.

This exercise shows that the trick is mathematically justified for a confined or semi-confined aquifer. For an unconfined aquifer the trick is not mathematically justifiable, but is still a reasonable ploy.

EXERCISE. Determine under which conditions the previous exercise can be extended to an unconfined aquifer.

Even in cases where conditions cannot in any sense be considered as steady-state, it is often handy still to compute under steady conditions as a 'worst possible case'.

Apply superposition if possible

The principle of superposition is well-known for linear differential equations. It therefore applies to the equation (1.11) for flow in a confined or semi-confined aquifer. This can therefore be fruitfully applied where appropriate. Basically, it means that in a predictive situation instead of simulating absolute groundwater levels, one can suffice with the computation of the change that will occur. This is an especially useful simplification in time-dependent simulations where the studied change occurs in the boundary conditions.

Besides computational advantage, superposition may also be applied to improve accuracy. Suppose for example (as in the case study in Subsection 2.2) that the first computations are considered accurate enough not to warrant further calibration, but are not wholly in agreement with measurement. Then instead of trying to predict the exact groundwater levels, one can compute the change in the groundwater levels in the future situation. The computed change can then be superimposed upon the current measured groundwater levels. It is to be expected that this leads to better results than would obtained with direct prediction.

Estimating recharge by feeding in the answer

This trick can best be explained by an example. Imagine an aquifer in an alluvial basin where a large but unknown amount of recharge is derived from surface or subsurface flow from surrounding hills. The task is to predict the effect of some change in groundwater withdrawal pattern in the aquifer. In the first instance, the imposition of known flux or source conditions to cover the natural recharge is excluded because the latter is unknown. On the other hand, the alternative of imposing known groundwater heads on the boundary of the aquifer cannot be used for the prediction, for it is precisely the groundwater levels which can be expected to change in the future as a result of the altered withdrawal pattern. The ruse is to impose the present known groundwater elevations for a simulation of the present circumstances. If it is necessary this model can be calibrated. Using the results of this simulation the recharge which is compatible with these results can be calculated. This gives no less than the recharge which has to be fed into the model to simulate the present conditions. Assuming that recharge does not change as a consequence of the altered withdrawal pattern, one can then use these data as the input for the predictive model.

For a transient simulation obtain good dynamic initial conditions

Transient groundwater studies are considerably more difficult than steady-state studies. This is one of the reasons why we considered a steady-state study in Subsection 2.2, and why the first 'trick' mentioned in this subsection is so important. The extra difficulty is partly technical in the sense that one has to solve a time-dependent extension of the steady-state equation. It is also practically more difficult, because one needs more data. One needs estimates of storage coefficients and one needs measurements of groundwater levels at different time intervals to calibrate or verify the model. In practice it is very difficult to calibrate such models in such a way that they yield good results at all times and places.

In attempting a successful transient simulation it is very important that the initial conditions be realistic. Rushton and Wedderburn [102,103] have investigated the effect of different choices of the initial data on the results of a transient simulation. One should not put too much weight on their conclusions in absolute terms, but they are instructive. The first alternative they investigated was imposing initial conditions which were computed by carrying out a steady-state stimulation with boundary conditions fixed as the transient boundary conditions at the initial time. This is an attractive option, because it can be automated. Their conclusion however was that this alternative gave such disappointing results as not to be worthy of serious consideration. It was even worse that the second alternative they considered. This was to impose a representative fixed uniform value for the initial groundwater elevation. They derived the rule of thumb, that the effects of the error in this assumed initial data was negligible after a time

$$t > 2{\cdot}5 \int_{\Omega} (T/S)dxdy.$$

Here, T denotes the transmissivity and S the storage coefficient in the aquifer with domain Ω. Their last alternative was to compute the initial data as a steady-state under time-averaged conditions (as suggested as the first trick of the trade in this subsection). Here, they observed that the transient simulation was acceptable after an elapsed time

$$t > \int_{\Omega} (T/S)dxdy.$$

In practice, if one is interested is seasonal fluctuations under a supposed meteorological cycle, it is handy to run the simulation for a number of successive cycles before analysing the results. In this way any effects of an initial bad guess can be ironed out, and one works with dynamic initial conditions as it were.

Conduct a sensitivity analysis

Play with a model. If one is unsure about a certain aspect of a model one can experiment with it. For instance, one can try different schematizations or network configurations. Perhaps more relevantly, one can experiment with data values of which one is unsure. Let us for example consider a model for which one has only scarce information on the hydraulic conductivity. One can begin by putting the best available estimate in the model. Then one can increase the hydraulic conductivity by a factor of two say. If this has little effect on the model results, one could quite happily accept the first estimate. On the other hand, if the results change drastically, one knows that for a reliable model one needs more information on the hydraulic conductivity. This is one of the reasons why it is important that in a geohydrological study with modelling, the modellers should be involved as early as possible. In an early stage it is possible to identify areas where it is important to collect more data or even where there is superfluous data. This is of course far better than coming to the conclusion that there is insufficient data to calibrate a model after the field work has been completed.

The ideal modeller would conduct a sensitivity analysis with every parameter in his model, and the ideal model would not be sensitive to more than reasonable changes within the uncertainties of the model. In practice analysing sensitivity to every parameter is too expensive both in time and man-power. Notwithstanding, by conducting a sensitivity analysis for one or more aspects of a model one can build up confidence in the model, one can identify its weak points. Both are useful sets of information.

3 Unsaturated groundwater flow modelling - general

3.1 Importance of the unsaturated groundwater flow zone

The unsaturated groundwater flow zone fulfils a key role in the hydrological cycle. It is the buffer between the atmosphere and the saturated groundwater flow regime [69]. For people concerned with agriculture, the unsaturated groundwater zone is the most critical component of the hydrological cycle, since it is from this zone that plants extract their water requirements. The state of the unsaturated zone is therefore important for the development of plants. If these people talk about losses from the unsaturated zone, they mean water which infiltrates into the deeper underground and is consequently no longer available for plant use. On the other hand, people concerned with drinking water supply are concerned about the unsaturated zone as the medium through which groundwater replenishment takes place. If these people talk about losses from the unsaturated zone they mean the water evaporated from the surface or transpired by plants which cannot then be transported to the exploitable saturated groundwater reservoirs.

For the understanding of the possibilities for mathematically modelling the unsaturated groundwater zone, it is a good idea to look at the essential question to be addressed. This is the question of constructing a suitable water-balance. With this, we shall consider models for the unsaturated zone at a given point ((x, y) in the notation of Section 1). This is largely for convenience. It is also related to the overriding consensus that for the purposes of hydrological modelling on a regional basis, unsaturated groundwater flow may be considered as an essentially one-dimensional vertical phenomenon [2,10,19,58,69,76,93]. Some shortcomings of this approach have been pointed out in [111].

The following factors are important to the construction of a water-balance for the unsaturated groundwater flow zone at any given point. Here the dimensions reflect the choice of one-dimensional modelling. Herewith:

N - precipitation [dim. L/T]

A - actual evapotranspiration [dim. L/T]

P - potential evapotranspiration [dim. L/T]

P_0 - open-water evaporation [dim. L/T]
Q - exchange with saturated groundwater zone [dim. L/T]
Θ - storage in the unsaturated zone [dim. L]
h - elevation of the water-table [dim. L]
d - thickness of the root zone [dim. L].

These factors require some further explanation.

Precipitation N is the rainfall, hail or snow which falls at the point. This is generally data supplied by the meteorological office in the terms of mm/day.

Actual evapotranspiration, A, is the cumulative name for the actual moisture transpiration by the plants and evaporation from the surface of the soil. These two are lumped together because in measurement in practice it is generally impossible to distinguish between the separate contributions. Potential evapotranspiration, P, is a hypothesized evapotranspiration which would have taken place from the soil under conditions of sufficient water supply. The basic thought behind this concept is that vegetation and other surface conditions do not influence atmospheric conditions. Temperature, sun radiation, and atmospheric moisture content determine the amount of moisture which a plant must transpire and which should evaporate from the surface. The actual transpiration occurs at the potential rate as long as the subsurface contains sufficient moisture. If the subsurface does not, the actual evapotranspiration is less than the potential evapotranspiration and in this case the plants are damaged. Conflicts around damage to agriculture subsequently often revolve around the question of whether the evapotranspiration was potential or not. When the actual evapotranspiration is less than potential, the discrepancy is sometimes referred to as the ***reduction in evapotranspiration.*** Strictly speaking the concept of potential evapotranspiration is not absolute [84]. One can imagine that the more plants transpire the greater the moisture content of the atmosphere. Thus the taxation of atmospheric conditions on the plants can be influenced. Notwithstanding, it is an extremely useful concept for modelling purposes. In this respect we mention the concept of open-water evaporation, P_0. This is the amount of water which would evaporate from an open water surface, under identical meteorological conditions. This can be measured by specially designed apparatuses known as evaporation pans.

The symbol Q denotes the exchange of the saturated groundwater zone with the unsaturated zone. In a coupling with models for saturated groundwater flow as described in Section 1 this is identical to the term Q_ς in equation (1.14).

The symbol Θ denotes the total amount of water stored in the unsaturated zone at any instant.

The objective of modelling the unsaturated groundwater zone is now to determine the interrelation between and thereby to quantify the factors described so far. The elevation of the water-table h and the depth d to which the plant roots penetrate the surface at the given point play a role in this. Strictly speaking, it is not so much the elevation of the water-table which is important, but its depth below the land surface. Since though we are only going to consider models at one point, without loss of generality we may take the land surface as being at zero (0) elevation. In this case, the thickness of the unsaturated zone is $0 - h = |h|$.

3.2 Classification of the different approaches

In principle, although there is considerable overlap, models for unsaturated groundwater flow in a vertical profile may be classified into five main categories, viz:

- empirical models,
- reservoir models,
- schematic solutions of the differential equation of unsaturated flow,
- numerical solutions of the differential equation of unsaturated flow,
- 'reproduction' models.

These categories will be reviewed in turn below. The references are by no means complete, but the intention is to indicate the various possibilities.

3.3 Empirical models

On the fringe of empirical techniques for mathematical modelling of the unsaturated zone we mention ***personal experience***. The farmer who has tilled a piece of ground for many years will have a good idea of how much he should irrigate by looking at the conditions in the field. In practice, it often turns out to be difficult for mathematical models to compete with such experience.

Another type of model with which mathematical models have to compete is the *lysimeter* [31]. This is a large box, generally constructed of concrete, with a horizontal dimension of 4×4 m say, which is closed in at a depth of approximately 2 m below the ground. The box is then filled with soil, which in view of the size of the box, can be cultivated as best as possible near to natural conditions. Under the box is an outlet through which natural discharge can be recorded. By placing a rain-gauge next to the box, it is possible to determine how much water enters the lysimeter. Any discrepancy between the calculated inflow and measured discharge can be attributed to evapotranspiration. In essence, the mathematical model is

$$A = N - Q. \tag{3.1}$$

Here, the effects of interception of precipitation by plants and its eventual evaporation are automatically included in the evapotranspiration. Furthermore, storage effects in the soil are neglected. However, if one looks at (3.1) averaged over a long period, the latter objection becomes insignificant.

By placing an evaporation pan next to the lysimeter, one can also determine what is referred to as the *crop factor*. If we denote this by c_f it is the empirical unknown in the relation

$$A = c_f P_0.$$

The Food and Agricultural Organization of the United Nations has compiled reference tables for the crop factors of a large number of crops and different types of vegetation under diverse climatic conditions. These are a useful medium for estimating the water use of crops in circumstances where no sophisticated apparatus is available, but where open water evaporation measurements are known.

For saturated groundwater modelling it is often not necessary to model the unsaturated zone in detail. It can be enough to begin with to assume that the saturated groundwater recharge is the difference between the precipitation and the potential evapotranspiration. The only mechanism which needs to be added is one which accounts for diminished evapotranspiration when the subsurface becomes dry, whereby the latter is coupled with the water-table level. One such model in widespread use in that from the United States Geological Survey [80,96,102,110] which is (3.1) where

$$A = PF(|h|)$$

and F is the cut-off function

$$F(\nu) = \begin{cases} 1 & \text{if} \quad \nu \le \nu_1 \\ (\nu_2 - \nu)/(\nu_2 - \nu_1) & \text{if} \quad \nu_1 < \nu \le \nu_2. \\ 0 & \text{if} \quad \nu > \nu_2 \end{cases} \tag{3.2}$$

Here, ν_1 and ν_2 are critical depths of the water-table elevation. Furthermore, P can be determine from open-water evaporation P_0 with the use of crop factors, for instance.

This last model can be refined in a number of ways. A few techniques in circulation are the following:

- Impose some physical constraints.
  For example, constrain the maximal value of Q by the assumed saturated hydraulic conductivity of the soil.

- Introduce the effect of waterlogging.
  The refinement here is based upon the observation that not only does the actual water use of a crop decrease as a result of drying-out of the soil, but also if the soil becomes too wet then plant roots may die and then also the actual evapotranspiration will decrease. This can be realized by a simple modification of the function in (3.2).

- Model effects of natural delay.
  One of the primary objections to (3.1) is that it does not really take the time-buffer effects of the unsaturated zone into account. This objection can be partially swayed by using the formula

  $$Q(t) = Q(0) + \alpha \int_0^t \{N(\tau) - A(\tau)\} \exp\{\alpha(\tau - t)\} d\tau$$

  in place of (3.1). Here α is a positive parameter. Under conditions of constant $N - A$ it yields

  $$Q(t) \to Q(\infty) := N - A \quad \text{as} \quad t \to \infty,$$

  and

  $$Q(t) = Q(0)\exp(-\alpha t) + Q(\infty)\{1 - \exp(-\alpha t)\}.$$

As α is an arbitrary constant, it could be weighted to account for a deep water-table, e.g.: $\alpha = K_s/|h|^2$ where K_s denotes the assumed saturated hydraulic conductivity of the soil.

– Combine some of the above.

3.4 Reservoir models

Conceptually reservoir models are the physically most simple. This is one of their most attractive features. It is easy to understand the thought behind them. They are also easy to control.

In the crudest form of a reservoir model, the unsaturated soil-moisture zone is considered as a single hydrostatic reservoir. In more sophisticated cases, the unsaturated groundwater profile is divided into a column of reservoirs situated one above the other as shown in Figure 3.1.

We suppose that there are I reservoirs and use a subscript i to denote an index for the reservoirs beginning with 1 at the top. Each reservoir can accommodate a minimum storage Θ_i^{min} and a maximum storage Θ_i^{max}, $i = 1, 2, \ldots, I$,. We begin at a time t when each of the reservoirs contains a storage $\Theta_i(t)$, $i = 1, 2, \ldots, I$. We consider a time-step Δt during which we assume that the precipitation N and potential evapotranspiration P can be represented by a constant value.

Now, let Θ_1^* denote the storage which would be found in the first reservoir at the new time $t + \Delta t$ if this reservoir were closed and there were no further constraints, i.e.

$$\Theta_1^* = \Theta_1(t) + \Delta t(N - P).$$

If $\Theta_1^{min} \le \Theta_1^* \le \Theta_1^{max}$ then we define $\Theta_1(t + \Delta t) = \Theta_1^*$, set the actual evapotranspiration for the time-step $A = P$, and define the flow to the second compartment $Q_1 = 0$. If though there is insufficient storage to permit this, i.e. $\Theta_1^* < \Theta_1^{min}$, then we define $\Theta_1(t + \Delta t) = \Theta_1^{min}$, we compute a reduction in the actual evapotranspiration by $A = N + \{\Theta_1(t) - \Theta_1^{min}\}/\Delta t < P$, and again give the flow to the second compartment is $Q_1 = 0$. Finally, if $\Theta_1^* > \Theta_1^{max}$ then we define $\Theta_1(t + \Delta t) = \Theta_1^{max}, A = P$ and can compute a flow to the next compartment by $Q_1 = N - P - \{\Theta_1^{max} - \Theta_1(t)\}/\Delta t$. The process is then repeated for the second compartment. Setting

$$\Theta_2^* = \Theta_2(t) + Q_1 \Delta t,$$

we define

$$\Theta_2(t+\Delta t) = \Theta_2^* \text{ and } Q_2 = 0 \text{ if } \Theta_2^* \leq \Theta_2^{max}$$

and

$$\Theta_2(t+\Delta t) = \Theta_2^{max} \text{ and } Q_2 = Q_1 - \{\Theta_2^{max} - \Theta_2(t)\}/\Delta t \text{ if } \Theta_2^* > \Theta_2^{max}.$$

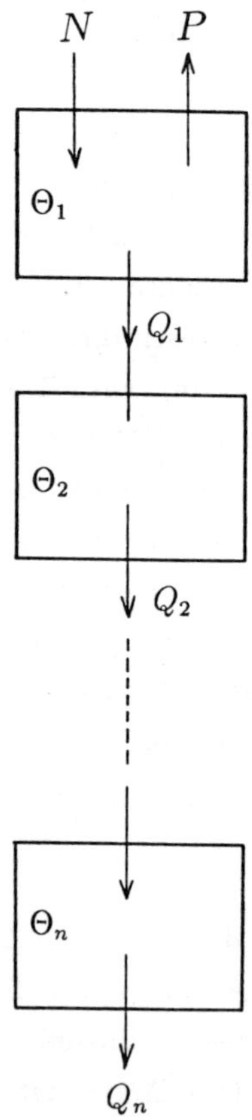

Figure 3.1: Reservoir model

Here Q_2 denotes a flow from the second to the third compartment. The last described process is repeated for all the compartments. The finally computed flow Q_I denotes the contribution to the saturated groundwater zone.

The major disadvantage of the scheme presented above is that it is incapable of predicting capillary rise. Water may only flow downwards from one reservoir to the next.

The method may be modified for example so that only precipitation is allocated to the uppermost layer and evapotranspiration occurs from the lower reservoirs. A few further adaptations which can be applied are the following:

- Allocate evapotranspiration to more than one reservoir, for instance with a weight related to the depth below the surface signified by the index i [107].

- Allocate evapotranspiration to more than one reservoir with a weight related to the moisture content in each reservoir [46].

- Define a capillary rise in the lowest reservoir Q_I as a function of the depth to the water-table $|h|$ [107].

- Define the maximum storage Θ_i^{max} in each reservoir as a function of the depth to water-table $|h|$ [107].

- Define the relationship between A and P as a function of Θ_1 which can be calibrated from field measurements [29].

- Relate the flow from one reservoir to the next Q_i to the storages in both reservoirs [46].

For the last alternative a sort of resistance relationship could be given. This can be physically founded by analogy to Darcy's law [46]. At this level of sophistication though, the reservoir models begin to metamorphose into models based upon discretization of a differential equation describing unsaturated groundwater flow. These are the subject of the next subsection but one.

We mention that a number of reservoir models have been well and successfully used [3,23]. An outstanding example is the Stanford Watershed Model [31], named after the university where it was developed, in which the whole hydrological cycle is modelled as a system of connected reservoirs.

3.5 Schematic solutions of the differential equation of unsaturated groundwater flow

Before we can talk about methods for modelling unsaturated groundwater flow based on the differential equation of unsaturated groundwater flow, it is necessary to say what we mean by the differential equation. The starting point is equation (1.3), which since we are only considering vertical flow, we may write as

$$\frac{\partial \theta}{\partial t} = \frac{\partial}{\partial z}\left(K\frac{\partial \Phi}{\partial z}\right) - w \tag{3.3}$$

where z denotes elevation. This is increasing upwards. For the rest, θ denotes moisture content, t time, K hydraulic conductivity, Φ potential in the form

$$\Phi = p + z \tag{3.4}$$

with p denoting water pressure head, and w water abstraction. Substituting (3.4) in (3.3) yields

$$\frac{\partial \theta}{\partial t} = \frac{\partial}{\partial z}\left(K\frac{\partial p}{\partial z} + K\right) - w. \tag{3.5}$$

In unsaturated groundwater flow, capillary suction retains moisture in the pores in the ground. For this reason, the pressure head is negative. Recall, that we have defined the transition from saturated to unsaturated flow as being given by $p = 0$. In magnitude the pressure head is equated with the terms *soil-moisture tension* or *soil-moisture suction.*

As the moisture content in a soil progressively increases it is generally conceived [30,108] that the smaller pores in which the capillary forces are the greatest are filled before the larger pores. Hence, the pressure head p in an increasing function of the soil-moisture content θ. Contemporaneously, since the larger pores in the soil are the most effective conductors, the hydraulic conductivity increases as the soil-moisture content increases and does so at a progressively greater rate. Consequently, hydraulic conductivity K can be viewed as an increasing function of θ. In practice these relations are subject to hysteresis. Neglecting hysteresis, typical examples of the relations between p and θ, and between K and θ as found in the literature [18,19,20,30,92,108] are shown in Figures 3.2 and 3.3.

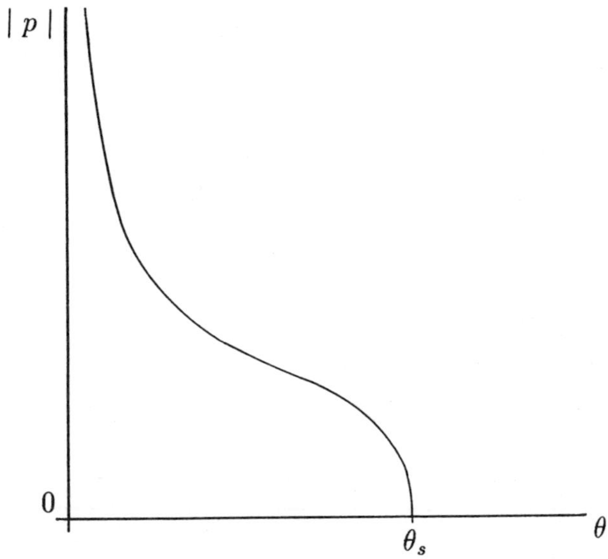

Figure 3.2: Typical $p - \theta$ relationship

Assuming that the soil is homogeneous and hysteresis is neglected, and thus that p is well-defined as a function of θ we may write (3.5) as

$$\frac{\partial \theta}{\partial t} = \frac{\partial}{\partial z}\left\{D(\theta)\frac{\partial \theta}{\partial z} + K(\theta)\right\} - w \tag{3.6}$$

where

$$D(\theta) = K(\theta)\frac{dp}{d\theta}(\theta)$$

is termed the ***soil-moisture diffusivity*** [18,20,30,92,108]. On the other hand assuming that at every point z in the medium we can write θ as a function of p and z, (3.5) may be formulated as

$$c(p,z)\frac{\partial p}{\partial t} = \frac{\partial}{\partial z}\left\{K(p,z)\frac{\partial p}{\partial z} + K(p,z)\right\} - w \tag{3.7}$$

where

$$c(p,z) = \frac{\partial \theta}{\partial p}(p,z).$$

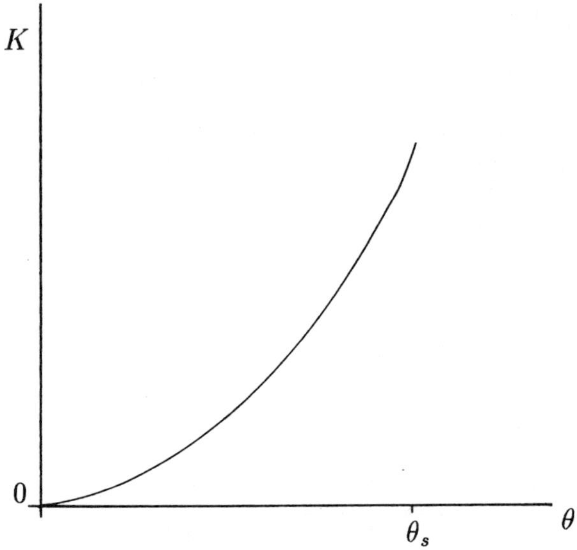

Figure 3.3: Typical $K - \theta$ relationship

The equations (3.6) and (3.7) are generally known as the *Richards' equation* [47,100,108], either formulated in terms of moisture content ((3.6)) or pressure head ((3.7)).

We remark that mathematically there is something to be said for both formulations. For instance, equation (3.6) cannot be used to describe the transition to saturated flow in a rigid soil. For then, in the saturated regime, the moisture content remains constant as pressure head increases and in this zone equation (3.6) becomes nothing more than the statement "$0 = 0$". On the other hand, equation (3.7) is undesirable for describing the transition from a wet to a dry soil, for then p becomes infinite in magnitude. Cf. Figure 3.2. We remark too that for describing the flow in a heterogeneous soil it is essential to use the formulation (3.7). This is evidently the case in layered soils where the moisture content θ may not even be a continuous function of z. This can be seen in Figure 3.4, where stationary moisture profiles have been computed in the indicated layered soil.

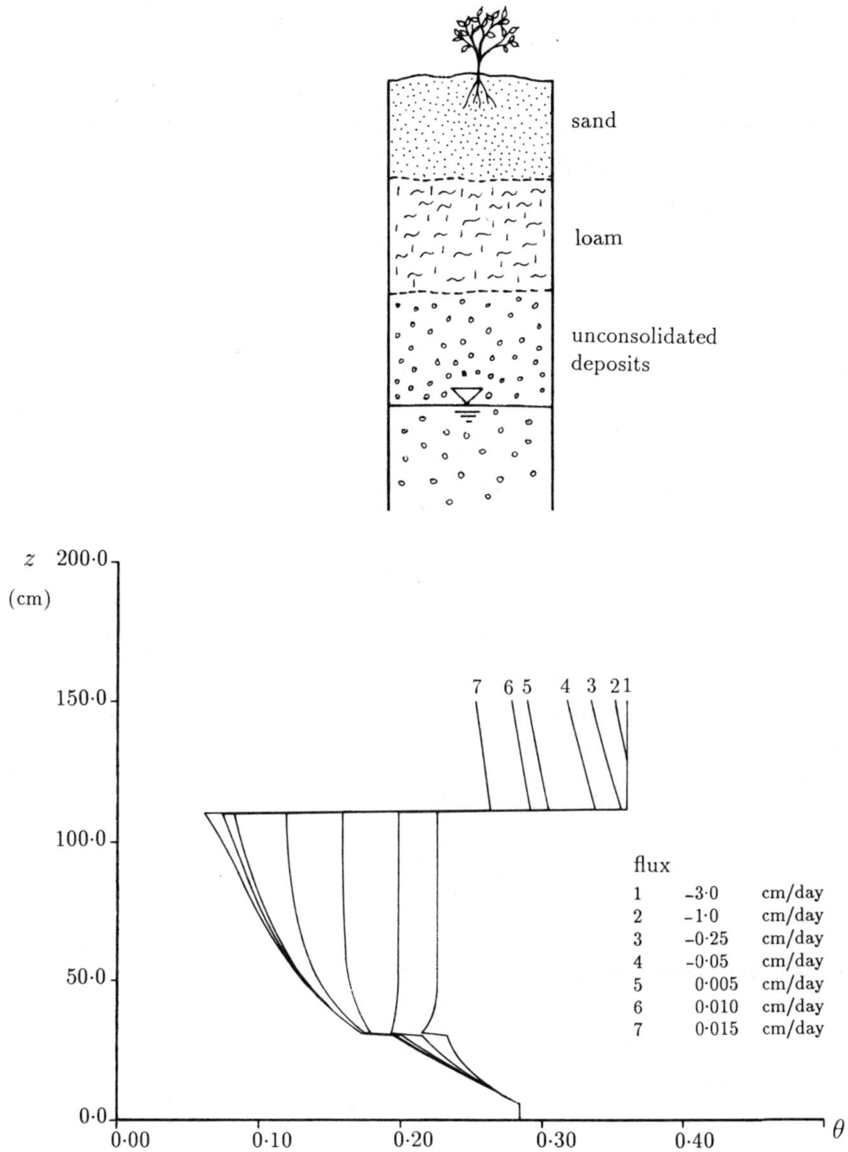

Figure 3.4: Stationary moisture profile in a layered soil

In the present context, equations (3.6) and (3.7) are valid for values of the elevation between the water-table and the surface elevation. Thus for

$$h(t) < z < 0,$$

where this domain can change with time. Since the lower boundary denotes the water-table the appropriate boundary condition here is

$$p(z,t) = 0 \quad \text{for} \quad z = h(t), \tag{3.8}$$

or for a homogeneous soil

$$\theta(z,0) = \theta_s \quad \text{for} \quad z = h(t) \tag{3.9}$$

where θ_s denotes the moisture content at saturation.

The appropriate upper surface boundary condition is a little more complex [43,58,86,91]. Transpiration by the plants will result in water-uptake by the plants and since this will only occur through roots in the soil, we can consider this as occurring through the sink term w in equations (3.6) and (3.7). We therefore only have to consider the component of the potential evapotranspiration which can be ascribed to evaporation of moisture from the soil surface for formulating the boundary condition at the upper surface. We suppose that this surface evaporation may also be considered as a potential flux and denote it by E. The other component which plays a role is the precipitation, N. Under ideal conditions the boundary condition at the surface is then

$$q_z(z,t) = E(t) - N(t) \quad \text{for} \quad z = 0,$$

where

$$q_z = -K\frac{\partial \Phi}{\partial z} = -\left(K\frac{\partial p}{\partial z} + K\right)$$

denotes the specific discharge as defined in Section 1. At the surface this discharge is positive if water is leaving the soil and negative if it is entering.

There are however two constraints on this boundary condition [49,58,86,91]. Firstly, if the soil becomes desiccated then moisture can no longer be transferred to the atmosphere. This constraint is realized by imposing a minimum admissible pressure head p_d. If this level is reached then less moisture is transferred to the atmosphere than is demanded by the potential evaporation possibly corrected for the precipitation. The value of the minimum pressure head can be determined physically by assuming that at this value the pressure head is in equilibrium with the atmosphere [45,86]. Often in practice [42,72] a value of $p_d = 160000$ mm is taken (for the initiated pF=4·2). The second constraint is that if the soil becomes saturated at the surface, the infiltration capacity is limited and water ponds on the surface. This is realized by imposing a maximal possible pressure head $p_s = 0$ at the surface. If this value is attained less water infiltrates than is given by the precipitation possibly corrected for the evaporation. In models for the complete hydrological cycle, the water which cannot infiltrate is subsequently routed to surface runoff or considered to pond on the surface until it can infiltrate. Incidentally, this constraint of a maximal pressure head is also necessary to violate our basic assumption that there is a well-defined unsaturated zone above a saturated regime.

In mathematical terms this leads to the surface boundary condition [49,58]

$$\begin{cases} -\left(K\frac{\partial p}{\partial z}+K\right)(z,t) = E-N & \text{and} \quad p_d \le p(z,t) \le 0 \\ -\left(K\frac{\partial p}{\partial z}+K\right)(z,t) < E-N & \text{and} \quad p(z,t) = p_d \\ -(K\frac{\partial p}{\partial z}+K)(z,t) > E-N & \text{and} \quad p(z,t) = 0 \end{cases} \tag{3.10}$$

for $z = 0$ for equation (3.7). For equation (3.6) it becomes

$$\begin{cases} -(D\frac{\partial \theta}{\partial z}+K)(z,t) = E-N & \text{and} \quad \theta_d \le \theta(z,t) \le \theta_s \\ -(D\frac{\partial \theta}{\partial z}+K)(z,t) < E-N & \text{and} \quad \theta(z,t) = \theta_d \\ -(D\frac{\partial \theta}{\partial z}+K)(z,t) > E-N & \text{and} \quad \theta(z,t) = \theta_s \end{cases} \tag{3.11}$$

where θ_d and θ_s denote respectively appropriate minimal and maximal moisture contents.

For coupling with saturated groundwater flow, once equation (3.7) for $0 < z < h(t)$ with boundary conditions (3.8) and (3.10) has been solved, the outflow to the saturated zone is given by

$$Q = -(K\frac{\partial p}{\partial z}+K)(z,t) \quad \text{with} \quad z = h(t).$$

For equation (3.6) with boundary conditions (3.9) and (3.11) it is

$$Q = -(D\frac{\partial \theta}{\partial z}+K)(z,t) \quad \text{with} \quad z = h(t).$$

In all the models based on the schematic solution of equations (3.6) and (3.7) the term w in the equations is neglected. Correspondingly, in (3.10) and (3.11) the variable $E - N$ is replaced by $P - N$.

The first model which we consider in this class is that related to the Green-Ampt equations. These equations are primarily directed towards predicting surface ponding. They are based on the assumption that a steep moisture profile develops in a soil after a rainfall event. This travels downwards at the maximal possible rate [5,32]. The Green-Ampt models neglect any interaction with the water-table. They can be grouped in this class though, since although the equations were obtained through other considerations, it has since been shown [92] that they are equivalent to taking a diffusion coefficient $D(\theta) = D_0\,\delta(\theta - \theta_0)$ in (3.6), where δ denotes the Dirac-delta 'function' and D_0 and θ_0 parameters. A more sophisticated variant has been proposed in [82,83].

The second schematic way [71] of solving (3.6) is to develop a solution in the form

$$z = \Psi_1(\theta)t^{1/2} + \Psi_2(\theta)t + \Psi_3(\theta)t^{3/2} + \Psi_4(\theta)t^2 + \dots$$

The form of the functions $\{\Psi_i\}$ can be determined by substitution in the equation (3.6). The first term in this expansion is derived from the well-known Boltzmann transformation [92]. Eagleson [35-41] has developed this idea considerably for predicting yearly water budgets with a stochastic element.

The next most important class of models based on the schematic solution of equations (3.6) or (3.7) are the so-called 'quasi-stationary' models. The idea is to solve the steady-state version of equation (3.6) or (3.7) with time-dependent boundary conditions, the argument being that this leads to a simpler problem whilst giving a realistic treatment of the conditions in the soil and being able to take time-dependence into account. In the simplest form, this model works as follows.

The steady-state version of (3.7) neglecting abstraction is

$$0 = \frac{\partial}{\partial z}\left\{K(p)\frac{\partial p}{\partial z} + K(p)\right\} \quad \text{for} \quad h < z < 0. \tag{3.12}$$

In the first instance this equation is considered with the boundary conditions

$$p(h) = 0 \tag{3.13}$$

and

$$-\left\{K(p)\frac{\partial p}{\partial z} + K(p)\right\} = P - N. \tag{3.14}$$

Cf. (3.8) and (3.10). Integration of (3.12) subsequently and imposition of (3.14) gives

$$-\left\{K(p)\frac{\partial p}{\partial z}+K(p)\right\}=q$$

where

$$q=P-N.$$

Whence,

$$K(p)\frac{\partial p}{\partial z}=-\{q+K(p)\}.$$

Further integration and imposition of (3.13) then yields

$$z-h=\int_p^0\frac{K(\nu)}{q+K(\nu)}d\nu \quad \text{for} \quad h<z<0. \tag{3.15}$$

Now, equation (3.15) is assumed to describe the pressure head profile, as long as the pressure head at the surface is not less that the minimum level. This is the case, if

$$0-h\leq\int_{p_d}^0\frac{K(\nu)}{q+K(\nu)}d\nu. \tag{3.16}$$

If (3.16) is violated then q is chosen such that

$$-h=\int_{p_d}^0\frac{K(\nu)}{q+K(\nu)}d\nu$$

and then the actual evapotranspiration A is defined by

$$q=A-N<P-N.$$

This idea applied to equation (3.6) can be found in [91]. A further extension in which the role of water uptake by plants is modelled with more sophistication can be found in [42].

A drawback of the quasi-stationary approach as explained above is that the effect of any change in atmospheric conditions is immediately apparent in the entire computed moisture profile. Thus the buffer effect of the unsaturated zone above the saturated zone is lost, and the model overreacts to strong changes in surface boundary conditions. A model which was designed to overcome this objection is the model of de Laat [76]. This model became the subject of considerable attention in the Netherlands at the time of its development, because it was seen as providing the ideal compromise for practical applications - it reflected physical conditions and was computationally efficient. This has the following idea behind it.

Let $\tilde{\theta}$ denote the moisture content function at a point z in the soil when the pressure head is p. The superscript *tilde* is introduced to emphasize that in contrast to the variable θ defined earlier, here the moisture content is a function of p and z. In fact, to be explicit, we have

$$\theta(z,t) = \tilde{\theta}(p(z,t), z).$$

Then, with an explicit solution $p(z)$ of (3.15) we can define the total moisture deficit in the soil by

$$\Lambda = \int_h^0 \{\tilde{\theta}(0,z) - \tilde{\theta}(p(z),z)\}dz.$$

This is plainly dependent only on h and q, i.e.

$$\Lambda = \Lambda(q,h).$$

Similarly, supposing that the roots of the plants occupy a zone to a depth d in the soil we can define a deficit in this root zone by

$$\Lambda_r = \int_{-d}^0 \{\tilde{\theta}(0,z) - \tilde{\theta}(p(z),z)\}dz$$

and again

$$\Lambda_r = \Lambda_r(q,h).$$

To proceed from a time t to $t+\Delta t$ one then considers the water-balance for the root-zone

$$\Lambda_r(q^{t+\Delta t}, h^{t+\Delta t}) = \Delta_r^t + \Delta t(P - N - q^{t+\Delta t}) \tag{3.17}$$

and for the entire soil-moisture profile

$$\Lambda(q^{t+\Delta t}, h^{t+\Delta t}) = \Delta^t + \Delta t(P - N + Q). \tag{3.18}$$

Here superscripts denote the variables at the appropriate time-levels. The equations (3.17) and (3.18) have three unknowns, viz. $q^{t+\Delta t}$, $h^{t+\Delta t}$ and Q. The latter two are linked however through models of the underlying saturated zone. Cf. equation (1.14). Thus under most conditions (3.17) and (3.18) can be solved, and in this case the actual evapotranspiration $A = P$. If no solution can be found then one replaces P in (3.17) and (3.18) by $A < P$ so that a solution is possible. The great advantage of this approach is that the functions $\Lambda(q,h)$ and $\Lambda_r(q,h)$ have only to be calculated once and can be tabulated. The approach is also easily adaptable for layered soils.

EXERCISE. Assuming that hydraulic conductivity K and moisture content θ are a strictly increasing functions of pressure head p in a homogeneous soil, show that the functions $\Lambda(q, h)$ and $\Lambda_r(q, h)$ defined above are strictly increasing in both variables q and h.

Various other analytical solutions of equations (3.6) and (3.7) which can be obtained under restricted circumstances may also be used to form the basis of models for the unsaturated groundwater flow regime [25,71,78,79,113,114].

3.6 Numerical solution of the differential equation

The most detailed models of unsaturated soil-moisture flow for purposes of hydrological modelling are based on the numerical solution of equation (3.7) with (3.8) and (3.10) as boundary conditions. Alternatively, for a homogeneous soil, (3.6) with (3.9) and (3.11), or, the problem which can be obtained by applying the transformation known as the Kirchhoff transformation [71,99] can be used. Diverse finite difference and finite element techniques can be applied [2,4,43-45,58,71,86,93,99]. Furthermore, many simplifications can be made, see for instance [101] for a model based on a numerical scheme which is essentially a sophisticated reservoir model.

There is one facet of these computations though which we have not yet described and that is the realization of the plant abstraction function w and its role in determining the actual (evapo)transpiration.

The general approach is to set $E(t) = 0$ and to define

$$w(p, z, t) = \rho(z, t) w_0(p, t) \tag{3.19}$$

where $\rho(z, t)$ is a function to describe the distribution of plant roots in the subsurface [4,43-45,81]. A typical form is shown in Figure 3.5. The time-dependence comes from the fact that roots grow or whither in season.

A form for the function w_0 which has been widely used [44,45,81] is

$$w_0(p, t) = \alpha P(t) F(p)$$

where F is as shown in Figure 3.6. Between two critical values p_1 and p_2 the function takes the value 1. It drops linearly or logarithmically to the value zero as p decreases to p_d. The factor α is defined by

$$\alpha = 1/\int_{h(t)}^{0} \rho(z,t)dz.$$

What does this mean? It means that if the pressure head in the root zone is between the levels p_1 and p_2, the total water abstraction by the plants

$$\int_{h(t)}^{0} w(p,z,t)dz = \int_{h(t)}^{0} \rho(z,t)\alpha P(t)dz = P(t).$$

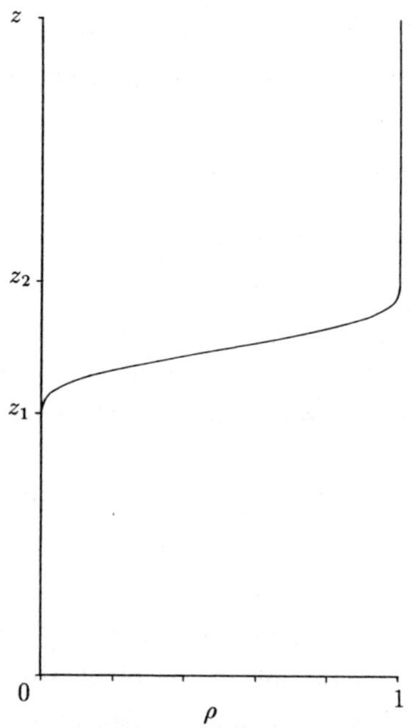

Figure 3.5: Root distribution function

So the plants transpire water at the potential rate. If though the pressure head drops too much, so that $p < p_2$, one gets

$$\begin{aligned} A(t) &= \int_{h(t)}^{0} w(p,z,t)dz = \int_{h(t)} \rho(z,t)\alpha P(t)F(p(z,t))dz \\ &< \int_{h(t)}^{0} \rho(z,t)\alpha P(t)dz = P(t). \end{aligned}$$

Often the value p_1 is set equal to zero. Introduction of the option $p_1 < 0$ allows one to take into account the fact that roots may become waterlogged if the soil is too wet, and thus for over-wet as well as over-dry soils the actual evapotranspiration may be less that the potential evapotranspiration.

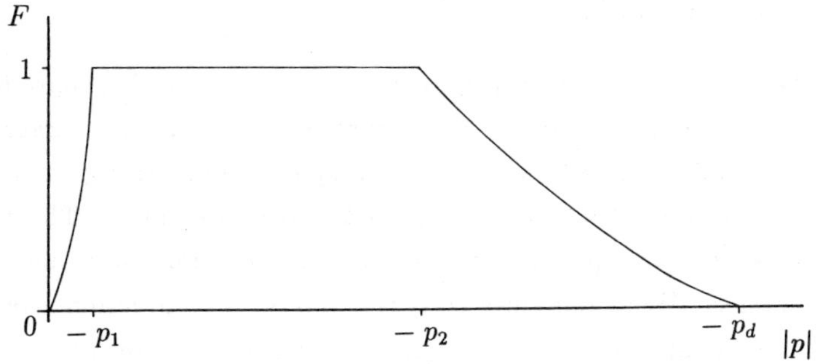

Figure 3.6: Plant abstraction function

Another form which has been widely used [43,81,86] is

$$w_0(p,t) = \begin{cases} rK(p,z)(p-p_r) & \text{if} \quad p \geq p_r \\ 0 & \text{if} \quad p < p_r. \end{cases} \tag{3.20}$$

The thought behind this function is that there is a pressure head p_r at the interface between plant roots and the soil. If the pressure head in the soil is greater than the interfacial pressure ($p \geq p_r$) water may flow to the roots. However, plants are incapable of putting moisture back into the soil. When $p \geq p_r$ the flow is governed by a resistance relationship where r denotes the resistance factor. There is some controversy as to whether or not the hydraulic conductivity function $K(p, z)$ should be included in this formula. Some say that it should, because the water in the soil has to flow (horizontally) to the roots and thus is subject to the influence of the conductivity. Others argue however that this is already modelled in equations (3.6) and (3.7). The interfacial pressure head p_r is a dynamic factor in formula (3.20). It is the maximal value which can be chosen under the constraint $p_r \geq p_d$ such that

$$\int_h^0 w(p(z,t), z, t; p_r)dz = P(t).$$

If no such choice is possible, one sets $p_r = p_d$ and then the actual evapotranspiration

$$A(t) = \int_h^0 w(p(z,t), z, t; p_d)dz < P(t).$$

In precise mathematical notation this dynamic parameter can be defined as

$$p_r(t) = \inf \{p' \geq p_d : \int_{h(t)}^0 w(p(z,t), z, t; p')dz \leq P(t)\}. \tag{3.21}$$

A review of different plant-uptake functions has been presented in [81].

3.7 Reproduction models

The last class of models to be considered here are the so-called reproduction models. The motivation for these models is that to model the unsaturated zone precisely one is required to fall back on a technique for numerically solving the differential equations as described in the last two subsections. The trouble though is that such techniques are costly and unwieldy for use in large scale hydrological models. What one does then is compute a large number of situations with a numerical model, and then try and capture the results in a table or in an empirical model. This is the what is known as the reproduction. For examples of such models we mention those described in [3,10,23].

4 Unsaturated groundwater flow modelling - case study

4.1 The comparison study

Throughout the world, many models of one or more of the types described in the previous section have been developed for modelling unsaturated groundwater flow. In the Netherlands, virtually every institute working in this field has developed its own model. In 1982, the Commission for Hydrological Research TNO and the Study Group for the Hulpselse Beek experimental catchment hit upon the idea of arranging a comparison test for all the models available in the Netherlands at that time [72]. These were:

- DEMGEN - a reservoir model from the Dutch Ministry of Transport and Public Works.
- MUST - the model of de Laat described in brief in the previous section.
- ONZAT - a numerical scheme for solving the Richards' equation developed by the Institute for Public Health.
- SOMOF - a numerical scheme for solving the Richards' equation developed at the Delft Hydraulics Laboratory.
- SWATRE - a numerical scheme for the Richards' equation developed by the Institute of Water Management.

The present author took part in this study as developer of the model SOMOF [58]. The name is an acronym of *So*il-*Mo*isture *F*low.

Each participant was provided with identical data on the soil characteristics, the rainfall N, computed potential evapotranspiration P, measured actual evapotranspiration A, and groundwater levels h, at a point in the experimental catchment for the period from 1 March 1976 to 24 December 1978. Using these data the participants were able to calibrate their models for the unsaturated groundwater flow at this point as far as this was possible or necessary. For the period from 1 January 1979 to 24 December 1982, the participants were subsequently provided with data on the rainfall, potential evapotranspiration and groundwater level only. For this period the objective was to use the model to predict the actual evapotranspiration. The results were subsequently submitted to an independent adjudicator who had access to the measured actual evapotranspiration data for this period. These had been hitherto withheld from the participants. All the models were then evaluated by comparison of their predicted actual evapotranspiration with that measured.

4.2 Setting up the model

The computer program SOMOF is based upon the numerical solution of the differential equation (3.7) with the boundary conditions (3.8) and (3.10) as described in the previous section. To deal with the difficulties of the moving lower boundary, a simple time-independent scaling transformation is applied. The numerical method itself is the Douglas-Jones predictor-corrector method [34,71,93,99]. It is not the intention here to go into details of the computational side of the exercise. The emphasis is on the practical aspects. For details of the computer program, see [58,69,72].

For application of the given model the following information is necessary:

- soil characteristics, i.e. the relations $K(p, z)$ and $\theta(p, z)$;
- boundary condition parameters, i.e. the value of p_d;
- time-dependent boundary conditions, i.e. the variables $h(t)$, $N(t)$, $E(t)$ and $P(t)$ for the given time-period;
- initial conditions for starting the simulations;
- a formulation for the plant abstraction function w.

Data for the first four of the above-mentioned classes of information were supplied by the organizers of the comparison exercise; and all participants agreed to interpret the data in the same fashion. For the model SOMOF, the plant abstraction function was the remaining medium open for calibration. Below the incorporation of each class of data in the model will be described.

Soil characteristics

Data on the soil characteristics at the study point were supplied by the Foundation of Soil Mapping (STIBOKA).

The subsurface can be viewed as being composed of four soil horizons, which to follow the nomenclature of the Foundation of Soil Mapping can be designated A_1, B_2, C_{11} and C_{12}. The uppermost horizon A_1 stretches from the surface to a depth of 15 cm below the surface and the subsequent horizon B_2 from this to 35 cm below the surface. The horizon C_{11} stretches from below the last-mentioned level to 90 cm below the surface. The last horizon classified, C_{12}, is to be found beneath a depth of 90 cm below the surface. For each soil type, laboratory experiments have been carried out to determine the relation between the hydraulic conductivity and pressure head in the soil, and between the moisture content and the pressure head. An exception is formed by the hydraulic conductivity function for the A_1 horizon. For this it can be assumed that the relationship is the same as that for the B_2 horizon. The results of the measurement are shown by the points in Figures 4.1 - 4.7.

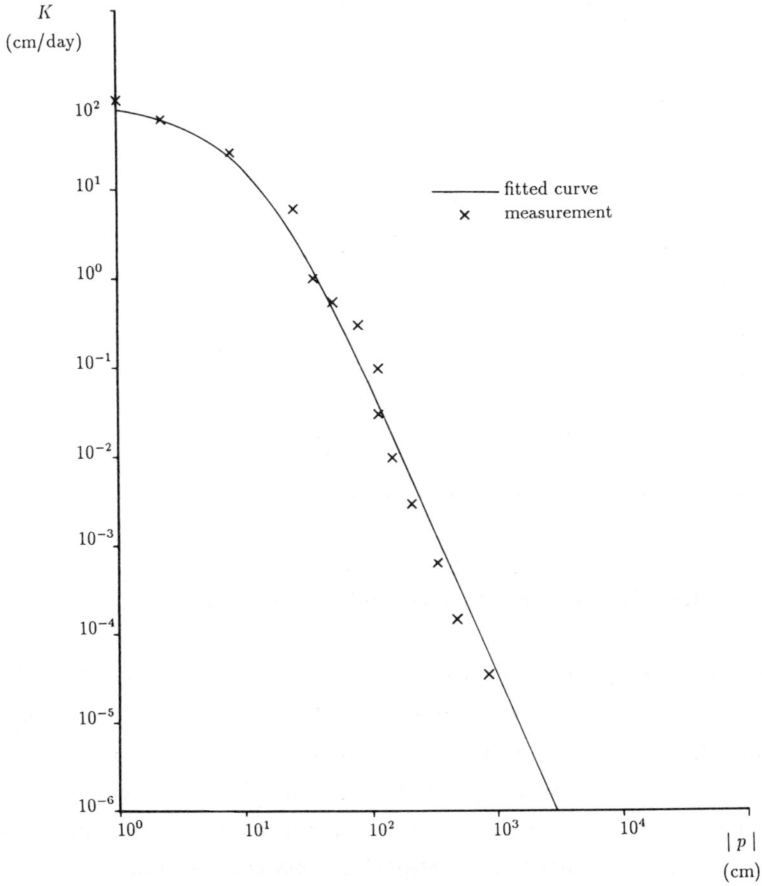

Figure 4.1: Hydraulic conductivity relationship for layers A_1 and B_2

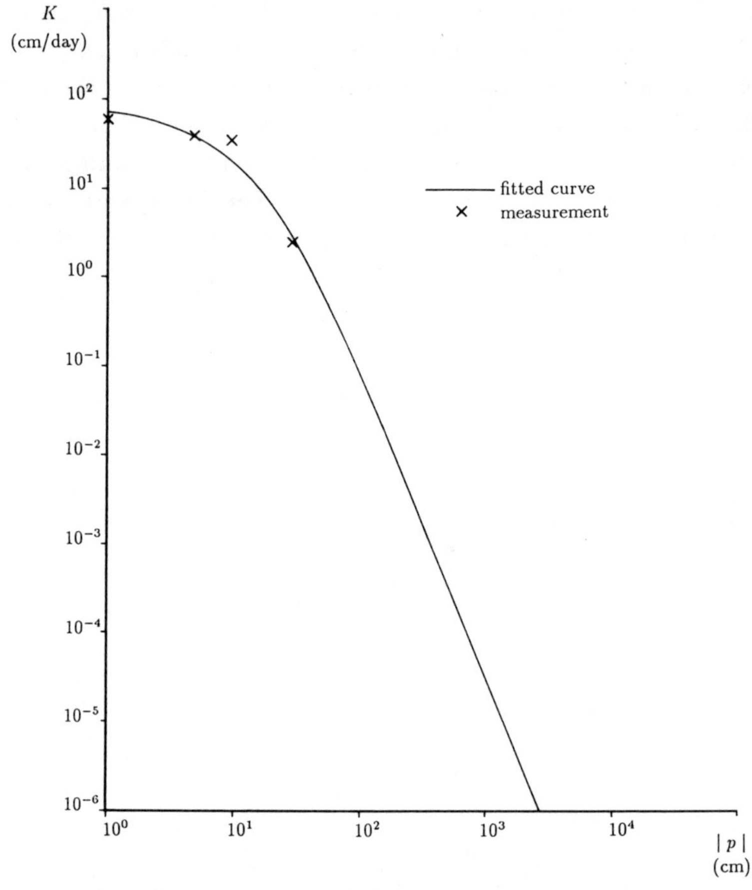

Figure 4.2: Hydraulic conductivity relationship for layer C_{11}

Here already the first difficulty in practical modelling can be seen. In the previous section, we described the theory of modelling soil-moisture flow using a differential equation. It was commented that K and θ were monotonic functions of p. The measurements do not altogether confirm this. Particularly illustrative are Figures 4.4 and 4.7. The data do not show any monotonicity. Indeed it is difficult to imagine a functional relationship between the points in these figures. In this respect, the reader should be aware that these measurements were carried out using the most modern techniques available, and care was taken to remove the effects of any hysteresis in the measurements. Here then is the first lesson: Theory is one thing; practice is something else.

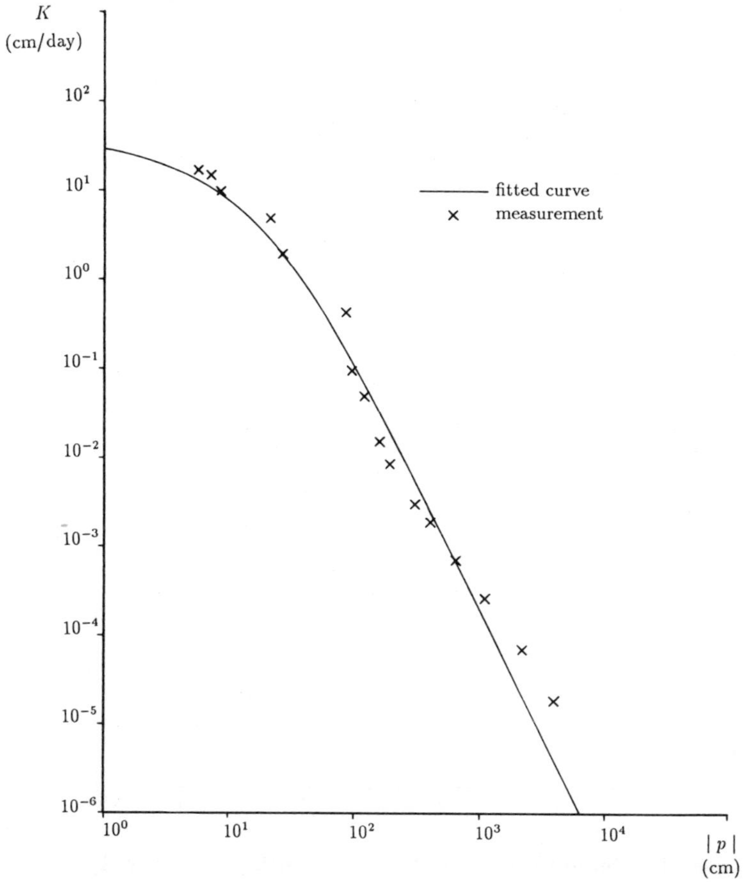

Figure 4.3: Hydraulic conductivity relationship for layer C_{12}

For substitution in equation (3.7), explicit functions for the representation of $K(p)$ and $\theta(p)$ have to be used. There are a number of formulae in circulation, some with more or less well-founded physical backgrounds. For this study a promising formula appeared to be that of van Genuchten [51,52]. This is

$$K(p) = K_s\{1+(-\alpha p)^m\}^{(1-m)/2m}[1-\{1+(-\alpha p)^{-m}\}^{(1-m)/m}]^2 \qquad (4.1)$$

with

$$\theta(p) = \theta_s\{1+(-\alpha p)^m\}^{(1-m)/m}. \qquad (4.2)$$

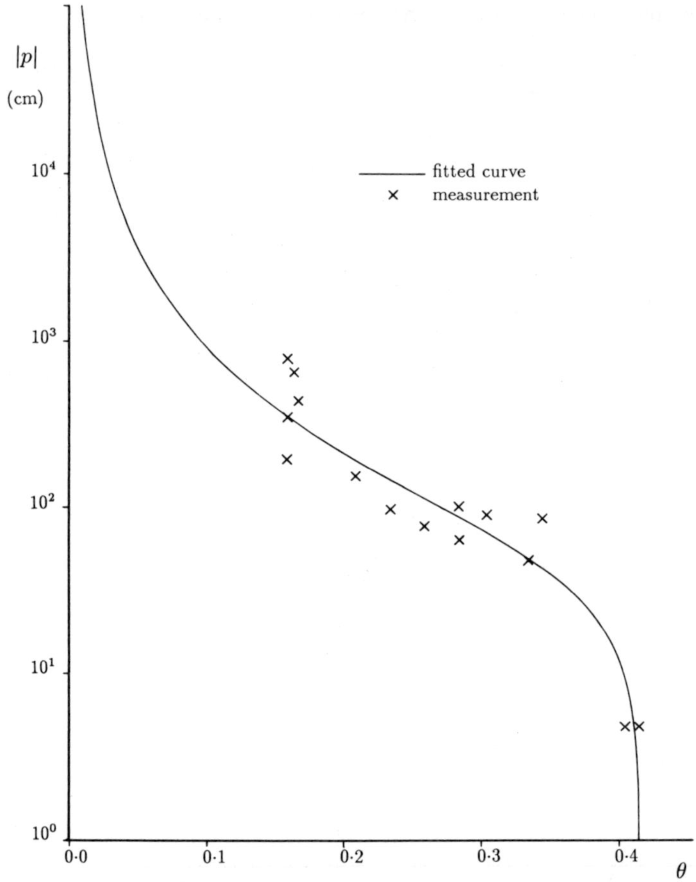

Figure 4.4: Soil-moisture content relationship for layer A_1

Here K_s and θ_s denote respectively the hydraulic conductivity and moisture content at saturation. The parameters $\alpha > 0$, $m > 1$ can be chosen freely. Functions described by these formulae have all the theoretical characteristics described in Subsection 3.5. The strength of the relationship lies in the fact that once the $K(p)$ or $\theta(p)$ relationship has been established, the other is known with respect to a scaling around the value at saturation.

What we did then was to plot the measured data on the log-log or semi-log scales as shown in Figures 4.1 - 4.7 and choose the parameters K_s, θ_s, α and m such that the resulting curves best-fitted the data. There are computer algorithms available for doing this work. However, we decided to do the work by trial and error with the subjective judgement of whether or not the curve was a good fit. This has the advantage that the geohydrologist can easily decide when a mathematically optimal fit is unreasonable. We found though that for each soil type, if the $\theta(p)$ curve represented a good fit of the data the corresponding $K(p)$ curve was a lousy fit. Vice versa, if the $K(p)$ data were well-fitted, the $\theta(p)$ data were not at all well matched. So much for the coupling of the two sets data through a few simple parameters! Lesson two: Be prepared for disappointment and be prepared to rethink the modelling concepts.

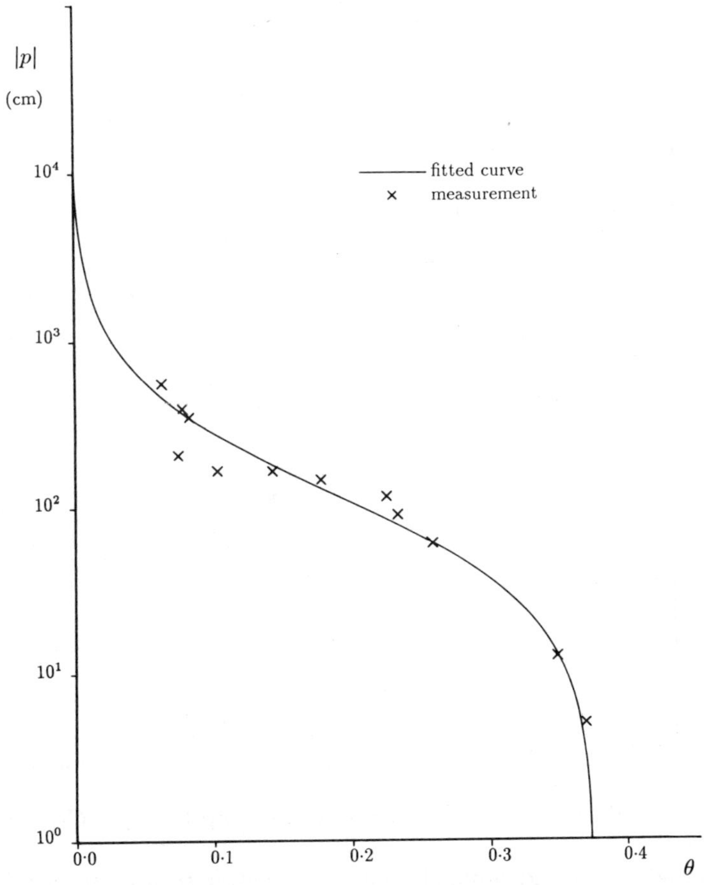

Figure 4.5: Soil-moisture content relationship for layer B_2

We solved the difficulty by uncoupling the parameter α and m in (4.1) and (4.2) and by replacing $(1-m)/m$ in (4.2) by $-m'$ where m' was a new positive parameter. We then fitted curves to the $K(p)$ and $\theta(p)$ data independently as best we could. One has to bear in mind that our goal here was merely to find suitable closed form-expressions for the $K(p)$ and $\theta(p)$ relationships. The results are contained in Figures 4.1 - 4.7.

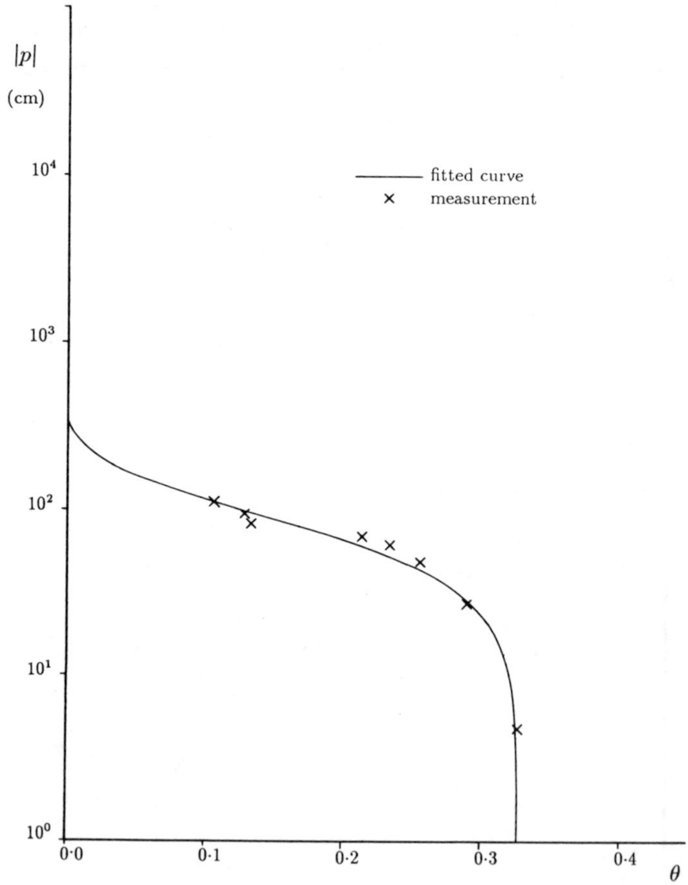

Figure 4.6: Soil-moisture content relationship for layer C_{11}

With hindsight, it would probably have been better to use the raw data in table form with a smooth interpolation, for instance one based on splines. Such an approach was adopted by some of the other participants in the modelling exercise. However, we can all be wiser after the event.

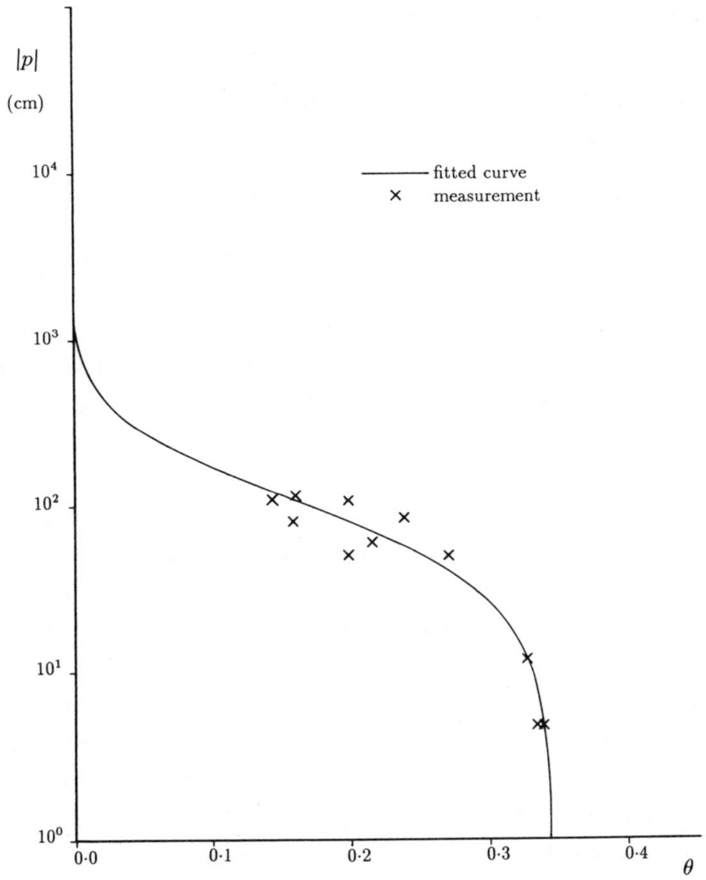

Figure 4.7: Soil-moisture content relationship for layer C_{12}

Boundary condition parameters

This came down to a suitable choice of the parameter p_d. For want of any other information, the generally adopted value of

$$p_d = -160000 \text{ mm}$$

was assumed. This illustrates a point made in the general discussion of saturated groundwater flow modelling. Sometimes, one just has to make do.

Time-dependent boundary conditions

Data on the depth of the groundwater table below the surface, the precipitation, the potential evapotranspiration, and on the actual evapotranspiration for the calibration period, were provided on a daily basis by the Hupselse Beek Study Group who maintain the experimental catchment.

For the interested reader, we mention that the potential evapotranspiration was calculated from available meteorological data by the formula of Thom and Oliver [106]. The actual transpiration was measured in the period from 1 April to 30 September in each year by the profile method [105]. For the remainder of each year, the actual evapotranspiration was assumed to be no less than the calculated evapotranspiration. This is a reasonable assumption for the colder (winter) months in the Netherlands.

For the purposes of this comparison, it was agreed among the participants that all the precipitation would be considered as effective precipitation, i.e. no rainfall would be intercepted by plants, and that there would be no surface runoff. Thus if at any stage ponding occurred the accumulated water would be added to the rainfall over the coming day. Furthermore, it was agreed that all the evapotranspiration would be considered as transpiration. In the present context, this means that the input variable E was identically zero.

For introduction in the model SOMOF, all the data were considered as applying at noon each day. For interpretation of the data at computational time-steps in between these periods, linear interpolation was used.

Initial conditions

For practical purposes the simulation was divided into subperiods. In fact though, we only had to consider the initial conditions for starting the simulation on 1 March 1976. For the remaining period, the last calculated profile at the end of one subperiod is the natural initial profile for the next. For lack of further information, it was assumed that on the 1 March 1976 the soil-moisture profile was in hydrostatic equilibrium, i.e.

$$p(z) = h - z \text{ for } h < z < 0,$$

and that there was no surface ponding.

Plant abstraction function

On the basis of favourable reports in the literature, the plant abstraction function (3.19) - (3.21) was selected for the application. The root distribution function was chosen to be

$$\rho(z) = \begin{cases} 1 & \text{for} \quad z \geq z_2 \\ \frac{z-z_1}{z_2-z_1} + \frac{1}{2\pi}\sin\left(2\pi\frac{z-z_1}{z_2-z_1}\right) & \text{for} \quad z_1 \leq z < z_2 \\ 0 & \text{for} \quad z < z_1 \end{cases} \tag{4.3}$$

where z_1 and z_2 are parameters. This is in some sense an arbitrary choice. It has the property though that it is monotonic and that there is an elevation z_2 above which the roots are fully-developed and an elevation z_1 below which there are no roots. Furthermore, it is smooth, which is advantageous for the numerical computation. It is shown in Figure 3.5.

There was considerable discussion among the participants about the depth to which the plant roots may penetrate. It would appear that this is a parameter to which the computations are extremely sensitive. This is especially true for the reservoir models. Eventually it was agreed that all modellers should use the same basic fact. This was that for the study point in the Hupselse Beek catchment the root zone had been measured to have a thickness of 25 cm. The root zone was here defined as that portion of the subsurface in which 80% of the plant roots may be found. For the chosen distribution (4.3) this meant that only one of the parameters z_1 and z_2 could be freely chosen. The other was then fixed by the relation

$$\int_{-25}^{0} \rho(z)dz \; / \int_{z_1}^{0} \rho(z)dz = 80\%. \tag{4.4}$$

4.3 First analysis

Having gathered all the information described in the previous subsection, the modelling exercise appeared straightforward. It merely required taking suitable values for the parameters z_2 and r in the root abstraction function and adjusting these until the computed actual evapotranspiration agreed with that measured. In reality this turned out to be anything but straightforward.

The first hitch came with the calibration of the parameter z_2. It was found that under the constraint (4.4) the best choice occurred when the corresponding parameter z_1 was of unrealistic proportions. For want of any guidelines, it was therefore eventually decided to take $z_2 = -25$ cm, this being the given thickness of the root-zone. The constraint (4.4) subsequently led to $z_1 = -37{\cdot}5$ cm.

The second hitch came when it was discovered that adjustment of the parameter r in the plant uptake function alone was insufficient to reproduce the measurements. This was specifically the case in 1976, which has entered the record books in Europe as an extremely dry year. In the simulation, everything went well whilst the actual evapotranspiration was potential, i.e. whilst the pressure head at the root soil interface p_r was larger than its minimum value p_d. When this was no longer the case, one had $p_r = p_d$ and the actual evapotranspiration was limited. At this stage though it made no difference if atmospheric conditions became even dryer, the actual evapotranspiration remained the same ($p_r = p_d$). Looking at the measurements though, it was clear that the actual evapotranspiration increased when the potential evapotranspiration did so, even if simultaneously the soil had become dryer. Talks with people with more knowledge of plant physiology indicated that this observation was quite plausible. Somehow plants can respond to more severe atmospheric conditions by increasing their transpiration rate. This was obviously a phenomenon not incorporated in the plant abstraction function (3.19) - (3.21). In this respect, there is therefore a shortcoming in this function.

The dilemma was resolved by choosing a base-value for r, r_0, and then setting

$$r = \begin{cases} r_0 & \text{if} \quad p_r > p_d \\ r_0(1 + \gamma \frac{P - P_c}{P_c}) & \text{if} \quad p_r = p_d. \end{cases}$$

Here γ is a fixed parameter, $0 \leq \gamma < 1$, and P_c is the critical level of evapotranspiration defined as the greatest possible potential evapotranspiration which does not correspond to a reduction in evapotranspiration. It is thus dependent upon the soil conditions and can be determined by

$$P_c(t) = \int_{h(t)}^{0} w(p(z,t), z, t; p_d)\, dz$$

with

$$w(p, z, t, p_d) = \rho(z) r_0 K(p, z) \max\{0, p - p_d\}.$$

Cf. (3.19) - (3.21). In total there are three different possible situations:
$P < P_c$ and then $p_r > p_d$ and $r = r_0$,
$P = P_c$ and then $p_r = p_d$ and $r = r_0$,
$P > P_c$ and then $p_r = p_d$ and $r > r_0$.
In the first two cases, the actual evapotranspiration is the same as was previously calculated, and in fact is potential. In the last case though, now

$$A = P_c + \gamma(P - P_c) < P,$$

instead of

$$A = P_c < P$$

which would have been obtained using the old scheme. By way of speaking, in this adapted plant uptake function, the parameter r_0 determines whether or not the actual evapotranspiration is potential, whilst the parameter γ determines the ratio of the actual evapotranspiration to the potential evapotranspiration when they are not equal. Of course, in reality this picture is oversimplified, since both parameters influence the moisture content of the soil which in turn influences the later evapotranspiration. However, it serves to illustrate the role of the different parameters.

The third real hitch came from the practicalities of carrying out the computational procedure. So far we have not discussed numerical aspects. Even now it is not the intention to dwell on this topic. Notwithstanding, they are worthy of some mention, because they again show that even when things are the theoretically sound there can be practical problems. We began the study with the optimism that for the numerical solution of the differential equation 10 vertical calculation points and 5 equal time-steps per day would be sufficient. Slowly, we learned that if the saturated groundwater level sunk too low, 20 vertical steps were better. Eventually, for extremely dry periods in which rain fell we ended up with 40 divisions. The same story can be told with respect to the time-steps. On occasions the computations were carried out with 10, 20 or 25 time-steps per day. In the most extreme case, a day in which the boundary conditions varied rapidly was begun with a time step of no more that 2×10^{-3} days. By using a weighting factor, this initial time step could be progressively increased though.

4.4 Calibration

The goal of the calibration in this project was the adjustment of the parameters r_0 and γ in the modified plant abstraction function described in the previous subsection, so that the best model of the real actual evapotranspiration was obtained. As material for judging this, the data from 1 March 1976 to 31 December 1978 was available. The judgement was carried out on the basis of two plots of the measured and calculated actual evapotranspiration versus time. In the one plot the total evapotranspiration over the foregoing ten days was set out. In the other plot the accumulated evapotranspiration over a larger time span was drawn. Examples are shown in Figures 4.8 - 4.13. The choice of using these variables for the calibration was based on the decision at the start of the exercise to use figures of the evapotranspiration over periods of ten days and of the accumulated evapotranspiration for comparing the models. In fact, there was no real decision in the latter, since the participating models of reservoir type could not deliver more-detailed results. The specific choice of using the visual judgement of plots was warranted by the considerations mentioned earlier with respect to the fitting of the $K(p)$ and $\theta(p)$ functions.

Because, during the colder winter periods in the Netherlands the actual evapotranspiration was always potential, the calibration of the unknown parameters primarily concerned the summer periods. The whole calibration period was therefore divided into the following seven sub-periods:

- 1 March to 31 March 1976 (winter),
- 31 March to 31 September 1976 (summer),
- 30 September 1976 to 31 March 1977 (winter),
- 31 March to 30 September 1977 (summer),
- 30 September 1977 to 31 March 1978 (winter),
- 31 March to 30 September 1978 (summer),
- 30 September to 31 December 1978 (winter).

With parameter values estimated from the first analysis, all the subperiods were subsequently simulated one after the other in order. The results from the winter periods were then left untouched for the time being. These served solely as providers of initial-values for computations of the summer periods. The actual calibration was then carried out on the summer periods.

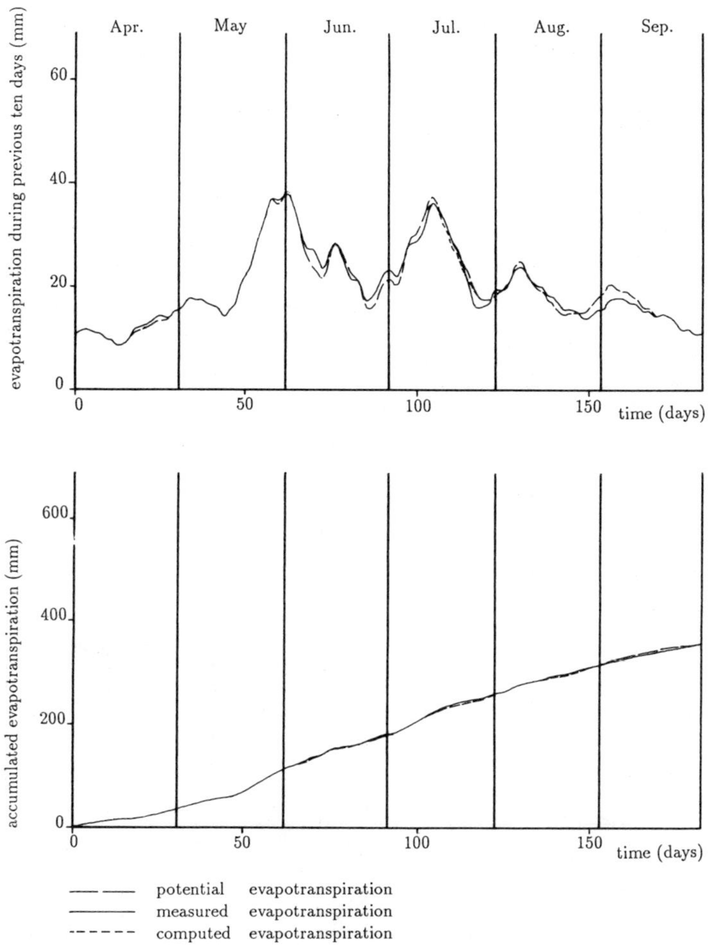

Figure 4.8: Calibration calculations for 1977

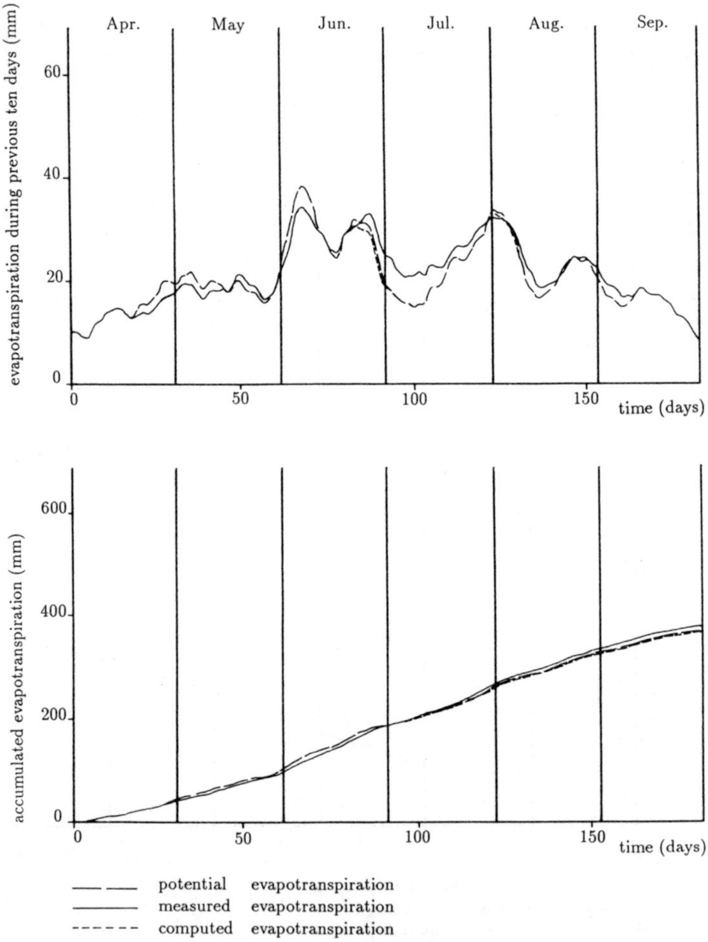

Figure 4.9: Calibration calculations for 1978

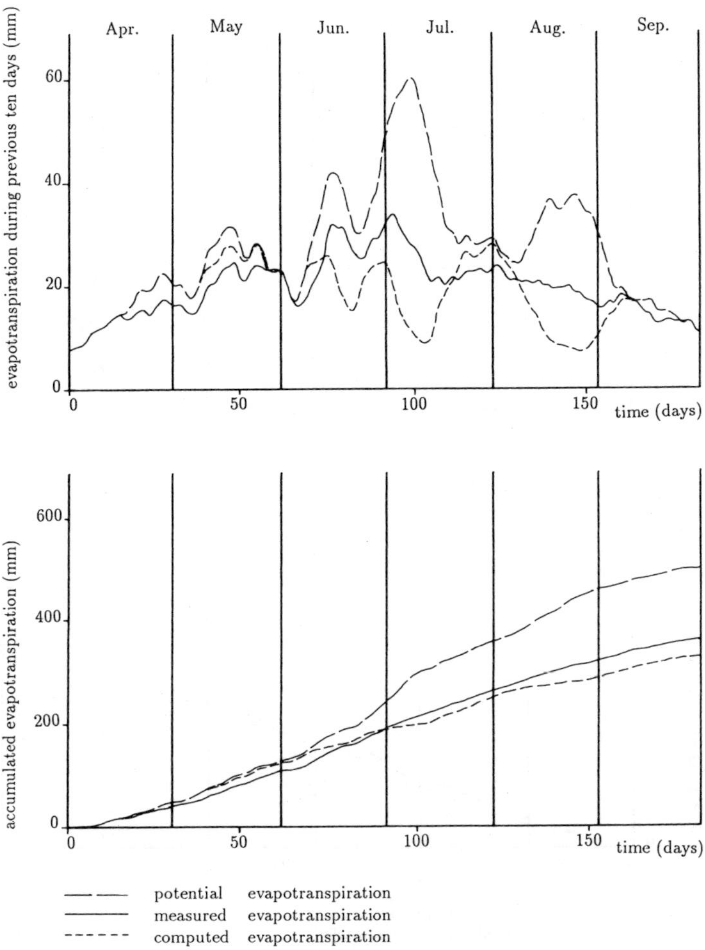

Figure 4.10: Calibration calculations for 1976 (with $\lambda = 0$)

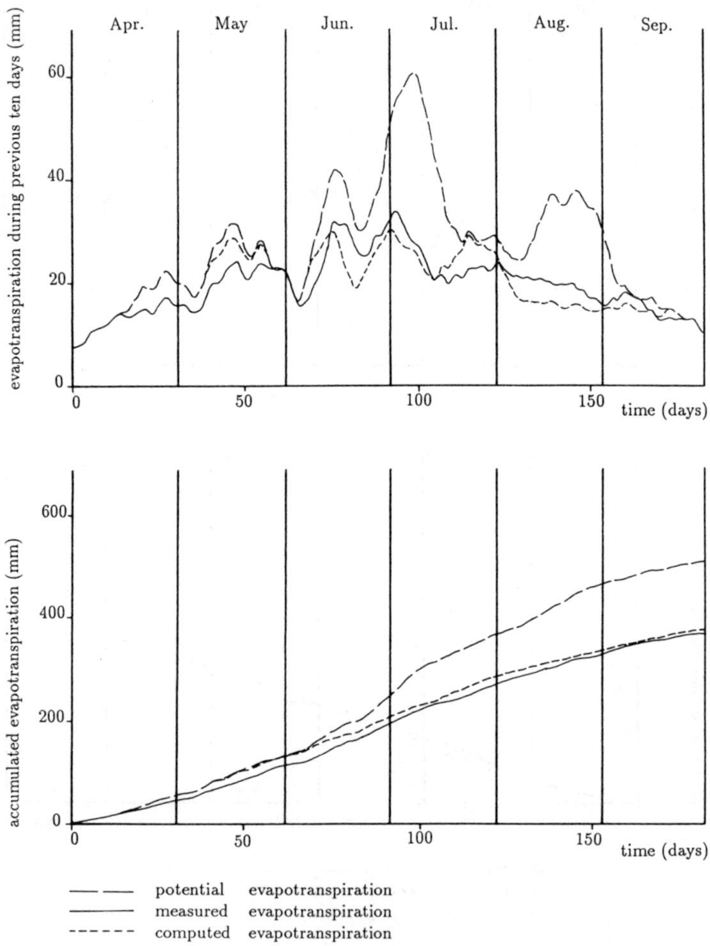

Figure 4.11: Calibration calculations for 1976 with λ=0·4

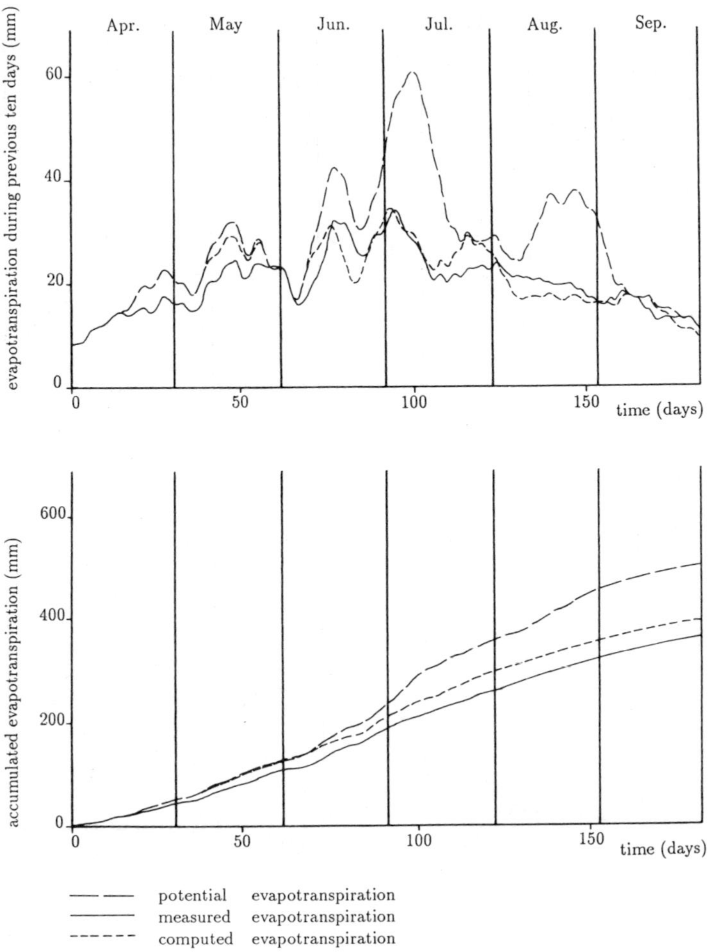

Figure 4.12: Calibration calculations for 1976 with λ=0·5

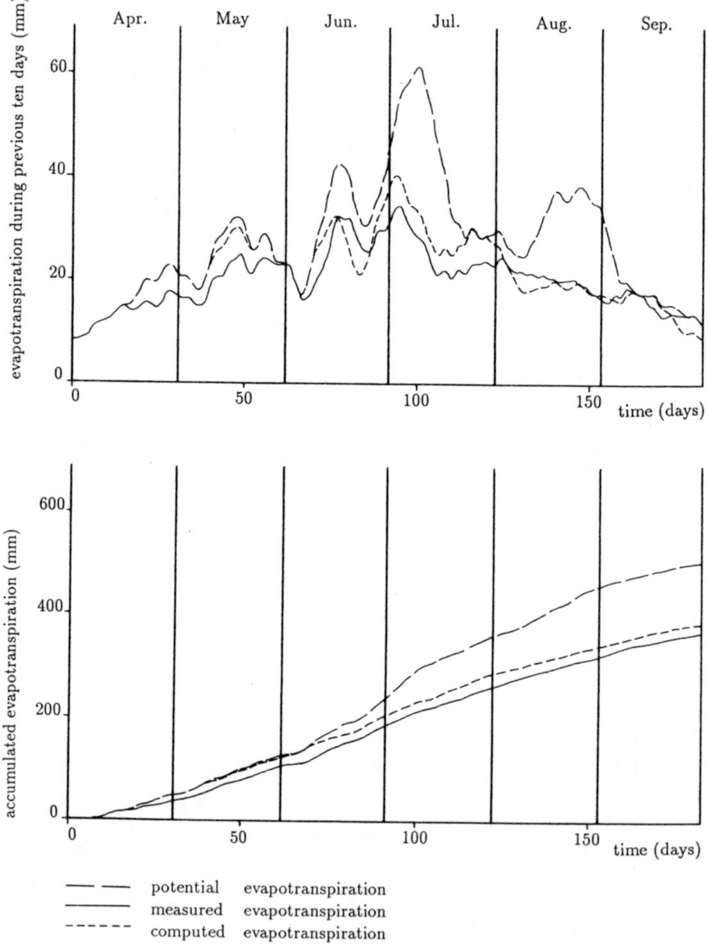

Figure 4.13: Calibration calculations for 1976 with $\lambda=0{\cdot}45$

To calibrate the parameters r_0 and γ the following strategy was adopted. First the parameter γ was set equal to zero. As mentioned earlier, the parameter r_0 has the greatest influence in determining in which the periods there is an evapotranspiration reduction. The parameter r_0 was then adjusted so that the periods in which the computed actual evapotranspiration was less than potential showed the greatest resemblance to those observed. Thereafter the parameter γ was calibrated so that in these periods the computed actual evapotranspiration best approximated that measured.

It was soon discovered that the value of the parameter r_0 which yielded the best results for 1976 led to poor results for 1977 and 1978. Actually, in 1977 and 1978 there was hardly any reduction of the actual evapotranspiration below its potential level measured in the field (see Figures 4.8 and 4.9). In fact, the observant reader will notice that in 1978 the accumulated measured actual evapotranspiration exceeds the accumulated potential evapotranspiration. Considering the inherent definition of these variables, there was not one model capable of emulating this. This led to much discussion among the participants about whether the data should be corrected or not. However, since it was uncertain which of the data sets was at fault, it was not possible to make an unbiased correction of this phenomenon and the proposal was dropped. This again illustrates the limitations of conducting a modelling study in practice, be it here that the limitation is in the data which do not comply to the modelling concepts.

It was also discovered that if a value of the parameter r_0 was so chosen that the measured reduction in evapotranspiration in March and April 1976 was reproduced by the model, the simulations of the remaining periods were totally unrealistic. Other modellers experienced the same difficulty. Eventually, by talking to various people, it became clear that the observed reduction in evapotranspiration in the spring of 1976 had an exceptional cause. Namely, there was some late frost. This killed some of the plants. As a result, the water-use was less than otherwise would have been the case.

Here too there are some lessons to be learnt. Firstly we see how a model can be thrown by phenomenon - such as frost action in this case - which are totally ignored. Secondly, we learn how important it is to talk to other people for finding explanations and developing insight.

As a consequence of the conclusion regarding the period of March to April

1976, we subsequently completely ignored this period in the remaining calibration.

We thus finally decided to choose the parameter r_0 so that the reduction in evapotranspiration in the summers of 1977 and 1978 was marginal. The discrepancies in 1976 were then accepted. By subsequently increasing r_0 with steps of $0{\cdot}5 \times 10^{-2}$ mm^{-2} we arrived at the choice of $r_0 = 2{\cdot}5 \times 10^{-7}$ mm^{-2}.

Because now there was virtually no reduction of the potential evapotranspiration in the summers of 1977 and 1978, the calibration of the parameter γ subsequently fell 'de facto' solely to the summer period of 1976. Furthermore, since the spring was excluded, this actually came down to the period of June to September 1976.

By trial and error it was judged that a value of $\gamma = 0{\cdot}4$ was too low and that a value of $\gamma = 0{\cdot}5$ was too high for acceptable results. For $\gamma = 0{\cdot}45$ the results were judged satisfactory.

Further refinement of the values of r_0 and γ was deemed pointless seeing all the uncertainties in the model and data, and so the above values were retained. With these values the whole period from 1 March 1976 to 31 December 1982 was then computed through again from start to finish in one sweep for presentation of the results of the calibration. The final figures obtained are Figures 4.8, 4.9 and 4.13. In Figure 4.13, one can clearly see the influence of the difficulty of the spring of 1976 working through.

4.5 Verification

With the calibration complete, the computations for the verification period from 1 January 1979 to 31 December 1982 could be carried out. The results are presented in Figures 4.14 - 4.17 in the same form as that used for the calibration. In Table 4.1 the accumulated evapotranspiration figures for the summer periods from 1 April to 31 September are presented. Here again in 1979 and 1981 one can see that the accumulated measured actual evapotranspiration exceeds the accumulated potential evapotranspiration.

For assessment of the results of our modelling in comparison to that of the others one should refer to Figures 4.18 - 4.21. These are taken from the contribution of Koopmans to [72]. Figure 4.18 presents the results of carrying out a linear regression between the evapotranspiration reduction in the verification

		accumulated potential evapotranspiration P in mm	measured accumulated actual evapotranspiration A in mm	computed accumulated actual evapotranspiration A in mm
calibration period	1976	505	368	384
	1977	363	361	362
	1978	373	383	371
verification period	1979	382	393	382
	1980	414	398	386
	1981	396	407	396
	1982	445	420	404

Table 4.1: Synopsis of results for 1 April to 31 September

period computed by the various models and that measured. Recall that the reduction of evapotranspiration is the difference between the potential evapotranspiration and the actual evapotranspiration, and is determinate for any possible crop damage. If the models had been perfect the lines would have run precisely from the bottom left-hand corner with coordinates $(0, 0)$ to the top right-hand corner with coordinates $(50, 50)$. As it turned out, in the verification period there were really only two years, viz. 1980 and 1982, which gave rise to a reduction in evapotranspiration. Moreover, only for 1982 was it possible to make a distinction between the results of the different models. Figure 4.19 presents graphs of the accumulated reduction in evapotranspiration plotted against time for each of the models together with the corresponding graph for the measured data. In Figures 4.20 and 4.21 one can see block-diagrams representing the origin of the actual evapotranspiration computed by each of the models. For clarity, "storage" represents the difference between the total amount of water in the unsaturated zone at the start and end of the growing season. "Capillary rise" refers to that fraction of the water used by the plants which was calculated to be taken from the saturated groundwater zone. "Precipitation" refers to the remainder. Here, precipitation which percolates through the unsaturated zone to the saturated regime to be used later appears

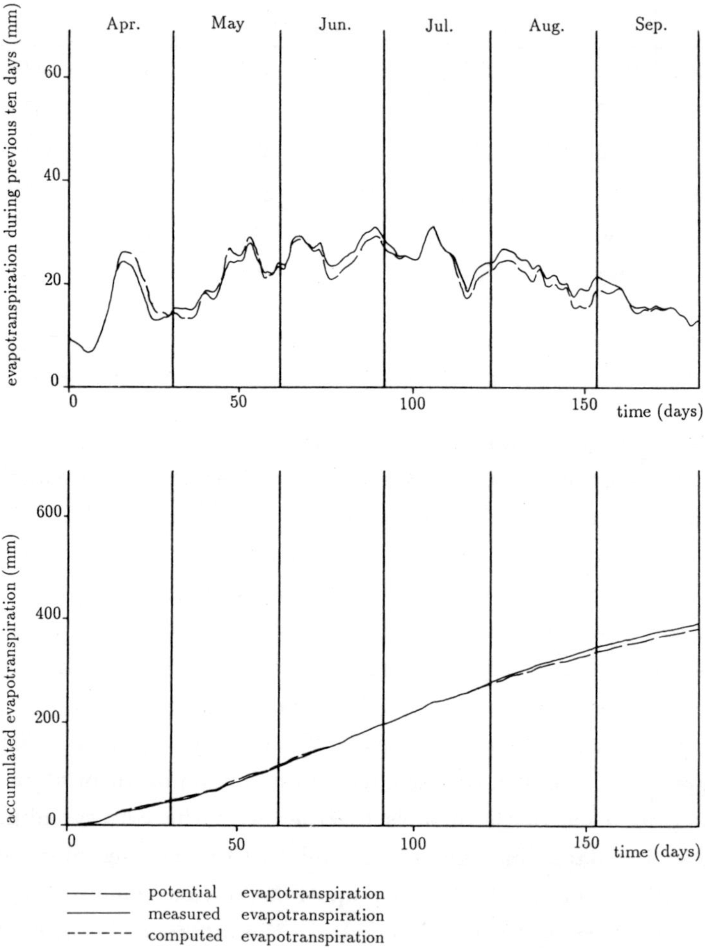

Figure 4.14: Verification calculations for 1979

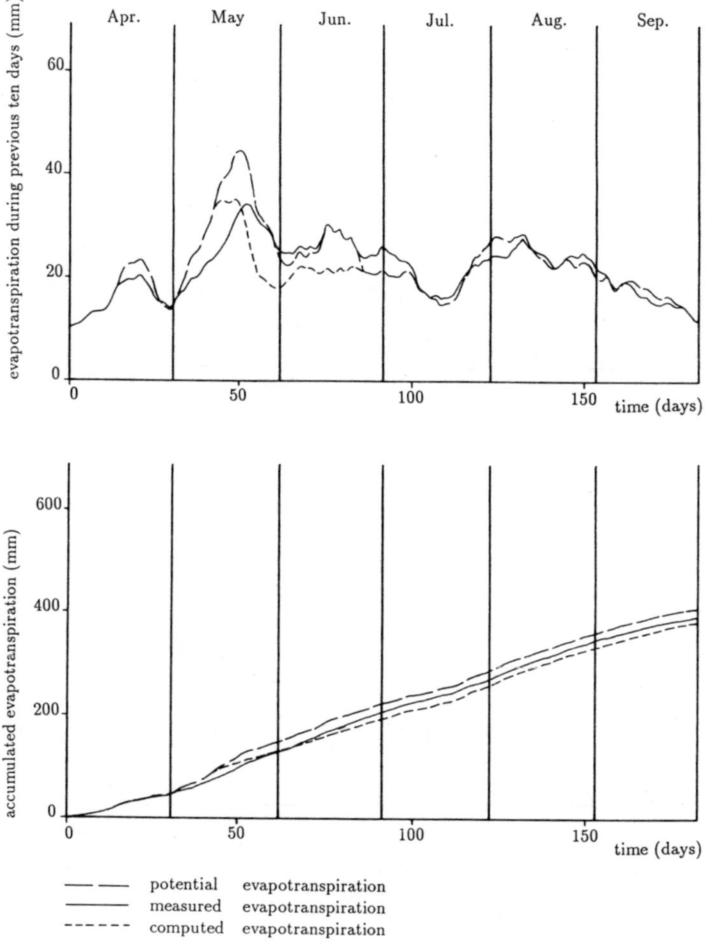

Figure 4.15: Verification calculations for 1980

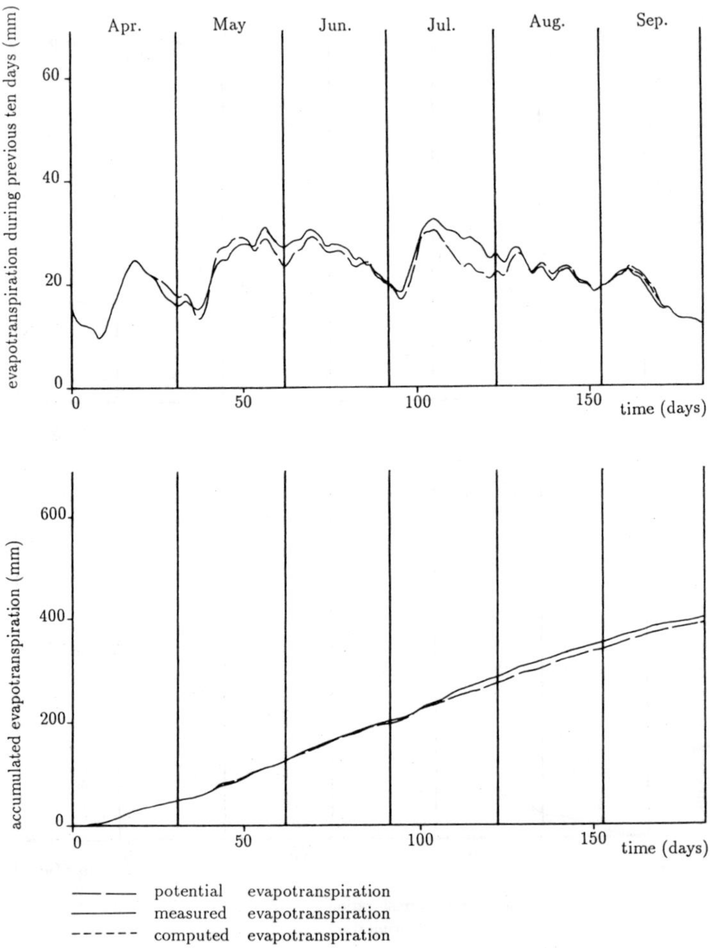

Figure 4.16: Verification calculations for 1981

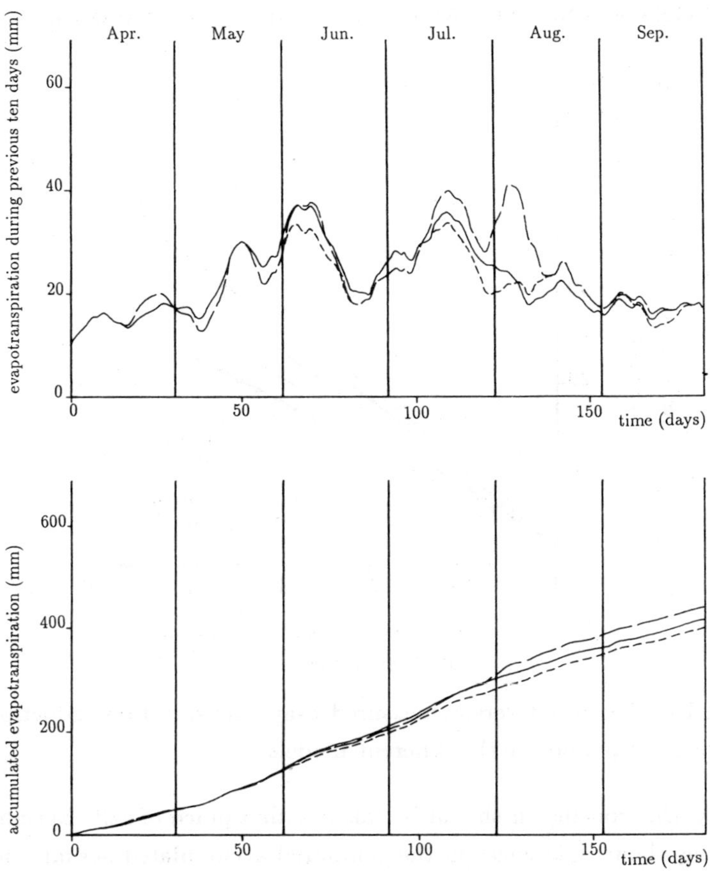

— — potential evapotranspiration
——— measured evapotranspiration
- - - - - computed evapotranspiration

Figure 4.17: Verification calculations for 1982

under "capillary rise". In the last two figures, the results of the different models do not differ significantly. Furthermore, it is not possible to say which is good.

From Figures 4.18 and 4.19 it can be construed that the best two models are DEMGEN and SOMOF. About this a number of remarks can be made.

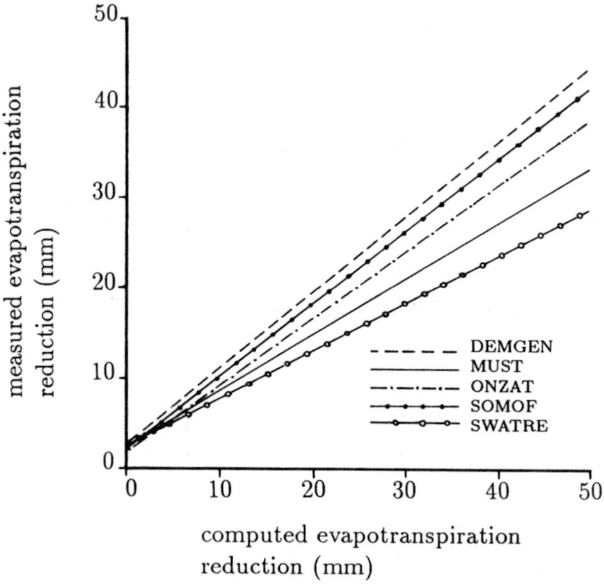

Figure 4.18: Computed versus measured evapotranspiration reduction during the verification period for the different models

Firstly, this conclusion should be taken with a pinch of salt. Specifically, we are talking about differences in the computed accumulated actual evapotranspiration of no more than three to four millimetres during a season. Whilst, the results all the models do not differ by more than 15 mm. These figures should be seen in the light of the fact that, by way of the discrepancy between the figures of potential and measured actual evapotranspiration, it is clear that these data cannot be determined at the present time with an accuracy of more than 10 mm.

Secondly, in the opinion of the author, the relatively good performance of SOMOF is not so much related to the qualities of the computer program but to the strategy of the modelling. To be specific, we chose to ignore the troublesome

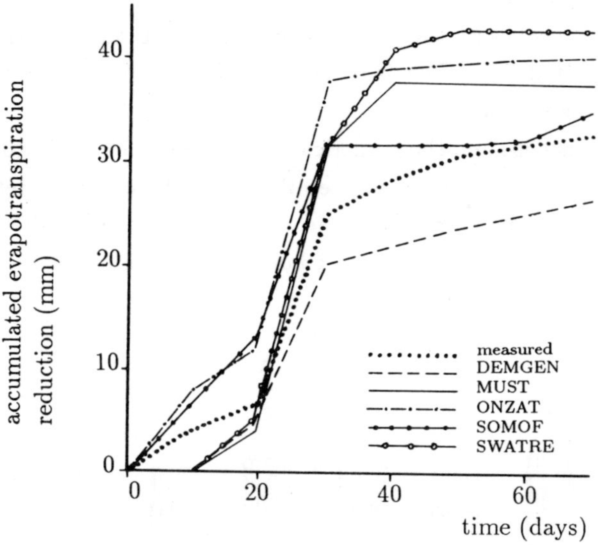

Figure 4.19: Accumulated evapotranspiration reduction in the summer of 1982 for the different models

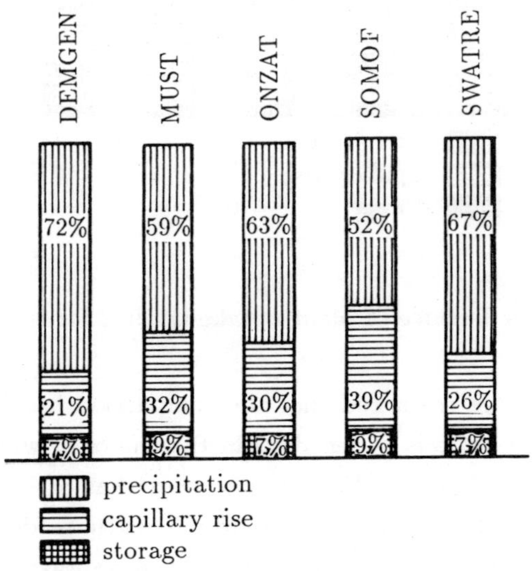

Figure 4.20: Water balance of plant uptake in 1981 for the different models

period of March and April 1976. In contrast, most of the other modellers took this into account. As a result their calibration on the summer of 1976 over-compensated the early deficit. This is reflected in their later over-prediction of the actual evapotranspiration.

Thirdly, SOMOF is based on the numerical solution of (3.7), (3.8), (3.10) and is subsequently relatively expensive to use. However, the other model singled out at the end is a relatively-simple reservoir model which is not at all expensive in computer time. Moreover, the results of both models are comparable.

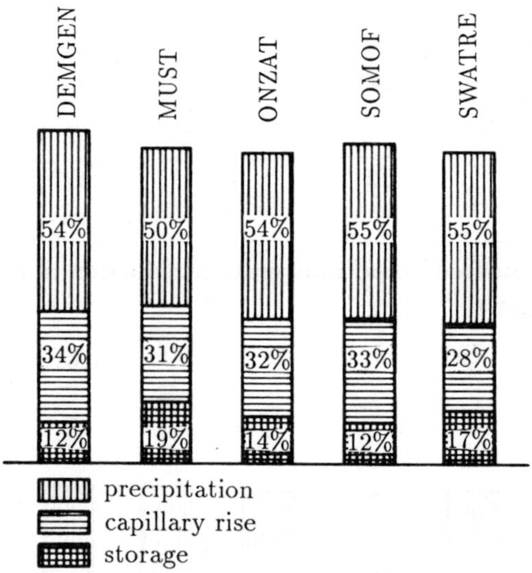

Figure 4.21: Water balance of plant uptake in 1982 for the different models

Finally it should be remarked that the comparison has been made on the basis of figures over periods of ten days in the choice of which there is a certain arbitrariness. In Figure 4.13, it can be seen that SOMOF can reproduce variations on a daily basis, which the more schematic models cannot. With a presentation of the data on a ten day basis, this advantage is not so apparent. With hindsight everyone was in agreement that a better test would have been made if the roles of the chosen calibration and verification periods had been reversed.

4.6 Conclusions

It was funny that at the end of the study everybody drew the conclusions that they wished. The organizers of the comparison study concluded that such an exercise was well worthwhile, because it deepened the insight of the participants and stimulated their contact with one another which could only benefit scientific contact in the Netherlands. The suppliers of the soil characteristics data concluded how useful these data were, and that more effort should be placed on deriving them more accurately, in particular for the purposes of unsaturated groundwater flow modelling. The managers of the experimental catchment concluded that their facility was indispensable for the verification of models. The hydrologists concerned with the determination of actual and potential evapotranspiration concluded that their techniques were reliable but also deserved more research. The modellers concluded that their models could be used with confidence but likewise needed further development [72].

In the opinion of the author there are a number of other conclusions which can be drawn:

- In some sense our ability to build computer programs for modelling unsaturated groundwater flow and its interaction with plants exceeds our understanding of the underlying physical/biological phenomena. This is particularly true with regard to water use by plants. (By way of illustration, we recall that the introduction of the parameter γ here was motivated by the need to obtain satisfactory results and not by any real plant-physiological considerations.)
- Experience in the use of models can play a much more important role in their satisfactory application than the inherent quality of the model itself.
- At the present time, for regional groundwater studies, sophisticated models do not deliver significantly better results than simple models.
- Models based on the solution of the differential equation of unsaturated groundwater flow are capable of providing more detailed results and are more suitable for developing insight into flow phenomena.
- As much as one could be disappointed by the quality of the results of the different models, one can be reassured by the fact that the methods discussed are the best there are!

5 Nonlinear diffusion-convection

5.1 Mathematical questions

In the previous sections, and particularly in Sections 2 and 3, we have discussed differential equations describing groundwater flow and the numerical solution of these equations without considering any mathematical theory for these equations. For the linear partial differential equations much is already known. However, for most of the nonlinear equations discussed, even today, a solid mathematical theory is far from being complete. Notwithstanding, in recents years, considerable progress has been made in the study of these equations. In this section we shall describe some of this progress. This is concentrated on the Richards' equation (3.6) in the particular case that plant abstraction is ignored, the medium is considered to be homogeneous, and only unsaturated flow is considered. To conform to the traditions in the mathematical literature on this subject, we write this equation as

$$\frac{\partial u}{\partial t} = \frac{\partial^2}{\partial x^2}(a(u)) + \frac{\partial}{\partial x}(b(u)). \tag{5.1}$$

To lay the link with the previous discussion of equation (3.6), here x denotes the place variable which may correspond to the elevation z or depth $-z$, u denotes the unknown in the equation corresponding to the soil-moisture content θ, $a(u)$ is the indefinite integral of the soil-moisture diffusivity coefficient D and the coefficient $b(u)$ is associated with $\pm K(\theta)$ in (3.6) with the sign according to the choice of the of the variable x.

Equation (5.1) may also be used to describe one-dimensional horizontal infiltration when the coefficient $b \equiv 0$. In this case x denotes distance in the horizontal direction, with the remaining variables as described above.

Equation (5.1) may be classified as being a nonlinear diffusion-convection equation with the second-order term on the right-hand side representing the diffusion component and the first-order term or the right-hand side representing the convection component. In some literature [19,62,92] this equation is referred to as the nonlinear Fokker-Plank equation. This is because of its resemblance to the (linear) Fokker-Planck equation known from statistical mechanics [28].

For the ensuing discussion of the mathematical analysis of (5.1) we shall assume that the functions a and b are defined for all nonnegative arguments.

Moreover

$$a, b \in C([0,\infty)) \cap C^2(0,\infty)$$

and are such that a'' and b'' are locally Hölder continuous on $(0,\infty)$,

$$a'(\nu) > 0 \quad \text{for} \quad \nu > 0 \tag{5.2}$$

and

$$a(0) = b(0) = 0. \tag{5.3}$$

With regard to the physical background to (5.1), the condition (5.2) is equivalent to the positivity of the soil-moisture diffusivity function at non-zero moisture contents. The condition (5.3) is merely for convenience and involves no loss of generality. Specifically, the hypothesis $a(0) = 0$ in (5.3) amounts to no more than a fixing of the constant of integration in the identification of a with the integral of the soil-moisture diffusivity coefficient. We note also that the assumption $b(0) = 0$ conveniently coincides with the observation that the hydraulic conductivity function tends to zero as the soil-moisture content approaches zero.

In the remainder of this section we shall discuss a number of topics concerned with the development of a rigorous mathematical theory for equation (5.1) and relate these to the physical background to the equation. Some important features of the mathematical techniques underlying the development will be stressed.

5.2 Generalized solutions

From a mathematical point of view the essential difficulty with the analysis of equation (5.1) is associated with its degeneracy. If $a, b \in C^1([0,\infty))$ and $a'(\nu) > 0$ for all $\nu \geq 0$ then the equation may be classified as a quasilinear equation of parabolic type. In this case there is a standard theory for the equation [48,77]. However, writing (5.1) as

$$\frac{\partial u}{\partial t} = a'(u)\frac{\partial^2 u}{\partial x^2} + \frac{\partial}{\partial x}(a'(u))\frac{\partial u}{\partial x} + \frac{\partial}{\partial x}(b(u))$$

we see that if we admit the possibility that $a'(\nu) \to 0$ as $\nu \to 0$ then this equation may revert to hyperbolic type as it were when $u = 0$. This is the implicit degeneracy of the equation.

The degeneracy of the equation has a number of far-reaching consequences [8,75,87,90]. In the first instance it represents an impediment to the application of the standard theory for parabolic partial differential equations. This means for the development of a suitable theory for this equation one has possibly to generalize (abstract) the notion of a solution of the equation. As we shall see later, in the main this is essential for equation (5.1).

Consider a domain Ω of the form

$$\Omega = (\eta_1, \eta_2) \times (\tau_1, \tau_2] \tag{5.4}$$

or

$$\Omega = (\eta_1, \eta_2) \times (\tau_1, \infty)$$

with

$$-\infty \leq \eta_1 < \eta_2 \leq \infty \text{ and } 0 \leq \tau_1 < \tau_2 < \infty. \tag{5.5}$$

Let R denote a bounded rectangle of the canonical form

$$R = (x_1, x_2) \times (t_1, t_2] \subseteq \Omega \tag{5.6}$$

and

$$\psi \in C^{2,1}(\overline{R}) \tag{5.7}$$

denote a function such that

$$\psi(x_1, t) = \psi(x_2, t) = 0 \quad \text{for all} \quad t \in [t_1, t_2]. \tag{5.8}$$

Then formally if we have a solution of equation (5.1) in Ω, multiplying equation (5.1) by ψ and integrating by parts over R there holds

$$\begin{aligned}
&\int_{t_1}^{t_2} \int_{x_1}^{x_2} \{a(u)\frac{\partial^2 \psi}{\partial x^2} - b(u)\frac{\partial \psi}{\partial x} + u\frac{\partial \psi}{\partial t}\} dx\, dt \\
&= \int_{x_1}^{x_2} \{u(x, t_2)\psi(x, t_2) - u(x, t_1)\psi(x, t_1)\} dx \\
&+ \int_{t_1}^{t_2} \{a(u(x_2, t))\frac{\partial \psi}{\partial x}(x_2, t) - a(u(x_1, t))\frac{\partial \psi}{\partial x}(x_1, t)\} dt.
\end{aligned} \tag{5.9}$$

Thus if u has a solution in Ω, (5.9) must hold for all rectangles R of the form (5.6) and functions ψ of the type (5.7), (5.8). We shall henceforth say that u is a *generalized solution* of (5.1) in Ω if u is continuous, nonnegative and bounded in Ω and satisfies (5.9) for all rectangles of the form (5.6) and functions ψ satisfying (5.7), (5.8). This means that any classical solution is a generalized solution. However the reverse is not necessarily true, and we have extended the concept of a solution.

5.3 Existence and uniqueness

We now examine the question of finding a generalized solution to equation (5.1) satisfying certain given initial and boundary conditions. For convenience here, we shall restrict attention to the *Cauchy problem*, i.e. the problem of solving (5.1) in the half-plane

$$\Omega = (-\infty, \infty) \times (0, \infty)$$

subject to the initial condition

$$u(x, 0) = u_0(x) \quad \text{for} \quad -\infty < x < \infty. \tag{5.10}$$

Here u_0 is a given nonnegative bounded continuous function. Results similar to those presented here for the Cauchy-Dirichlet and first boundary value problems can be found in [61].

We have already mentioned that if the equation (5.1) would be not degenerate, then standard theory available for treating it. We take advantage of this for getting a grip on solutions of (5.1) in general.

Let

$$M > \sup\{u_0(x) : -\infty < x < \infty\}.$$

Then [61], we can construct a sequence of functions $\{u_{0,k}\}_{k=1}^{\infty}$ with the following properties:

$$u_{0,k} \in C^2([-k, k]),$$

$$1/k \le u_{0,k}(x) \le M \quad \text{for all} \quad x \in [-k, k], \tag{5.11}$$

$$u_{0,k}(\pm\, k) = M,$$

$$((a(u_{0,k}))'' + (b(u_{0,k}))')(\pm k) = 0, \; u_{0,k+1}(x) \le u_{0,k}(x) \quad \text{for all} \quad x \in [-k, k] \tag{5.12}$$

and

$$u_{0,k}(x) \downarrow u_0(x) \text{ as } k \uparrow \infty \text{ for all } x \in (-\infty, \infty).$$

See Figure 5.1. Next, we modify the functions a and b by introducing the functions a_k and b_k for any $k \geq 1$ with the following properties:

$$a_k(\nu) = a(\nu) \text{ for } 1/(k+1) \leq \nu \leq M \tag{5.13}$$

and

$$b_k(\nu) = b(\nu) \text{ for } 1/(k+1) \leq \nu \leq M. \tag{5.14}$$

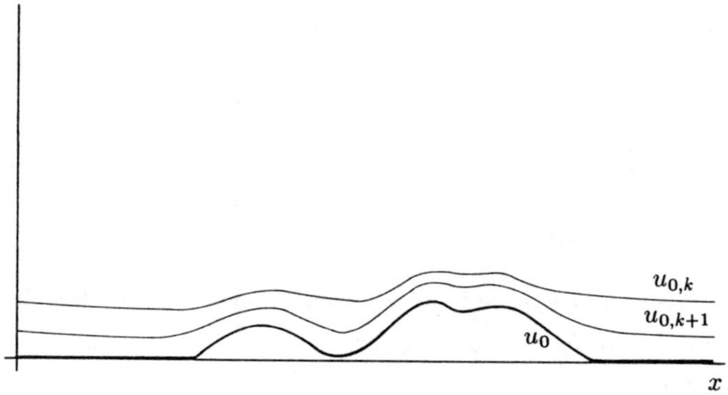

Figure 5.1: Approximating sequence for initial data functions

Moreover $a_k, b_k \in C^2(-\infty, \infty)$ and are such that a_k'' and b_k'' are uniformly Hölder continuous on $(-\infty, \infty)$ and

$$a_k'(\nu) > 0 \text{ for all } \nu \in (-\infty, \infty).$$

See Figure 5.2. For details of these constructions we refer to [61].

Now, by the standard theory [48,77] the equation

$$\frac{\partial u}{\partial t} = \frac{\partial^2}{\partial x^2}(a_k(u)) + \frac{\partial}{\partial x}(b_k(u))$$

has a classical solution $u_k(x,t)$ in the domain

$$\Omega_k := (-k, k) \times (0, \infty)$$

satisfying

$$u_k(\pm k, t) = M \text{ for all } t \in [0, \infty) \tag{5.15}$$

and

$$u_k(x,0) = u_{0,k}(x) \quad \text{for all} \quad x \in [-k, k]. \tag{5.16}$$

Furthermore, by application of the maximum principle from the standard theory [48,97] we know that by (5.11), (5.15), (5.16)

$$1/k \le u_k(x,t) \le M \quad \text{for all} \quad (x,t) \in \Omega_k \tag{5.17}$$

and by (5.12)

$$u_{k+1}(x,t) \le u_k(x,t) \quad \text{for all} \quad (x,t) \in \Omega_k.$$

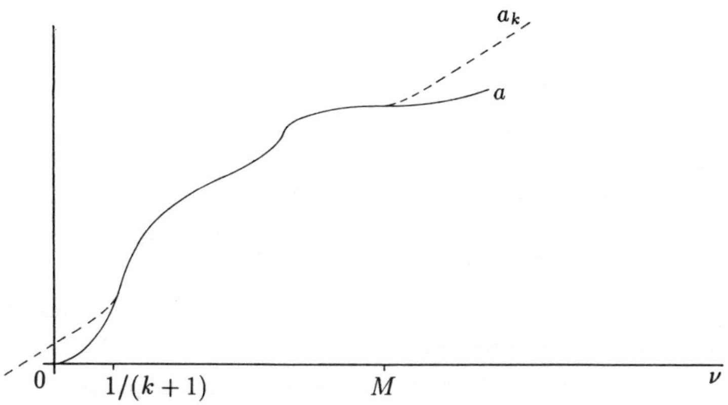

Figure 5.2: Modification of coefficients

The key is now that since (5.17) holds, and, by (5.13) and (5.14) the coefficients a_k and a, and, b_k and b, are identical for the range of values in (5.17), this means that u_k actually solves (5.1) in Ω_k.

Thus, we have constructed a decreasing sequence of positive classical solutions $\{u_k\}_{k=1}^{\infty}$ of (5.1) on a sequence of increasing domains $\{\Omega_k\}_{k=1}^{\infty}$. Our contention is that the limit

$$u(x,t) := \lim_{k \to \infty} u_k(x,t)$$

defines a solution of problem (5.1), (5.10).

By the construction, the limit $u(x,t)$ is nonnegative and bounded. Moreover, since as we have seen any classical solution is a generalized solution, each u_k satisfies (5.9) for all $R \subseteq \Omega_k$ and appropriate ψ. Subsequently taking the limit $k \uparrow \infty$ it is easily verified that the limit u satisfies (5.9) of the definition of a generalized solution of equation (5.1) in Ω. Plainly (5.10) holds too. The crunch comes in showing that u is continuous in $\overline{\Omega}$.

The earliest proposed method of showing that u is continuous was to obtain an estimate of $\frac{\partial}{\partial x}(a(u_k))$ in Ω_k which is independent of k [54,66,75,87]. This leads to an estimate of the equicontinuity of the sequence $a(u_k)$ with respect to x and if $a'(\nu)$ is bounded on $[0, M]$ also to an estimate of the equicontinuity of $\{u_k\}_{k=1}^{\infty}$ with respect to t. However, some restrictions, either on the coefficients a and b, or on the initial data function u_0 were traditionally needed in this approach. Recently though [61] there has been considerable progress in this direction. Curiously enough this has come down to replacing the search for estimates of $\frac{\partial}{\partial x}(a(u_k))$ by estimates of

$$q^{(k)} = -\left\{\frac{\partial}{\partial x}(a(u_k)) + b(u_k)\right\}.$$

By considering $q^{(k)}$ the optimal results [61] can be obtained. With hindsight this has some physical significance for $q^{(k)}$ can be equated with the flux in the derivation of equation (3.6). So, the technique comes down to uniformly estimating the flux of the approximating sequence of solutions $\{u_k\}_{k=1}^{\infty}$. This leads to the required continuity of the limit function u and an estimate in the mathematical sense of distributions of the flux term for this solution. The whole displays a remarkable symbiosis of mathematics and physics.

Taking advantage of the construction process, uniqueness of the generalized solution of problem (5.1), (5.10) may also be established [61].

We note that the assumptions on the functions a and b are the minimal required for the construction of the sequences of functions $\{a_k\}_{k=1}^{\infty}$ and $\{b_k\}_{k=1}^{\infty}$ with the stated properties. Thus, within the confines of the construction method these assumptions cannot be further relaxed.

5.4 Travelling waves

From the previous analysis it is of course still not clear that it was actually necessary to define a generalized solution of problem (5.1), (5.10). All that we have remarked up to now is that because of the implicit degeneracy of the equation, we could not consider classical theory. To see that we have not been taking unnecessary precautions, we exhibit now some generalized solutions of equation (5.1) which are not classical solutions. Having remarked that it is the degeneracy of the equation at $u = 0$ which is the cause of the difficulties we naturally look for solutions which become zero at some point.

A suitable starting point for finding tangible solutions of a differential equation is to search for similarity solutions. These are solutions which in general depend on a smaller number of variables. Thus in this case, reducing from the variables x and t, solutions which depend on only one variable. In the next section we shall discuss similarity solutions in a general setting and in a specific setting of equation (5.1) in more detail. A very obvious similarity solutions is a travelling-wave solution. In fact for (5.1) with arbitrary coefficients we do not have any other candidate for a similarity solution. This is a solution of the form

$$u(x,t) = f(\xi) \quad \text{where} \quad \xi = \lambda t - x \tag{5.18}$$

for some parameter λ.

Formally, if equation (5.1) has a travelling-wave solution of the type (5.18), substituting in (5.1) the function f must satisfy

$$\lambda f' = (a(f))'' - (b(f))'.$$

Subsequently integrating

$$(a(f))' = \varepsilon + \lambda f + b(f) \tag{5.19}$$

where ε is a constant of integration. For the further analysis of (5.19) we have to make a distinction between whether $(a(f))' = 0$, $(a(f))' > 0$, or $(a(f))' < 0$. For simplicity, let us consider the case that $a(f)$ is nondecreasing. In this case, if we are to permit a solution which becomes zero we must have

$$\varepsilon + \lambda f + b(f) \geq 0, \tag{5.20}$$

and in particular when $f = 0$, $\varepsilon \geq 0$. Then, integrating (5.19) once more yields

$$\int_{f_0}^{f} a'(\nu)/\{\varepsilon + \lambda\nu + b(\nu)\}d\nu = \xi \tag{5.21}$$

for some constant f_0. Looking for solutions which become zero formally this means that one must have

$$\left|\int_{0}^{f} a'(\nu)/\{\varepsilon + \lambda\nu + b(\nu)\}d\nu\right| < \infty. \tag{5.22}$$

Subsequently one can rewrite (5.21) as

$$\int_{0}^{f} a'(\nu)/\{\varepsilon + \lambda\nu + b(\nu)\}d\nu = \xi - \xi_0 \quad \text{for} \quad \xi > \xi_0 \tag{5.23}$$

for some value ξ_0. For completion, if f is to be nondecreasing there must hold

$$f(\xi) = 0 \quad \text{for} \quad \xi \leq \xi_0. \tag{5.24}$$

Now, if $\varepsilon > 0$ then (5.20) is automatically satisfied. However, checking (5.23), (5.24) in the definition of a generalized solution it can be ascertained that in this case the function $u(x,t)$ defined by (5.18) cannot be a generalized solution of (5.1). Actually it constitutes a generalized subsolution. Cf. [60,65]. It is therefore necessary to take $\varepsilon = 0$.

Taking $\varepsilon = 0$ in (5.20) and (5.22) means that for a travelling-wave solution of (5.1) which can become zero there must hold

$$\lambda\nu + b(\nu) \geq 0 \quad \text{for} \quad 0 \leq \nu \leq \delta \tag{5.25}$$

and

$$\int_0^\delta a'(\nu)/\{\lambda\nu + b(\nu)\}d\nu < \infty \tag{5.26}$$

for some $\delta > 0$.

This is a formal argument. Notwithstanding, retracing it, it can be checked that if (5.25) and (5.26) hold, then

$$\int_0^{u(x,t)} a'(\nu)/\{\lambda\nu + b(\nu)\}d\nu = \lambda t - x - \xi_0 \quad \text{for} \quad x < \xi_0 + \lambda t \tag{5.27}$$

$$u(x,t) = 0 \quad \text{for} \quad x \geq \xi_0 + \lambda t \tag{5.28}$$

really does define a generalized solution of (5.1). Moreover, differentiating (5.23) and (5.24) which we can do as long as $x \neq \xi_0 + \lambda t$, we find that the constructed solution satisfies

$$u_t(x,t) = \begin{cases} \lambda[\{\lambda u + b(u)\}/a'(u)](x,t) & \text{for } x < \xi_0 + \lambda t \\ 0 & \text{for } x > \xi_0 + \lambda t \end{cases}.$$

Thus if

$$\{\lambda\nu + b(\nu)\}/a'(\nu) \not\to 0 \quad \text{as} \quad \nu \downarrow 0 \tag{5.29}$$

the time derivative of this solution has a discontinuity along the line $x = \xi_0 + \lambda t$ and subsequently cannot be a classical solution of (5.1).

As an example which illustrates that the above argument is not hollow we may take $a(\nu) = \nu^m$ and $b(\nu) = -\nu^n$ with $m > 1$ and $n > 1$. We find that for any $\lambda > 0$, (5.25) and (5.26) hold and (5.27), (5.28) defines a generalized solution of (5.1) which is bounded above by $\lambda^{1/(n-1)}$ in $(-\infty,\infty) \times (0,\infty)$. Moreover, (5.29) holds. Thus there are indeed generalized solutions of the problem (5.1), (5.10) which are not classical solutions.

5.5 Regularity

For the mathematician the situation is now clear-cut. For problem (5.1), (5.10) there is one and only one generalized solution, and in general this solution is not a classical solution. For the practitioner though, this is far from satisfactory. What does it mean to have a generalized solution? Does this mean that equation (5.1) (or (3.6)) is useless for modelling unsaturated groundwater flow? To see that this is not the case we have to examine the mathematical question of *regularity* [75].

By definition, any generalized solution $u(x,t)$ of problem (5.1), (5.10) has the following property:

- u is bounded and continuous in $\overline{\Omega}$.

To this we can add the following list [60,61]:

- The quantity $-\{(a(u))_x + b(u)\}$ is bounded in the sense of distributions in any set of the form $(-\infty,\infty) \times [\tau,\infty)$ with $0 < \tau < \infty$.
- If $-\{(a(u_0))' + b(u_0)\}$ is bounded in the sense of distributions then the quantity $-\{(a(u))_x + b(u)\}$ is actually bounded in the sense of distribution in $\overline{\Omega}$.
- If there is a constant $\mathcal{C}$ such that in the sense of distributions $-\{(a(u_0))' + b(u_0)\}(x) \leq \mathcal{C}u_0(x)$ for all $x \in (-\infty,\infty)$ then in the sense of distributions $-\{(a(u))_x + b(u)\}(x,t) \leq \mathcal{C}^*u(x,t)$ for all $(x,t) \in \overline{\Omega}$ for some constant $\mathcal{C}^* \geq \mathcal{C}$.
- Under the assumptions of the previous item, $-\{(a(u))_x + b(u)\}$ is continuous in $\overline{\Omega}$.
- If $a'(\nu)/\nu$ is integrable in a neighbourhood of zero then given any $\tau \in (0,\infty)$ there exists a constant $\mathcal{C}^*$ such that in the sense of distributions $-\{(a(u))_x + b(u)\}(x,t) \leq \mathcal{C}^*u(x,t)$ for all $(x,t) \in (-\infty,\infty) \times [\tau,\infty)$.
- Under the assumptions of the previous item, $-\{(a(u))_x + b(u)\}$ is continuous in $(-\infty,\infty) \times [\tau,\infty)$ for any $\tau \in (0,\infty)$.
- u is a classical solution of (5.1) at any point $(x,t) \in \Omega$ where $u(x,t) > 0$.

To interpret these results for the physical modelling of unsaturated groundwater flow, note that the quantity $-\{(a(u))_x + b(u)\}$ may identified with soil-moisture flux q, and that subsequently the quantity $-\{(a(u))_x + b(u)\}/u$ may be identified with the soil-moisture particle velocity $v = q/\theta$. In this light the above-mentioned regularity results may be explained for equation (3.6) as follows.

- The soil-moisture content θ is continuous and bounded in the domain under consideration.
- The flux is bounded from any positive time onwards.
- If the initial flux is bounded, the flux is bounded from time zero.
- If the initial particle velocity is bounded then the particle velocity is bounded at all later times.
- In this case the flux is continuous.
- If soil-moisture diffusivity approaches zero fairly quickly as soil-moisture content θ goes to zero, then soil-moisture particle velocity is bounded from any positive time onwards.
- In this case also the flux is continuous.
- The equation (3.6) is solvable after-all in the classical sense at any point where the moisture content is not zero.

Thus albeit that one has to consider solutions of problem (5.1), (5.10) defined in a generalized sense, these solutions turn out to possess all the properties which one would physically consider necessary [63]. Furthermore, since trivially (5.1) is satisfied in the classical sense in the interior of any subdomain of Ω where a solution u is identically zero, the only points in which (5.1), (5.10) is not solvable in the classical sense and which therefore actually necessitate the introduction of the notion of a generalized solution are points on an interface between subdomains where a solution is positive and where it is zero. With hindsight we note that this is wholly consistent with the physical derivation of (3.6), since the expression $p(\theta)$ and $(dp/d\theta)(\theta)$ are not defined at $\theta = 0$.

5.6 Necessary and sufficient conditions for interfaces

We have just remarked that the only points (x, t) in which a generalized solution $u(x, t)$ of problem (5.1), (5.10) may fail to be a classical solution are points on an interface between regions where u is positive and where it is zero. Incidentally, this property was exhibited by the travelling waves constructed in Subsection 5.4. The question may now be asked: Under what conditions do such interfaces occur?

This question is important with regard to the background to equation (3.6) [53,63]. It can be couched as: Do wetting-fronts occur? This is to say, given a dry medium into which moisture is introduced in a small area, can one detect a front demarking the extent of infiltration of the moisture at any later time. If this is not the case, the moisture must penetrate the medium at infinite speed. In mathematical terms, one talks of *infinite speed of propagation* for solutions of (5.1). This question can be epitomized as follows. Consider a generalized solution $u(x, t)$ of problem (5.1), (5.10) and set

$$\mathcal{P}(t) = \{x \in (-\infty, \infty) : u(x, t) > 0\} \tag{5.30}$$

for any $t \geq 0$. Define

$$\zeta(t) = \sup \mathcal{P}(t). \tag{5.31}$$

Then if $-\infty < \zeta(0) < \infty$, is it true that $-\infty < \zeta(t) < \infty$ for all $t \in (0, \infty)$? Note that we could equally well consider the equivalent interface defining the infimum of $\mathcal{P}(t)$. However by the simple transformation of coordinates $x \to -x$ we can transform this question into the one to be studied.

It turns out that the answer to the question is intricately bound up with the study of travelling-wave solutions of (5.1) [60,65]. As we have seen if we formally wished to construct a decreasing travelling-wave solution which became zero we needed the conditions (5.25) and (5.26). Noting that if these conditions hold for a certain parameter $\lambda = \lambda^*$, then choosing $\lambda > \lambda^*$ we have

$$\lambda\nu + b(\nu) > 0 \quad \text{for all} \quad s \in (0, \delta) \tag{5.32}$$

and

$$\int_0^\delta a'(\nu)/\{\lambda\nu + b(\nu)\} ds < \infty \tag{5.33}$$

for some $\delta \in (0, \infty)$. The precise answer to the question of finite speed of propagation is [60]: Equation (5.1) displays finite speed of propagation if and only if there exists a $\lambda \in (-\infty, \infty)$ such that (5.32) and (5.33) hold for some $\delta \in (0, \infty)$. This boils down to the statement that (5.1) admits interfaces if and only if there are travelling-waves with interfaces.

To prove the general result we need two tools. The first tool we require is the following identity for a solution of problem (5.1), (5.10):

$$\int_{-\infty}^{\infty} u(x,t)dx = \int_{-\infty}^{\infty} u_0(x)dx \tag{5.34}$$

for all $t \in (0, \infty)$. This identity can be interpreted in physical terms as the statement of *conservation of mass* for solutions of (3.6), since from the beginning we have assumed that density is constant and equal to unity. The identity (5.34) may be proved formally by integrating equation (5.1) with respect to x from $-\infty$ to ∞. It can be proved rigorously by taking a sequence of rectangles $R_k = (-k, k) \times (0, T]$ and a sequence of functions $\psi_k \in C^{2,1}(R_k)$ which approximate 1 at every point in Ω in the limit $k \to \infty$, in the definition of a generalized solution [53,60].

The second tool is a *comparison principle* [61]. We say that a function $\overline{u}(x,t)$ is a generalized supersolution of equation (5.1) in a domain Ω of the form (5.4), (5.5) if it conforms to the definition of a generalized solution with " $=$ " replaced by " $\leq$ " in (5.9). Likewise, we say that a function $\underline{u}(x,t)$ is a generalized subsolution of equation (5.1) in Ω if it conforms to the definition of a generalized solution with " $=$ " replaced by " $\geq$ ". The comparison principle states that given a generalized supersolution $\overline{u}(x,t)$ and a generalized solution $u(x,t)$ of equation (5.1) in a domain Ω of the form (5.4), (5.5), if $\overline{u}(x,t) \geq u(x,t)$ for all $(x,t) \in \overline{\Omega}\backslash\Omega$ there holds $\overline{u}(x,t) \geq u(x,t)$ for all $(x,t) \in \overline{\Omega}$. Similarly, given a generalized subsolution $\underline{u}(x,t)$ of equation (5.1) in Ω such that $\underline{u}(x,t) \leq u(x,t)$ for all $(x,t) \in \overline{\Omega}\backslash\Omega$ there holds $\underline{u}(x,t) \leq u(x,t)$ for all $(x,t) \in \overline{\Omega}$. This comparison principle may be interpreted in physical terms by saying that given two identical soil samples, if one is initially wetter than the other and is continually maintained wetter on the perimeter, then at any given time it will also be wetter everywhere in the interior of the sample. Also if one injects water into the first sample, or abstracts it from the second, the conclusion remains valid.

Using these two tools the general result can be proved [60]. The identity

(5.34) ensures that a solution of problem (5.1), (5.10) cannot vanish. The comparison principle firstly allows us to compare any solution of (1.1), (1.8) with a travelling wave with an interface to deduce that there is finite speed of propagation if (5.32) and (5.33) hold. On the other hand, in the derivation of travelling waves in Subsection 5.4 we remarked that for any $\varepsilon > 0$ and λ we could construct a generalized subsolution of equation (5.1). It is these generalized subsolutions that provide the key for the proof of the necessity of (5.32) and (5.33). For details, see [60,65].

The conditions (5.32) and (5.33) tell us precisely what is necessary for the existence of a wetting-front in the solution of equation (3.6). If one believes in wetting-fronts in upward and downward directions there must hold

$$K(\theta)/\theta \leq \lambda \text{ for some } \lambda < \infty \tag{5.35}$$

and

$$D(\theta)/\theta \text{ is integrable on } (0, \delta) \tag{5.36}$$

for some $\delta > 0$. Noting though that (5.35) is automatically satisfied if K is convex as shown in Figure 3.3 and as argued in Subsection 3.5, this comes down to the constraint (5.36) on admissible diffusivity coefficients for modelling soil-moisture flow.

Note that from a mathematical point of view if (5.32) and (5.33) and their counterparts for the interface $\inf \mathcal{P}(t)$ do not hold, even if equation (5.1) is formally of degenerate type this degeneracy has no real effect. For in this case every solution will be necessarily positive at any positive time, and a positive solution is always a classical solution.

5.7 Further properties of the interfaces

Having come to the conclusion that the admission of an interface of the type (5.30), (5.31) by a solution of equation (5.1) and the corresponding degeneracy of the equation is the essential difficulty with the mathematical analysis of the equation, one may be interested in specific properties of such an interface. From a physical point of view, these are the properties of the *wetting-front* [53,60,63].

The first property we mention which has been proven is that there is a minimum speed at which an interface may move. This is

$$\lambda_0 = \limsup_{\nu \to 0} -b(\nu)/\nu. \tag{5.37}$$

The precise mathematical result [60] is that for any $t_1 \in [0, \infty)$ and $t_2 \in (t_1, \infty)$ there holds

$$\zeta(t_2) \geq \zeta(t_1) + \lambda_0(t_2 - t_1).$$

Note that by the constraint (5.32) for the existence of an interface, necessarily $\lambda_0 < \infty$, although we cannot exclude the possibility that $\lambda_0 = -\infty$. Curiously, if we consider the Richards' equation (3.6) serving as a prototype for (5.1) we find that for the movement of a wetting-front in a vertical profile with x directed downwards

$$\lambda_0 = \limsup_{\theta \to 0} K(\theta)/\theta = ? \geq 0, \tag{5.38}$$

for the movement of water in a horizontal profile

$$\lambda_0 = 0$$

by default, whilst for the movement of water in a vertical profile with x directed upwards

$$\lambda_0 = \limsup_{\theta \to 0} -K(\theta)/\theta = ? \tag{5.39}$$

Noting though that since the indications are that hydraulic conductivity is a convex function (cf. Subsection 3.5 and Figure 3.3 in particular), the question marks in (5.38) and (5.39) may be replaced by zero. Thus, for the application of (5.1) in soil-moisture theory $\lambda_0 = 0$ always. This means that a wetting-front can actually only advance. This in turn has the consequence in physical terms that once moisture is introduced at a point in a soil there will always be moisture present at that point at any later time. Of course, in the underlying continuum model, this may occur is increasingly smaller concentrations in increasingly narrower pores.

The second property of interfaces we mention is that for an arbitrary solution of (5.1) which exhibits an interface (5.30), (5.31) if λ_0 defined by (5.37) satisfies $\lambda_0 > -\infty$, then [60] the interface $\zeta(t)$ is continuous on $[0, \infty)$. Thus, for soil-moisture flow where we have seen that $\lambda_0 = 0$ a wetting-front moves continuously. If $\lambda_0 = -\infty$ however the continuity of the interface $\zeta(t)$ remains as yet an open question. In general, all we know is that the interface is lower semi-continuous and continuous from the right on $[0, \infty)$. However, we cannot exclude the possibility that the interface jumps from left to right. That there are relatively simple forms of (5.1) for which this question is still open can be seen by examining $a(\nu) = \nu^m$ and $b(\nu) = \nu^n$ with $0 < n < 1$ and $m > n$.

The third property of the interface $\zeta(t)$ worth examining is that of its description of the interface as a mathematical free boundary. This property becomes clear if one returns to the use of (5.1) as a possible model for unsaturated soil-moisture flow. For such a flow, the interface is a material surface. On one side there is no moisture, and on the other side there is. The movement of the interface must subsequently be determined by the velocity of motion of the water particles constituting it. Recalling that in physical terms, the flux $q = -\{(a(u))_x + b(u)\}$ and the soil-moisture particle velocity v is given by $v = -\{(a(u))_x + b(u)\}/u$, one would expect therefore that the speed of the interface ζ at any time $t \in (0, \infty]$ be given by

$$\zeta'(t) = \lim_{x \uparrow \zeta(t)} -[\{(a(u))_x + b(u)\}/u](x, t).$$

That this property is true has been proven in a weak sense [53,60]. In this respect though there is still a lot of mathematical work to be done. For the specific case $a(\nu) = \nu^m$ with $m > 1$ and $b \equiv 0$ strong results have been obtained [8], but in general as yet there is little theory.

5.8 Localization

The final question to be studied in this survey of the mathematical analysis of (5.1) is that of *localization* [62]. For this it is more convenient to turn from the Cauchy problem for (5.1) to the *Cauchy-Dirichlet problem*. This is the problem of solving (5.1) in the quarter-plane

$$\Omega = (0, \infty) \times (0, \infty)$$

subject to the initial conditions

$$u(x, 0) = u_0(x) \quad \text{for} \quad 0 < x < \infty \tag{5.40}$$

and

$$u(0, t) = \Psi(t) \quad \text{for} \quad 0 < t < \infty. \tag{5.41}$$

Here u_0 and Ψ are both continuous functions on $[0, \infty)$ which satisfy the compatibility condition $u_0(0) = \Psi(0)$. Furthermore, it will be assumed that (5.32) and (5.33) hold and that u_0 has compact support in $[0, \infty)$. In this case existence, uniqueness and interface property results similar to those elaborated for the Cauchy problem can be obtained by analogy [62]. Thus setting

$$\mathcal{P}(t) = \{x \in (0, \infty) : u(x,t) > 0\}$$

we can define

$$\zeta(t) = \sup \mathcal{P}(t)$$

for all $t \geq 0$ in the knowledge that $\zeta(t) < \infty$ for all $t \in [0, \infty)$.

The question of localization is: Is

$$\sup\{\zeta(t) : 0 \leq t < \infty\} < \infty \tag{5.42}$$

or, if this is not true, is there is a sequence of values $\{t_i\}_{i=1}^{\infty} \subseteq (0, \infty)$ such that $\zeta(t_i) \to \infty$ as $t_i \to \infty$. If (5.42) holds *localization* is said to occur. From the view-point of modelling unsaturated groundwater flow, this question is that of soil-moisture penetration. Having observed that a wetting-front always advances one may ask if this advancement can occur without bound. The most topical case is that of the advancement of moisture in a vertical profile where on the one hand one has capillary suction continually trying to draw moisture upwards into empty pores whilst gravity tends to pull the moisture downwards. Factors which come into play are the behaviour of the $D(\theta)$ and $K(\theta)$ relationships and the behaviour of the boundary data function $\Psi(t)$ as $t \to \infty$. The latter corresponds to the behaviour of a controlled moisture content at the boundary of the moisture profile.

A large number of results on this question have been obtained. We refer to [62] for a complete overview. Here we shall only present a few which are probably of the most interest with regard to the physical background, and which also give a taste of the others. We concentrate on the case that

$$\Psi(t) = 0 \quad \text{for all} \quad t \geq 0. \tag{5.43}$$

This corresponds to the physically worst case possible, where the soil boundary is desiccated on the boundary. If one likes, one may think of a soil-surface which is dried out by the heat of the sun.

The comparison principle discussed in Subsection 5.6 is again an important tool in the obtainment of the results we review for the considered question. Another major tool is the identity

$$\int_0^\infty xu(x,\tau)dx = \int_0^\infty xu_0(x)dx + \int_0^\tau a(\Psi(t))dt - \int_0^\tau \int_0^\infty b(u(x,t))dx\,dt$$

for any $\tau > 0$. This is very similar to the conservation of mass formula (5.34) for solutions of (5.1), (5.10) discussed in Subsection 5.6. For solutions of problem (5.1), (5.40), (5.41), it describes ***conservation of momentum.*** It may be proved formally by multiplying (5.1) by x and integrating by parts. Alternatively one may derive it [62] by using a truncation of the function $\psi = x$ as a test function in (5.9) of the definition of a generalized solution.

Using the above mentioned tools the following has been proven:

– If $\lambda_0 > 0$ and there exists a real number σ such that

$$\sigma\nu - b(\nu) > 0 \quad \text{for all} \quad \nu \in (0,\delta)$$

and

$$\int_0^\delta a'(\nu)/\{\sigma\nu - b(\nu)\}d\nu < \infty$$

for some $\delta \in (0,\infty)$, then there is no localization for problem (5.1), (5.40), (5.41), (5.43).

– If

$$b(\nu) \le 0 \quad \text{for all} \quad \nu \in (0,\delta)$$

some $\delta > 0$, then there is no localization for problem (5.1), (5.40), (5.41), (5.43).

– If

$$\int_0^\delta a'(\nu)/\nu d\nu < \infty$$

and

$$\sup\{a'(\nu)/\int_0^\nu a'(\omega)/\omega d\omega : 0 < \nu < \delta\} < \infty$$

and

$$\sup\{|b'(\nu)|\int_0^\nu a'(\omega)/\omega d\omega : 0 < \nu < \delta\} < \infty$$

for some $\delta \in (0, \infty)$; then there is no localization for problem (5.1), (5.40), (5.41), (5.43).

– If

$$b(\nu) > 0 \text{ for } 0 < \nu < \delta$$

and

$$\int_0^\delta a'(\nu)/b(\nu)d\nu < \infty \tag{5.44}$$

for some $\delta \in (0, \infty)$, there is localization for problem (5.1), (5.40), (5.41), (5.43).

Once again we refer to [62] for further details and in particular to results which apply when (5.43) is violated.

The first of the above four results is not so relevant for the modelling of unsaturated groundwater flow, since we had observed in the previous subsection that $\lambda_0 = 0$ for this physically relevant case. In general though the condition $\lambda_0 > 0$ means that there is a positive minimum speed of propagation, whilst the remaining conditions infer that there is also finite speed of propagation to the left. So, as it were, water cannot flow through the fixed boundary $x = 0$ too fast.

In contrast the second result mentioned above is of great interest for soil-moisture flow modelling. Recalling that the term $b(u)$ in (5.1) corresponds with $K(\theta)$ in (3.6) for vertical flow directed downwards, and that $b(u)$ corresponds with the zero function for horizontal soil-moisture flow, this result states that for one-dimensional horizontal or vertically downward moisture infiltration, penetration cannot be limited. This is independent of the actual soil-moisture characteristic functions D and K and of course with the worst possible boundary conditions. Thus, for one-dimensional flow in a horizontal soil profile, once water is introduced in the soil it will eventually reach every other point. Similarly in a vertical profile, moisture will also eventually reach every other point below the area of its introduction.

As examples of functions $a(\nu)$ and $b(\nu)$ to which the third result applies we may once more take $a(\nu) = \nu^m$ and $b(\nu) = \nu^n$, now with $m > 1$ and $n \geq m$. The result implies that for certain soils even in vertical flow in a direction against the pull of gravity, moisture penetration cannot be limited. Thus for soils with $D(\theta)$ and $K(\theta)$ relationships which comply to the given criterion, upward capillary suction is a more dominant mechanism than gravitational pull.

The fourth and last result mentioned above is the most thought-provoking with regard to soil-moisture modelling. Recalling that in the parallel between equations (5.1) and (3.6), the function a' correspond to the soil-moisture diffusivity function $D(\theta) = K(\theta)(dp/d\theta)(\theta)$ and for upward soil-moisture flow the function b corresponds with the hydraulic conductivity function $K(\theta)$, the combination a'/b represents the derivative of the pressure head function $dp/d\theta$. In this light (5.44) requires that $p(\theta)$ be bounded as $\theta \to 0$. This is contradictory with the generally accepted picture. Cf. Subsection 3.5, and in particular Figure 3.2. Should soil-moisture pressure head be bounded as moisture content approaches zero, then the fourth result above implies that moisture penetration in an upward direction may be limited. However, seeing as mentioned, that this does not appear to be the case, for soils whose characteristics do not conform to the constraints implied in the third criterion above, the question of upward penetration in a soil with a desiccated lower boundary must necessarily remain open.

5.9 Postscript

We close this chapter with a moral which can be discerned from the previous analysis [63]. The interchangeable consideration of both mathematical and physical aspects of the problem in hand has been indispensable for the current understanding of both the mathematics and the physics involved. It is curious to note that among many mathematicians (cf. [54]) it was long thought that the inclusion of the convection term in equation (5.1) only represented a technical difficulty with regard to the extension of the theory for the pure diffusion equation without the convection term. However, it is now clear that this is not the case [60,64].

6 Similarity solutions of the porous media equation

6.1 What similarity solutions are and why we study them

In the previous chapters we have seen something of the role of differential equations in the study of groundwater hydrology, and the present state of some of the mathematical analysis of one such equation. From the latter it can be discerned that the current mathematical theory can sometimes be lagging behind the practical applications one has in mind. The question of what one should do when first confronted with a differential equation for which little or no theory is available can then be raised. In this section one hopeful line of attack will be illustrated. This is that of searching for similarity solutions of the equation in hand, and subsequently using elementary techniques from the theory of ordinary differential equations to investigate their properties.

In general a solution of a differential equation is said to be a similarity solution if it is invariant under a continuous transformation depending on some parameter for which the equation is also invariant [6,7,22]. As an example consider the equation

$$\frac{\partial u}{\partial t} = \frac{\partial^2 u^m}{\partial x^2} \quad \text{with} \quad m > 1 \tag{6.1}$$

which will actually be the centre of attention here. A function $u(x,t)$ is a similarity solution of (6.1) if there is a continuous transformation

$$t \to t'(x,t,u;\chi) \tag{6.2}$$

$$x \to x'(x,t,u;\chi) \tag{6.3}$$

$$u \to u'(x,t,u;\chi) \tag{6.4}$$

depending on the parameter χ such that

$$u'(x,t,u;\chi) = u(x',t')$$

and

$$\frac{\partial u'}{\partial t'} = \frac{\partial^2 (u')^m}{(\partial x')^2} \quad \text{for all } \chi.$$

In the study of partial differential equations one often comes across the concept of a similarity solution as a solution of the equation which is expressible in a form which leads to a reduction of the number of dependent variables. For example, for equation (6.1) we may consider a solution $u(x,t)$ of the form

$$u = f(\xi) \text{ with } \xi = xt^{-1/2}. \tag{6.5}$$

This is the famous Boltzmann transformation. Substituting in (6.1) we find that the function f must satisfy

$$\tfrac{1}{2}\xi \frac{df}{d\xi} = \frac{d^2 f^m}{d\xi^2}.$$

Thus the solution u dependent on the variables x and t has a form for which the partial differential equation becomes an ordinary differential equation depending only as the variable ξ. We note that a similarity solution of the form (6.5) is a similarity solution in the sense of the first-presented definition, for it is invariant under the transformation

$$t' = \chi^2 t, \; x' = \chi x, \; u' = u$$

with χ an arbitrary parameter. We refer the interested reader to [6,7,22,27,88] for a fuller discussion of the link between the two definitions.

Similarity solutions of partial differential equations are most often studied in cases when no explicit solutions of the equation in hand can be found, either in closed form or in series form. They are subsequently generally of the greatest interest for nonlinear problems. In the first instance, similarity solutions are studied solely because, depending on fewer variables, they are easier to analyse. They subsequently provide an easier means of obtaining insight into the properties to be expected of arbitrary solutions. This is naturally of the greatest benefit when it is not apparent how arbitrary solutions can be analysed. Similarity solutions have however also been shown to describe the asymptotic behaviour of certain classes of arbitrary solutions [6,15,16,55]. Furthermore, in this light, it is now also apparent that they are useful for the possession of insight needed for the development of appropriate numerical methods for solving the original differential equation. Lastly, we mention that not only do similarity solutions provide insight into the properties of arbitrary solutions, but under certain circumstances they can actually be used to characterize the behaviour of arbitrary solutions [8,65].

6.2 Transformations for the porous media equation

In this section we shall be concerned with the mathematical analysis of similarity solutions of equation (6.1). This equation has an adequate degree of complication to be able to illustrate the points we wish to make without being so complicated that one becomes bogged down in too many details. Equation (6.1) may be regarded as a special case of equation (5.1). Considering horizontal soil-moisture flow, so that the first-order gravity or convection term on the right-hand side of (5.1) vanishes, and taking a diffusion coefficient of power-law type, one obtains (6.1). Equation (6.1) is generally known as the ***porous media equation***, because of its occurrence as a model for the flow of an adiabatic gas in a porous medium [85]. This equation also arises though in many other fields [6,8,70,90]. See [6,8,90] for a general discussion, and [57] for a specific discussion of further hydrological applications of (6.1).

Here we shall explain the ideas of studying similarity solutions of (6.1) by looking for similarity solutions of the form

$$u(x,t) = X(t)f(\xi) \quad \text{with} \quad \xi = xY(t) \tag{6.6}$$

where $X > 0$, $f > 0$ and $Y > 0$. There are many methods of deriving admissible similarity solutions. A commonly used method is that of the algebraic analysis of infinitesimal transformations of the type (6.2)-(6.4) [6,7,22,27,88]. Here though we shall apply a straightforward extension of the separation of variable techniques [57] to determine all possibilities of the form (6.6).

Specifically, letting u denote a solution of equation (6.1) of the form (6.6), substituting in (6.1) we deduce that necessarily

$$X'f + XY^{-1}Y'\xi f' = X^m Y^2 (f^m)''.$$

Hence

$$(f^m)'' = X^{1-m}Y^{-3}Y'\xi f' + Y^{-2}X^{-m}X'f. \tag{6.7}$$

Differentiating with respect to t, now yields

$$0 = (X^{1-m}Y^{-3}Y')'\xi f' + (Y^{-2}X^{-2}X')'f. \tag{6.8}$$

Subsequently dividing by f and differentiating with respect to η, we find

$$(X^{1-m}Y^{-3}Y')'(\xi f^{-1}f')' = 0. \tag{6.9}$$

Suppose now that

$$(\xi f^{-1} f')' = 0. \tag{6.10}$$

Then $\xi f^{-1} f' = \kappa$ for some constant κ. Whence

$$f = \sigma \xi^{\kappa}$$

for some further constant σ. However in this case we can write

$$u = X(t)\sigma\xi^{\kappa} = X(t)\sigma(xY(t))^{\kappa},$$

or

$$u = X_1(t)\sigma(xY_1(t))^{\kappa}$$

where

$$X_1(t) = X(t)Y^{\kappa}(t) \quad \text{and} \quad Y_1(t) = 1.$$

Thus without loss of generality we make take $Y' = 0$ if (6.10) holds. Either way, from (6.9) we may conclude that

$$(X^{1-m}Y^{-3}Y')' = 0. \tag{6.11}$$

without loss of generality. Whence, by (6.8)

$$(Y^{-2}X^{-m}X')' = 0. \tag{6.12}$$

Solving (6.11) and (6.12) one then derives the following three alternatives for combinations of X and Y:

$$X = (t+\tau)^{\alpha},\ Y(t) = (t+\tau)^{-\{1+(m-1)\alpha\}/2}$$
$$X = (\tau - t)^{\alpha},\ Y(t) = (\tau - t)^{-\{1+(m-1)\alpha\}/2}$$

and

$$X = \exp(\alpha t),\ Y(t) = \exp\{-\tfrac{1}{2}(m-1)\alpha t\}.$$

EXERCISE. Solve (6.11) and (6.12) to derive the above three alternatives. Show that any constants other than α and τ which may appear may be absorbed in the function f.

Substitution of the above combination in (6.7) finally leads to the following three possibilities for similarity transformations of (6.1) of the type (6.6).

Transformation I.

$$u = (t+\tau)^{\alpha} f(\xi) \quad \text{with} \quad \xi = x(t+\tau)^{-[1+(m-1)\alpha]/2}$$

which gives

$$(f^m)'' + \tfrac{1}{2}\{1+(m-1)\alpha\}\xi f' = \alpha f. \tag{6.13}$$

Transformation II.

$$u = (\tau - t)^{\alpha} f(\xi) \quad \text{with} \quad \xi = x(\tau - t)^{-[1+(m-1)\alpha]/2}$$

which gives

$$(f^m)'' - \tfrac{1}{2}\{1+(m-1)\alpha\}\xi f' = -\alpha f. \tag{6.14}$$

Transformation III.

$$u = \exp(\alpha t) f(\xi) \quad \text{with} \ \xi = x \ \exp\{-\tfrac{1}{2}(m-1)\alpha t\}$$

which gives

$$(f^m)'' + \tfrac{1}{2}(m-1)\alpha\xi f' = \alpha f. \tag{6.15}$$

In [6,7,22,88] one may read how these transformations can be derived from algebraic considerations. One may also like to refer to [57] for transformations which are not of the type (6.6).

6.3 Weak solutions

The identification of admissible similarity transformations, such as has been carried out in the previous subsection, only provides us with a number of possibilities for reducing the original partial differential equation to an ordinary differential equation. This does not mean that we actually have solutions. To determine if this is so we must study the ordinary differential equation we have derived. This is then the next step in our analysis, which we shall cover in this and coming subsections.

Equations (6.13)-(6.15) may all be cast in the form

$$(f^m)'' + p\xi f' = qf \tag{6.16}$$

with p and q constants. We shall now study bounded nonnegative solutions of equation (6.16) in the domain $0 < \xi < \infty$ with the initial condition

$$f(0) = U \quad \text{and} \quad (f^m)'(0) = \beta \tag{6.17}$$

for some $U \geq 0$. We note that with regard to equation (6.1), any solution of (6.13) leads via the Transformation I to a solution of (6.1) on the half-strip

$$H = (0, \infty) \times (0, t^*]$$

when $\tau > 0$, whilst any solution of (6.14) leads via the Transformation II to a solution of (6.1) in the half-strip H when $\tau > t^*$, and finally any solution of (6.15) leads via the Transformation III to a solution of (6.1) in the half-strip H without provisos. Thus, formally, if we have a solution of (6.16) in one of the forms (6.13) - (6.15) we have an appropriate solution of (6.1) in H with dependence on only one essential variable.

In Section 5 we have seen that for the study of equations of the type (5.1) of which (6.1) may be regarded as a special case, it was necessary to introduce the notion of a generalized solution. One should then intuitively feel that for the study of similarity solutions cast in the form (6.16), some abstraction of the concept of a solution should also be necessary. However, whereas the theory for the partial differential equation is still in a phase of early growth, for ordinary differential equations a substantial theory has already been developed. We can profit from this. Indeed, this corresponds with our idea that the study of similarity solutions is easier than the study of the original equation.

As standard reference let us take the book by Fuchik and Kufner [50]. Consider the equation

$$(F(\xi, f(\xi), f'(\xi))' = g(\xi) \tag{6.18}$$

where

$$\{F(\xi, f, \nu_1) - F(\xi, f, \nu_2)\}\{\nu_1 - \nu_2\} \geq \delta\{\nu_1 - \nu_2\}^2 \quad \text{for all} \quad \nu_1, \nu_2$$

and some $\delta > 0$. A function $f \in W^{1,2}(\xi_1, \xi_2)$ is said to be a weak solution of (6.18) if

$$\int_{\xi_1}^{\xi_2} \{F(\xi, f(\xi), f'(\xi))\psi'(\xi) + g(\xi)\psi(\xi)\}d\xi = 0 \quad \text{for all} \quad \psi \in C_0^\infty(\xi_1, \xi_2).$$

(The reader, recalling the regularity results for the porous media equation (6.1) which can be deduced from Section 5, may like to note that this definition for a weak solution of (6.16) is thoroughly consistent with that of a generalized solution of (6.1).) Now, it follows from the Sobolev Imbedding Theorem [50] that any such weak solution $f \in C([\xi_1, \xi_2])$. Furthermore, one has the following regularity results:

- If $F \in C([\xi_1, \xi_2] \times \mathbf{R}^2)$ and $g \in C([\xi_1, \xi_2])$ then any weak solution $f \in C^1([\xi_1, \xi_2])$.
- If $F \in C^1([\xi_1, \xi_2] \times \mathbf{R}^2)$ and $g \in C([\xi_1, \xi_2])$ then any weak solution $f \in C^2([\xi_1, \xi_2])$ and f is a classical solution of (6.18) on $[\xi_1, \xi_2]$.

Writing equation (6.16) in the form $((f^m)' + p\xi f)' = (p+q)f$ the first of the afore-mentioned results tells us that for any weak solution of (6.16) in an interval (ξ_1, ξ_2),

$$f^m \in C^1([\xi_1, \xi_2]).$$

Whilst, then if

$$f(\xi) > 0 \text{ for all } \xi \in [\xi_1, \xi_2].$$

then the second of the afore-mentioned results tells us that

$$f \in C^2([\xi_1, \xi_2])$$

and f is a classical solution of (6.16) on $[\xi_1, \xi_2]$. What does this mean? It means that given any weak solution of (6.16) on $0 < \xi < \infty$ such that $f(\xi) > 0$ for $\xi_1 < \xi < \xi_2$ this solution f is actually a classical solution of (6.16) on (ξ_1, ξ_2). Moreover, $(f^m)'$ is continuous on $[0, \infty)$ and

$$(f^m)'(\xi) = 0 \text{ if } f(\xi) = 0 \text{ for any } 0 < \xi < \infty. \tag{6.19}$$

The last observation (6.19) is a consequence of the continuity of $(f^m)'$, because any point where a nonnegative solution f is zero must constitute a minimum of the solution. In summary, to study weak solutions of (6.16) it is actually sufficient to study positive classical solutions of the equation and if necessary extend them or patch them together using the condition (6.19). Recall that in the previous section, we established that any generalized solution of (5.1) was a classical solution at a point where it was positive and also under appropriate conditions that the flux q was continuous. The above results are no more than the analogy to this situation for equation (6.16).

6.4 Characterization

We assert now that given any bounded nonnegative weak solution $f(\xi)$ of equation (6.16) on $0 < \xi < \infty$, if there exists a point $\xi_0 \in (0, \infty)$ such that $f(\xi_0) = 0$ then $f(\xi) = 0$ for all $\xi \geq \xi_0$. Taking into consideration the observations of the previous subsection, this infers that in fact, weak solutions of (6.16) on $0 < \xi < \infty$ which are not identically zero can be divided into two types:

Type A (with compact support). There exists an $\Xi \in (0, \infty)$ such that

$$f(\xi) > 0 \text{ for all } 0 < \xi < \Xi \tag{6.20}$$

and f is a classical solution of (6.16) on $(0, \Xi)$, whilst

$$f(\xi) = 0 \text{ for all } \xi \geq \Xi. \tag{6.21}$$

Type B (positive everywhere). There holds

$$f(\xi) > 0 \text{ for all } 0 < \xi < \infty$$

and f is a classical solution of (6.16) on $(0, \infty)$.

The three tools we use to prove this and other results [56,67,68] are elementary. All three relate to the observation that any weak solution is a classical solution where it is positive. Suppose that $f(\xi) > 0$ for all $\xi \in [\xi^-, \xi^+]$ some $0 \leq \xi^- < \xi^+ < \infty$. Then, integrating equation (6.16) from ξ^- to ξ^+ there holds

$$(f^m)'(\xi^+) - (f^m)'(\xi^-) + p\xi^+ f(\xi^+) - p\xi^- f(\xi^-) = (p+q) \int_{\xi^-}^{\xi^+} f(\nu) d\nu. \tag{6.22}$$

Secondly, multiplying equation (6.16) first by ξ and then integrating

$$\begin{aligned} &\xi^+(f^m)'(\xi^+) - \xi^-(f^m)'(\xi^-) - f^m(\xi^+) + f^m(\xi^-) \\ &\quad + p(\xi^+)^2 f(\xi^+) - p(\xi^-)^2 f(\xi^-) = (2p+q) \int_{\xi^-}^{\xi^+} \nu f(\nu) d\nu. \end{aligned} \tag{6.23}$$

Finally, multiplying equation (6.16) by $(f^m)'$ and integrating

$$\begin{aligned} \tfrac{1}{2}\{(f^m)'\}^2(\xi^+) \; - \; \tfrac{1}{2}\{(f^m)'\}^2(\xi^-) &= \frac{qm}{m+1}\{f^{m+1}(\xi^+) - f^{m+1}(\xi^-)\} \\ &\quad - \; pm \int_{\xi^-}^{\xi^+} \nu f^{m-1}(\nu)(f')^2(\nu) d\nu. \end{aligned} \tag{6.24}$$

All three of these integral identities can easily be checked by differentiation. Moreover, recalling that $(f^m)'$ is continuous on $(0, \infty)$, from (6.22) - (6.24) by continuous extension, we may deduce that these inequalities actually hold for any $0 \le \xi^- < \xi^+ < \infty$ such that

$$f(\xi) > 0 \ \text{ for } \ \xi^- < \xi < \xi^+. \tag{6.25}$$

The identities (6.22) - (6.24) when (6.25) hold are the tools we use to prove that the only possible weak solutions of (6.16) are of the two earlier-mentioned types. We conduct the proof in two steps. First we eliminate the possibility that there exists a $\xi_1 > 0$ and a $\xi_2 > \xi_1$ such that a weak solution f satisfies

$$f(\xi_1) = f(\xi_2) = 0 \tag{6.26}$$

and

$$f(\xi) > 0 \ \text{ for } \ \xi_1 < \xi < \xi_2. \tag{6.27}$$

Thereafter we eliminate the possibility that there exists a $\xi_1 > 0$ such that

$$f(\xi_1) = 0 \tag{6.28}$$

and

$$f(\xi) > 0 \ \text{ for } \ \xi_1 < \xi < \infty. \tag{6.29}$$

We eliminate both of these possibilities by arguing by contradiction.

Suppose firstly that there exists a $0 \le \xi_1 < \xi_2 < \infty$ such that (6.26) and (6.27) hold for some weak solution f. Then by the analysis of the previous section there must hold

$$(f^m)'(\xi_1) = (f^m)'(\xi_2) = 0.$$

Moreover, plainly f must have a positive maximum on (ξ_1, ξ_2). So there also exists a ξ^* with

$$\xi_1 < \xi^* < \xi_2$$

such that

$$(f^m)'(\xi^*) = 0 \ \text{ and } \ f(\xi^*) > 0. \tag{6.30}$$

Now, taking $\xi^- = \xi_1$ and $\xi^+ = \xi_2$ in (6.24) we have

$$0 = pm \int_{\xi_1}^{\xi_2} \nu f^{m-1}(\nu)(f')^2(\nu)d\nu$$

which in view of (6.26) and (6.27) can only be the case if $p = 0$. However, then taking $\xi^- = \xi_1$ and $\xi^+ = \xi^*$ in (6.24) this means

$$0 = q\frac{m}{m+1}f^{m+1}(\xi^*)$$

which can now only be the case if $q = 0$. In this last event though, (6.24) becomes

$$\{(f^m)'(\xi^-)\}^2 = \{(f^m)'(\xi^+)\}^2 = \{(f^m)'(\xi^*)\}^2 = 0$$

for all $\xi_1 \le \xi^- \le \xi^* \le \xi^+ \le \xi_2$. So f must be a constant, and this is in contradiction with (6.26) and (6.27). Thus, by contradiction, the possibility that (6.26) and (6.27) hold is eliminated.

Our next goal is the elimination of (6.28), (6.29). To attain this goal, we shall first show that given any interval $(\xi_1, \xi_2) \subseteq (0, \infty)$ such that

$$f(\xi) > 0 \text{ for } \xi \in (\xi_1, \xi_2)$$

then one of the following five possibilities must hold:

(i) $(f^m)'(\xi) = 0$ for all $\xi \in (\xi_1, \xi_2)$,

(ii) $(f^m)'(\xi) > 0$ for all $\xi \in (\xi_1, \xi_2)$,

(iii) $(f^m)'(\xi) < 0$ for all $\xi \in (\xi_1, \xi_2)$,

(iv) $(f^m)'(\xi) > 0$ for all $\xi \in (\xi_1, \xi^*)$, $(f^m)'(\xi^*) = 0$, and $(f^m)'(\xi) < 0$ for all $\xi \in (\xi^*, \xi_2)$, for some $\xi^* \in (\xi_1, \xi_2)$,

(v) $(f^m)'(\xi) < 0$ for all $\xi \in (\xi_1, \xi^*)$, $(f^m)'(\xi^*) = 0$, and $(f^m)'(\xi) > 0$ for all $\xi \in (\xi^*, \xi_2)$, for some $\xi^* \in (\xi_1, \xi_2)$.

Suppose to begin with that there is a point $\xi^* \in (\xi_1, \xi_2)$ such that (6.30) holds. Then from equation (6.16) we have

$$(f^m)''(\xi^*) = qf(\xi^*).$$

So if $q > 0$ then ξ^* must be a strict maximum of f, and if $q < 0$ then ξ^* must be a strict minimum of f. If follows that if $q \neq 0$ then only the last four possibilities given above are admissible. On the other hand if $q = 0$ and there is a $\xi^* \in (\xi_1, \xi_2)$ such that (6.30) holds, from (6.24) we have

$$\tfrac{1}{2}\{(f^m)'\}^2(\xi^-) = pm\int_{\xi^-}^{\xi^*} \nu f^{m-1}(\nu)(f')^2(\nu)d\nu$$

and

$$\tfrac{1}{2}\{(f^m)'\}^2(\xi^+) = -pm\int_{\xi^*}^{\xi^+} \nu f^{m-1}(\nu)(f')^2(\nu)d\nu$$

for any $\xi < \xi^- < \xi^* < \xi^+ < \infty$. Whence

$$(f^m)'(\xi) = 0 \text{ for all } \xi \in (\xi_1, \xi_2).$$

So in the case $q = 0$, only the first three possibilities are admissible. (Note that in fact we have shown that the first possibility can only hold when $p = q = 0$, whilst the fourth possibility can only hold when $q < 0$, and the last possibility can only hold when $q > 0$.)

Using the above result, we deduce that if (6.28) and (6.29) hold then

$$f^* := \lim_{\xi\to\infty} f(\xi)$$

exists. Furthermore, subsequently letting $\xi^+ \uparrow \infty$ in (6.24) this implies that $\lim_{\xi\to\infty}(f^m)'(\xi)$ exists, and to obtain a bounded solution of (6.16) this must mean

$$\lim_{\xi\to\infty}(f^m)'(\xi) = 0.$$

Subsequently we can set $\xi^- = \xi_1$ and let $\xi^+ \uparrow \infty$ in (6.24) to deduce

$$\frac{qm}{m+1}(f^*)^{m+1} = pm\int_{\xi_1}^{\infty} \nu f^{m-1}(\nu)(f')^2(\nu)d\nu. \tag{6.31}$$

Likewise, we can set $\xi^- = \xi_1$, and let $\xi^+ \uparrow \infty$ in (6.22). If moreover we first divide by ξ^+, this yields

$$\begin{aligned} pf^* &= \lim_{\xi\to\infty}(p+q)\textstyle\int_{\xi_1}^{\xi} f(\nu)d\nu/\xi \\ &= \lim_{\xi\to\infty}(p+q)\textstyle\int_{\xi_1}^{\xi} f^* d\nu/\xi \\ &= \lim_{\xi\to\infty}(p+q)f^*(\xi-\xi_1)/\xi \\ &= (p+q)f^*. \end{aligned}$$

Hence

$$qf^* = 0. \tag{6.32}$$

Combining (6.31) and (6.32), necessarily $p = 0$. However, in this case, setting $\xi^- = \xi_1$ in (6.24) the latter becomes

$$\tfrac{1}{2}\{(f^m)'\}^2(\xi) = \frac{qm}{m+1}f^{m+1}(\xi) \text{ for all } \xi > \xi_1.$$

So $q \geq 0$, and

$$(f^m)'(\xi) = \sqrt{\frac{2qm}{m+1}} f^{(m+1)/2}(\xi) \quad \text{for all} \quad \xi > \xi_1.$$

However, the latter can now be rewritten as

$$\frac{2m}{m-1}(f^{(m-1)/2})'(\xi) = \sqrt{\frac{2qm}{m+1}}.$$

Whence integrating from ξ_1 to $\xi > \xi_1$

$$f^{(m-1)/2}(\xi) = \sqrt{\frac{q(m-1)^2}{2m(m+1)}}(\xi - \xi_1)$$

for all $\xi > \xi_1$. Dividing by ξ and letting $\xi \uparrow \infty$ yields $0 = q$. Here though, $p = q = 0$ can also only be the case if in addition

$$(f^m)'(\xi) = (f^m)'(\xi_1) = 0 \quad \text{for all} \quad \xi > \xi_1$$

by (6.24). This provides the desired contradiction of (6.28), (6.29).

In summary, using the identities (6.22) and (6.24), we have shown that if a weak solution f satisfies $f(\xi_0) = 0$ for some $\xi_0 \in (0, \infty)$, then $f(\xi) = 0$ for all $\xi \geq \xi_0$. Hence, any weak solution of (6.16) must be of one of the two earlier-mentioned types, Type A or Type B.

6.5 Shooting arguments for solutions with compact support

A very useful technique in the study of ordinary differential equations is a shooting argument [9,21]. In a shooting argument one studies the solution of an initial-value problem, which through some reasoning or other one knows exists locally. Then ones tries to extend this solutions onto as large a domain as possible. One hopes to find that this solution is defined on the whole interval under consideration and this solution fulfils all the given boundary conditions. If it does not, one attempts to modify the initial-value problem so that the solution does exist on the given interval and moreover does satisfy all the given boundary conditions. This is the idea of shooting. Starting from the initial value one shoots a solutions. One hopes that by suitably adjusting the aim, one can breach the given interval and hit the required target in the form of all the specified boundary conditions.

This technique is ideal for studying (weak) solutions of (6.16). We apply it first to study weak solutions of Type A. In the next subsection we apply it to the study of weak solutions of Type B.

For a weak solution of Type A we know that there exists a point $\Xi \in (0, \infty)$ and an $\varepsilon \in (0, \Xi)$ such that

$$f(\xi) > 0 \text{ for } \xi \in (\Xi - \varepsilon, \Xi) \tag{6.33}$$

and

$$f(\Xi) = (f^m)'(\Xi) = 0 \tag{6.34}$$

by (6.19) - (6.21). The (backwards) initial value problem (6.16), (6.33), (6.34) is the basis for our shooting.

To start with, we identify conditions which are necessary for (6.16), (6.33), (6.34) to have a solution. Once more we use the simple integral identities (6.22) - (6.24) as the only tools of our analysis. Supposing that (6.16) has a solution satisfying (6.33) and (6.34) setting $\xi^- = \xi$ and $\xi^+ = \Xi$ in (6.22) there holds

$$(f^m)'(\xi) + p\xi f(\xi) = -(p+q)\int_{\xi}^{\Xi} f(\nu)d\nu \tag{6.35}$$

for all $\xi \in (\Xi - \varepsilon, \Xi)$. Moreover, by the argument in the previous subsection regarding the monotonicity of f on $(\Xi - \varepsilon, \Xi)$ there must be a $\xi^* \in (\Xi - \varepsilon, \Xi)$ such that

$$(f^m)'(\xi) < 0 \text{ for all } \xi \in (\xi^*, \Xi). \tag{6.36}$$

However, in (6.35) this means

$$|(f^m)'(\xi) + p\xi f(\xi)| \le |p+q| \int_{\xi}^{\Xi} f(\xi)d\nu = |p+q|(\Xi - \xi)f(\xi). \tag{6.37}$$

Whence dividing by $f(\xi)$ we find

$$|\frac{m}{m-1}(f^{m-1})'(\xi) + p\xi| \le |p+q|(\Xi - \xi).$$

Letting $\xi \uparrow \Xi$ this gives

$$\lim_{\xi \to \Xi} \frac{-m}{m-1}(f^{m-1})'(\xi) = p\Xi.$$

By (6.36) this can only be the case if $p \ge 0$. Furthermore, if $p = 0$ then by (6.35)

$$(f^m)'(\xi) = -q\int_{\xi}^{\Xi} f(\nu)d\nu$$

for any $\xi \in (\xi^*, \Xi)$. So necessarily $q > 0$. Our conclusion is that the initial value problem (6.16), (6.33), (6.34) can only have a solution if

$$p > 0, \quad \text{or,} \quad p = 0 \text{ and } q > 0. \tag{6.38}$$

It turns out [67] that if (6.38) holds then (6.16), (6.33), (6.34) does have a solution for any Ξ provided ε is chosen small enough. This can be proven by a contraction mapping principle argument, and we refer to [67] for details. So (6.38) is actually both necessary and sufficient for a solution of the given initial-value problem.

The question is now given a solution of (6.16), (6.33), (6.34) does this yield a positive solution of (6.16) on $(0, \Xi)$. By the standard theory of ordinary differential equations, the given local solution of (6.16), (6.33), (6.34) can be extended backwards as long as it remains bounded and positive. Considering the monotonicity of any positive solution of (6.16) described in the previous subsection, there are then two things which can go wrong. Firstly, there could be a $\xi_0 \in (0, \Xi)$ such that for (6.16), (6.33), (6.34) the solution can be extended onto (ξ_0, Ξ) but

$$f(\xi) \uparrow \infty \quad \text{as} \quad \xi \downarrow \xi_0. \tag{6.39}$$

Secondly, there could exist a $\xi_0 \in (0, \Xi)$ such that the solution can be extended onto (ξ_0, Ξ) but

$$f(\xi) \downarrow 0 \quad \text{as} \quad \xi \downarrow \xi_0. \tag{6.40}$$

To see that the first possibility cannot be admitted, we use (6.35) and the monotonicity argument again. These tell us that

$$-(f^m)'(\xi) \leq p\xi f(\xi) + |p+q|(\Xi - \xi)f(\xi)$$

for $\xi \in (\xi_0, \Xi)$. Cf. (6.37). Hence dividing by $f(\xi)$

$$\frac{-m}{m-1}(f^{m-1})'(\xi) \leq p\xi + |p+q|(\Xi - \xi)$$

for all $\xi \in (\xi_0, \Xi)$. Whence integrating from $\xi \in (\xi_0, \Xi)$ and using (6.34)

$$\frac{m}{m-1} f^{m-1}(\xi) \leq \frac{p}{2}(\Xi^2 - \xi^2) + |\frac{p+q}{2}|(\Xi - \xi)^2 \tag{6.41}$$

for all $\xi \in (\xi_0, \Xi)$. However if (6.39) holds, when $\xi \downarrow \xi_0$ in (6.41) the left-hand side tends to infinity whilst the right-hand side is bounded. This contradiction subsequently implies that the possibility (6.39) can be excluded.

We now have to examine the possibility of whether or not (6.40) occurs for some $\xi_0 \in (0, \Xi)$. To do this we fall back on the integral inequality we have used the least up to now, namely (6.23). Setting $\xi^+ = \Xi$ and $\xi^- = \xi$ in (6.23) we have

$$f^m(\xi) - \xi(f^m)'(\xi) = p\xi^2 f(\xi) + (2p+q)\int_\xi^\Xi \nu f(\nu)d\nu. \tag{6.42}$$

Hence, if there exists a point $\xi_0 \in (0, \Xi)$ such that (6.40) holds,

$$\lim_{\xi \downarrow \xi_0} -\xi(f^m)'(\xi) = (2p+q)\int_{\xi_0}^\Xi \nu f(\nu)d\nu.$$

The limit on the left-hand side must be nonnegative though, by the monotonicity argument. In fact the limit must be strictly positive. For, if it is not, then we have the first possibility we excluded in the characterization of weak solutions. This is (6.26), (6.27) with $(\xi_1, \xi_2) = (\xi_0, \Xi)$. This implies that if (6.40) holds for some $\xi_0 \in (0, \Xi)$ then $2p + q < 0$. On the other hand if there is no point $\xi_0 \in (0, \Xi)$ such that (6.40) holds then we may take $\xi = 0$ in (6.42) to deduce

$$f^m(0) = (2p+q)\int_0^\Xi \nu f(\nu)d\nu \tag{6.43}$$

which implies $2p + q \geq 0$. We conclude that the possibility (6.40) holds if and only if $2p + q < 0$. In other words, the condition $2p + q \geq 0$ is both necessary and sufficient for the solution of the initial value problem (6.16), (6.33), (6.34) to yield a positive solution of (6.16) on $(0, \Xi)$.

This shooting procedure is illustrated in Figure 6.1 where for a fixed value of $m > 1$ and $p > 0$ the behaviour of the solution is indicated for different values of q.

Recapitulating and combining (6.38) with the condition $2p+q \geq 0$, we have found that given any $\Xi \in (0, \infty)$ the initial value problem (6.16), (6.33), (6.34) has a positive solution on $(0, \Xi)$ if and only if $p > 0$ and $2p + q = 0$, or, $p \geq 0$ and $2p + q > 0$. Subsequently there is a weak solution of (6.16) of Type A if and only if $p > 0$ and $2p + q = 0$, or, $p \geq 0$ and $2p + q > 0$. Moreover, there is a solution with support with least upper bound Ξ for every $\Xi \in (0, \infty)$, and substituting (6.17) in (6.35) and (6.43),

$$\beta = -(p+q)\int_0^\Xi f(\nu)d\nu \tag{6.44}$$

and

$$U^m = (2p+q)\int_0^\Xi \nu f(\nu)d\nu. \tag{6.45}$$

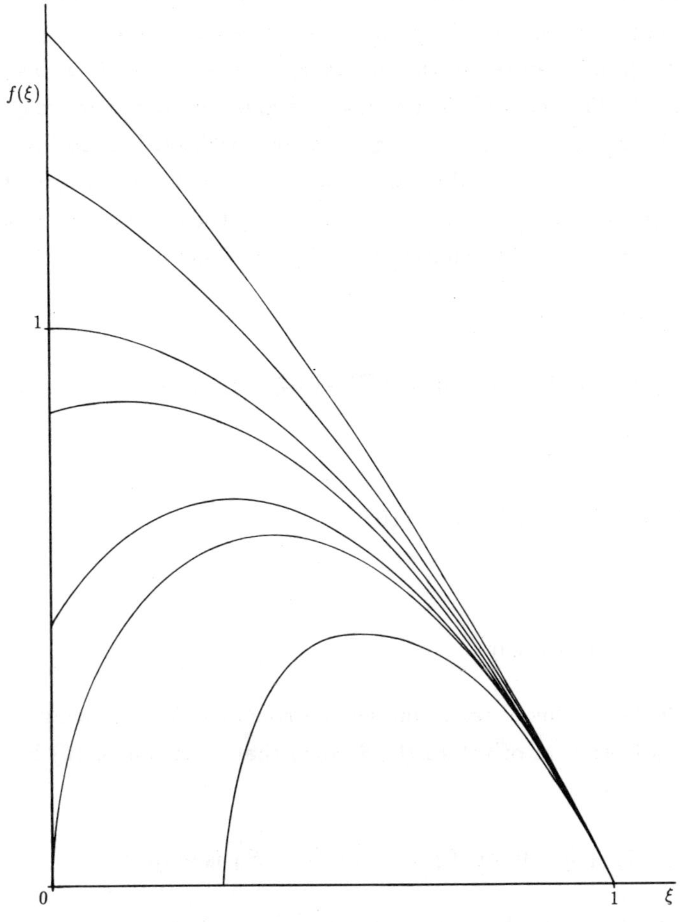

Figure 6.1: Backwards shooting from $\xi = 1$ for $(f^2)'' + 4\xi f' = qf$ with, from top to bottom, $q = 0, -2, -4, -5, -7, -8, -12$

It follows already that if $p > 0$ and $2p + q = 0$ then one can only have $U = 0$, whilst if $p \geq 0$ and $2p + q > 0$ then necessarily $U > 0$. Similarly if $p + q > 0$ then $\beta < 0$, if $p + q = 0$ then $\beta = 0$, and if $p + q < 0$ then $\beta > 0$.

With this remark we are almost, but not quite, finished. Although we have established exactly when given Ξ we can find a weak solution of Type A whose support is $(0, \Xi)$, and can relate the boundary-values U and β to this solution via (6.44) and (6.45), we still do not know if given U and β we may find an appropriate Ξ. In other words, although with our backwards shooting we know that we reach the origin, we do not know if we can hit an arbitrary target placed there. This though, can be taken care of by the observation that if $f(\xi)$ is a solution of (6.16), (6.33), (6.34) on $(0, \Xi)$, the function

$$\tilde{f}(\xi) = \chi^{-2/(m-1)} f(\chi\xi)$$

is a positive solution of (6.16) on $(0, \chi\Xi)$ satisfying

$$\tilde{f}(0) = \chi^{-2/(m-1)} f(0),$$

$$(\tilde{f}^m)'(0) = \chi^{-(m+1)/(m-1)} (f^m)'(0)$$

and

$$\tilde{f}(\chi\Xi) = (\tilde{f}^m)'(\chi\Xi) = 0$$

for any $\chi > 0$. With this observation we can conclude the following.

Let S_A denote the set of values (U, β) such that equation (6.16) has a weak solution of Type A.

(a) If $p < 0$, if $2p + q < 0$, or if $p = q = 0$ then S_A is empty.

(b) $p > 0$ and $2p + q = 0$ then $S_A = \{(0, \beta) : 0 < \beta < \infty\}$.

(c) $p \geq 0$ and $2p + q > 0$ then $S_A = \{(U, U^{(m+1)/2}\beta^*) : 0 \leq U < \infty\}$ for a unique value $\beta^* \in (-\infty, \infty)$ with $\beta^* < 0$ if $p + q > 0$, $\beta^* = 0$ if $p + q = 0$, and $\beta^* < 0$ if $p + q > 0$.

Herewith, all solutions of (6.16) of Type A have been characterized in terms of their initial values (6.17).

6.6 Shooting arguments for solutions which are positive everywhere

Having determined all possible solutions of Type A using a shooting argument, we consider now the similar determination of solutions of Type B. Whereas though for solutions of Type A we shot backwards from the least upper bound of the support of the solution (defined by the point Ξ), for the analysis of solutions of Type B we shoot forwards from the point $\xi = 0$. Thus we consider the initial value problem (6.16), (6.17) and ask whether this yields a bounded positive classical solution of (6.16) on the whole of $(0, \infty)$.

Necessarily for the existence of a local nonnegative solution of (6.16), (6.17), $U \geq 0$ and if $U = 0$ then $\beta \geq 0$. In fact if $U = 0$ we must have $\beta > 0$ for the option $\beta = 0$ can be excluded by the argument used to exclude (6.28), (6.29) in the characterization of solutions. Thus if (6.16), (6.17) is to yield a positive solution of (6.16) on $(0, \infty)$

$$U \geq 0, \quad \text{and,} \quad \beta > 0 \text{ if } U = 0. \tag{6.46}$$

To complement this, by using the standard theory [9,21] of ordinary differential equations it can be shown that if (6.46) holds then the initial-value problem (6.16), (6.17) does have a positive local solution. The question of the shooting argument is now whether or not the solution of this initial-value problem can be extended so that it yields a bounded positive solution of (6.16) on $(0, \infty)$.

Given a positive solution of the initial-value problem (6.16), (6.17) by the monotonicity argument in Subsection 6.4 there are four different types of behaviour with the shooting of this solution which can occur:

(i) There is a $\xi_0 \in (0, \infty)$ such that f is defined on $(0, \xi_0)$ but $f(\xi) \uparrow \infty$ as $\xi \uparrow \xi_0$.

(ii) There is a $\xi_0 \in (0, \infty)$ such that f is defined on $(0, \xi_0)$ but $f(\xi) \downarrow 0$ as $\xi \uparrow \xi_0$.

(iii) The function $f(\xi)$ extends to $(0, \infty)$ but $f(\xi) \uparrow \infty$ as $\xi \uparrow \infty$.

(iv) The function f extends to a bounded positive solution on $(0, \infty)$ as desired.

We shall not present the arguments which can be used to determine when these four possibilities occur. They are similar in style to those we have already seen for the characterization of solutions of Type A. We leave the major ones as exercises. Those who do not wish to tackle the exercises may refer to [68].

EXERCISE. Use the equality (6.23) to show that the possibility (i) given above cannot occur for any value of p and q.

EXERCISE. Use the equality (6.23) to show that if $q < 0$ and $2p + q \leq 0$ then possibility (ii) occurs.

EXERCISE. Use the equality (6.23) to show that if $p \geq 0$ and $2p + q \geq 0$ given two solutions of (6.16) f_1 and f_2 respectively with different sets of initial values (U_1, β_1) and (U_2, β_2) which satisfy (6.46) and with

$$U_1 \geq U_2 \quad \text{and} \quad \beta_1 \geq \beta_2,$$

then

$$f_1(\xi) \geq f_2(\xi) \text{ for all } \xi \in (0, \infty) \text{ such that both solutions are defined.}$$

EXERCISE. Use the outcome of the previous exercise and equality (6.23) to show that if $p \geq 0$, $2p + q > 0$ and $\beta \leq U^{(m+1)/2}\beta^*$ where β^* is the value in the characterization of solutions of Type A then possibility (ii) occurs. Whilst if $\beta > U^{(m+1)/2}\beta^*$ then possibilities (iii) or (iv) occur.

EXERCISE. Use the outcome of the previous exercises and equality (6.22) or (6.23) to show that if $q > 0$ and $p \geq 0$ then possibility (iii) occurs, whilst if $q \leq 0$ and $2p + q > 0$ and $\beta > U^{(m+1)/2}\beta^*$ where β^* is the value in the characterization of solutions of Type A then possibility (iv) occurs.

The conclusions of these exercises cover the characterization of all solutions of Type B except the case $q > 0$ and $p < 0$, and the case $q = 0$ and $p \leq 0$. The study of these last cases is a particularly good illustration of the application of a shooting argument. When $q > 0$ and $p < 0$ or when $q = 0$ and $p \leq 0$ it can be shown [68] that there is a value β_1 such that if $\beta < U^{(m+1)/2}\beta_1$ then any solution of the initial-value problem conforms to possibility (ii). Whilst there is also a value $\beta_2 \geq \beta_1$ such that if $\beta > U^{(m+1)/2}\beta_2$ then any solution of the initial-value problem (6.16), (6.17), (6.46) confirms to possibility (iii). Moreover, given U the set of values β such that possibility (ii) occurs is open and connected, and likewise the set of values such that possibility (iv) occurs is open and connected. Basically this means as it were, that if for fixed $U > 0$ one shoots too low one hits the ground, whilst if one shoots too high one shoots in the air. Continuity considerations subsequently tell us that there must be an angle (β^*) somewhere in between such that neither of the above occur. However, the only possibility which is open for this value of β^* is possibility (iv), i.e. that there is a solution of Type B. With some effort [56] it can also be shown that given any U there is at most one β with the desired effect.

The effect of all this shooting from the origin is illustrated in Figures 6.2 - 6.11. These figures consider a fixed value of m with different values of p and q on the unit-circle $p^2 + q^2 = 1$. Here, all the possibilities as discussed above can be discerned.

Summarizing, and letting S_B denote the set of values (U, β) such that (6.16), (6.17) yields a bounded positive solution of (6.16) on $(0, \infty)$ the following characterization of weak solutions of (6.16) of Type B emerges.

(a) If $q < 0$ and $2p + q \leq 0$ then S_B is empty.

(b) If $q \leq 0$ and $2p + q > 0$ then $S_B = \{(U, \beta) : 0 \leq U < \infty, U^{(m+1)}\beta^* < \beta\}$ where β^* is derived from the characterization of solutions of Type A for this range of parameter values.

(c) If $q > 0$ and $p \geq 0$ then S_B is empty.

(d) If $q > 0$ and $p < 0$, or if $q = 0$ and $p \leq 0$ then $S_B = \{(U, U^{(m+1)/2}\beta^*) : 0 < U < \infty\}$ for a unique value β^* with $\beta^* < 0$ if $q > 0$ and $\beta^* = 0$ if $q = 0$.

Herewith, we have completed our discussion of the characterization of all possible (weak) solutions of (6.16) on $(0, \infty)$. The only tools we have used to obtain this characterization are three integral identities coupled with shooting arguments. For further details, and in particular for details of the asymptotic behaviour of solutions of Type B as $\xi \to \infty$ we refer to [56,67,68].

A study of similarity solutions of equation (6.1) of the form $u(x, t) = X(t)f(x)$ which leads to the analysis of equation (6.16) with $p = 0$ on a bounded domain may be found in [59].

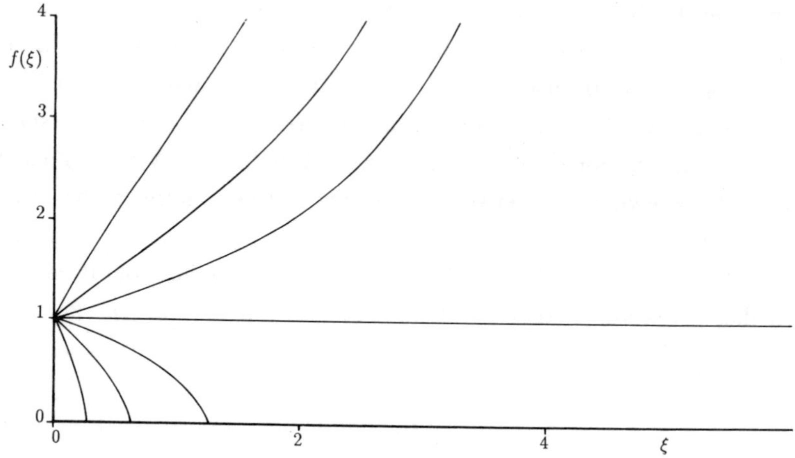

Figure 6.2: Forwards shooting for $(f^{3/2})''+p\xi f' = qf$, $f(0) = 1$ and $(f^{3/2})'(0) = \beta$ with $\beta = (3/2)\tan(i\pi/8)$ for, from top to bottom, $i = 3, 2, 1, 0, -1, -2, -3$. Here $p = -1$ and $q = 0$

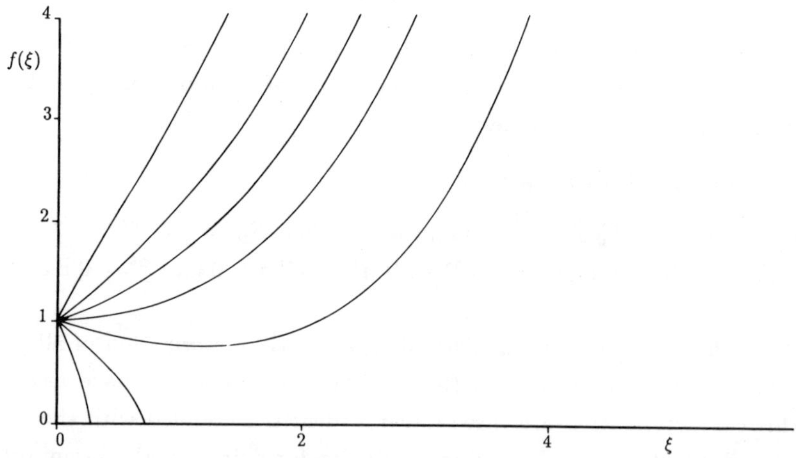

Figure 6.3: Forwards shooting for $(f^{3/2})''+p\xi f' = qf$, $f(0) = 1$ and $(f^{3/2})'(0) = \beta$ with $\beta = (3/2)\tan(i\pi/8)$ for, from top to bottom, $i = 3, 2, 1, 0, -1, -2, -3$. Here $p = -0{\cdot}7071$ and $q = 0{\cdot}7071$

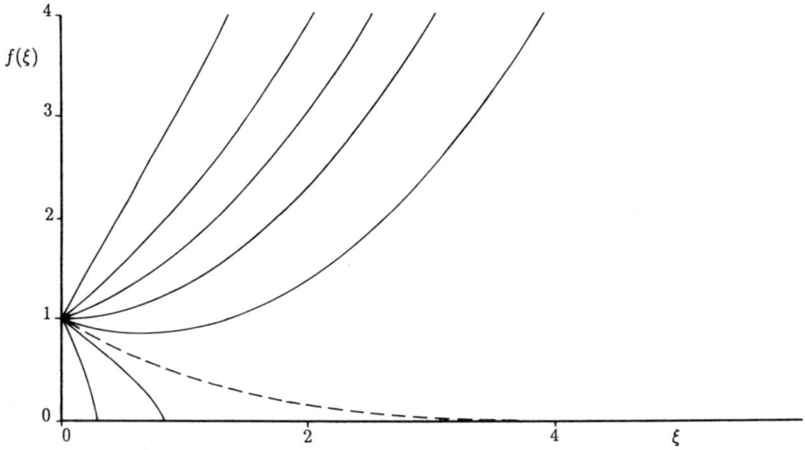

Figure 6.4: Forwards shooting for $(f^{3/2})''+p\xi f' = qf$, $f(0) = 1$ and $(f^{3/2})'(0) = \beta$ with $\beta = (3/2)\tan(i\pi/8)$ for, from top to bottom, $i = 3, 2, 1, 0, -1, -2, -3$. Here $p = 0$ and $q = 1$. The broken line indicates the explicit solution obtained with $\beta = -\sqrt{6/5}$

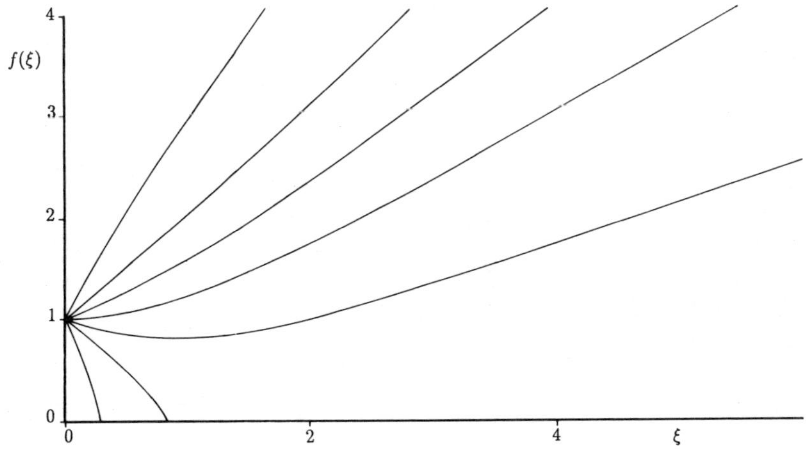

Figure 6.5: Forwards shooting for $(f^{3/2})''+p\xi f' = qf$, $f(0) = 1$ and $(f^{3/2})'(0) = \beta$ with $\beta = (3/2)\tan(i\pi/8)$ for, from top to bottom, $i = 3, 2, 1, 0, -1, -2, -3$. Here $p = 0{\cdot}7071$ and $q = 0{\cdot}7071$

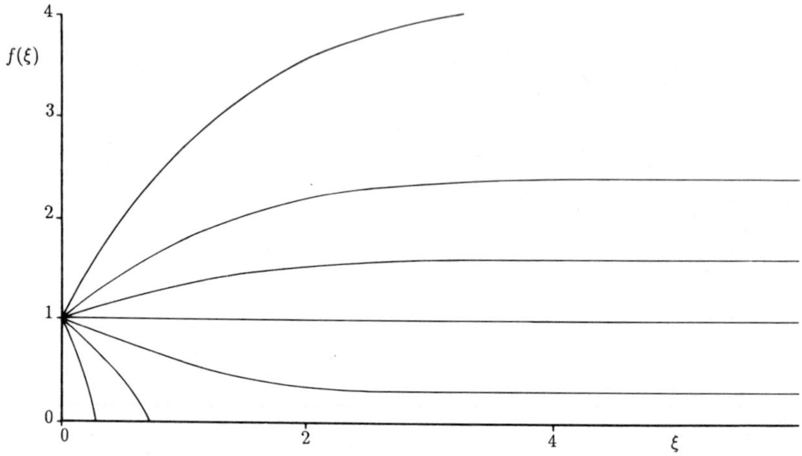

Figure 6.6: Forwards shooting for $(f^{3/2})''+p\xi f' = qf$, $f(0) = 1$ and $(f^{3/2})'(0) = \beta$ with $\beta = (3/2)\tan(i\pi/8)$ for, from top to bottom, $i = 3, 2, 1, 0, -1, -2, -3$. Here $p = 1$ and $q = 0$

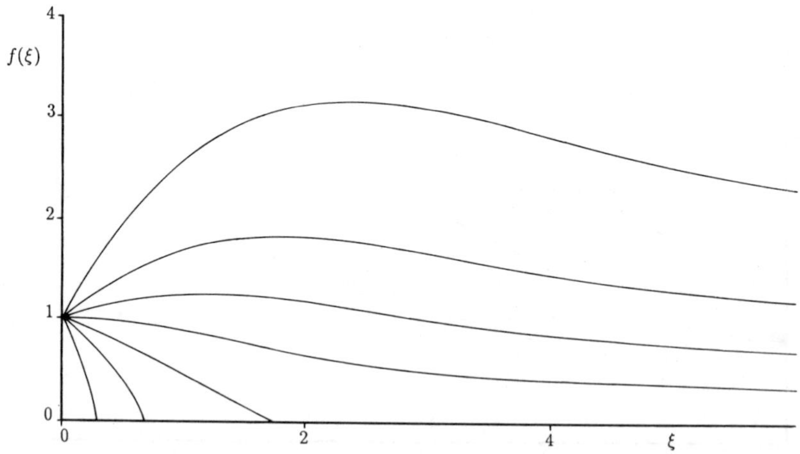

Figure 6.7: Forwards shooting for $(f^{3/2})''+p\xi f' = qf$, $f(0) = 1$ and $(f^{3/2})'(0) = \beta$ with $\beta = (3/2)\tan(i\pi/8)$ for, from top to bottom, $i = 3, 2, 1, 0, -1, -2, -3$. Here $p = 0{\cdot}9239$ and $q = -0{\cdot}3827$

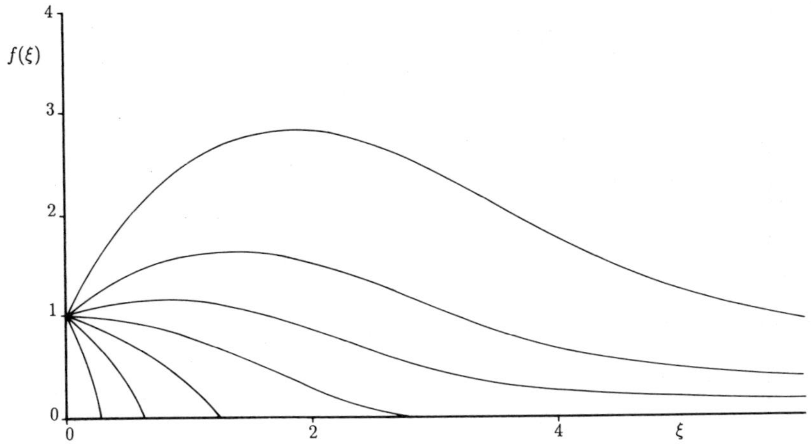

Figure 6.8: Forwards shooting for $(f^{3/2})''+p\xi f' = qf$, $f(0) = 1$ and $(f^{3/2})'(0) = \beta$ with $\beta = (3/2)\tan(i\pi/8)$ for, from top to bottom, $i = 3, 2, 1, 0, -1, -2, -3$. Here $p = 1/\sqrt{2}$ and $q = -1/\sqrt{2}$

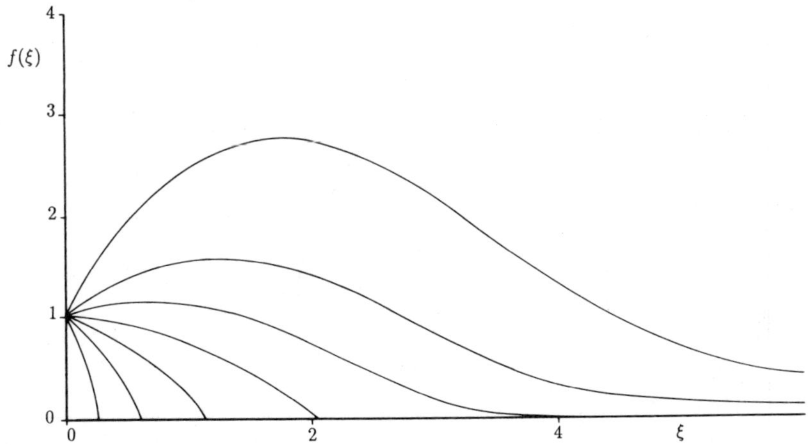

Figure 6.9: Forwards shooting for $(f^{3/2})''+p\xi f' = qf$, $f(0) = 1$ and $(f^{3/2})'(0) = \beta$ with $\beta = (3/2)\tan(i\pi/8)$ for, from top to bottom, $i = 3, 2, 1, 0, -1, -2, -3$. Here $p = 0{\cdot}5847$ and $q = -0{\cdot}8112$

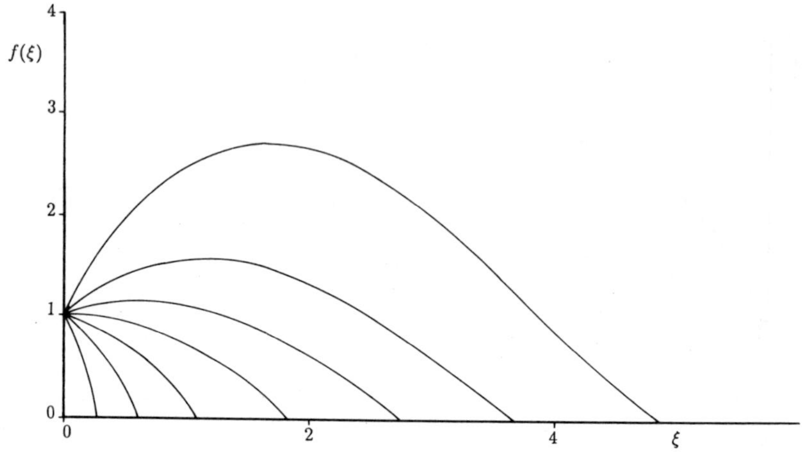

Figure 6.10: Forwards shooting for $(f^{3/2})'' + p\xi f' = qf, f(0) = 1$ and $(f^{3/2})'(0) = \beta$ with $\beta = (3/2)\tan(i\pi/8)$ for, from top to bottom, $i = 3, 2, 1, 0, -1, -2, -3$. Here $p = 1/\sqrt{5}$ and $q = -2/\sqrt{5}$

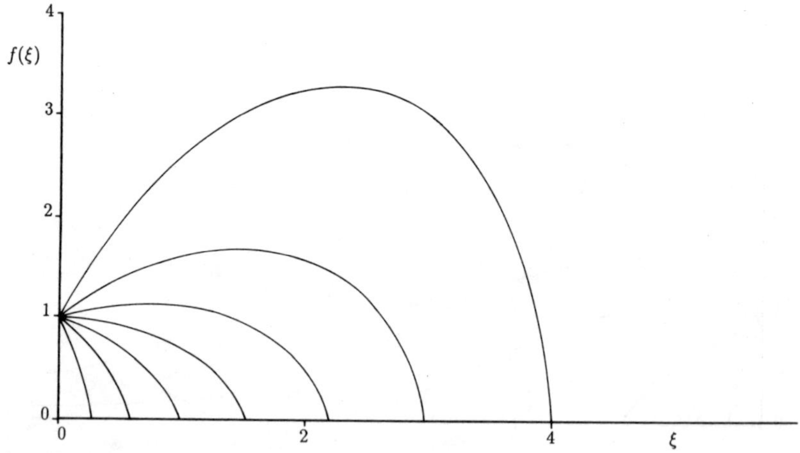

Figure 6.11: Forwards shooting for $(f^{3/2})'' + p\xi f' = qf, f(0) = 1$ and $(f^{3/2})'(0) = \beta$ with $\beta = (3/2)\tan(i\pi/8)$ for, from top to bottom, $i = 3, 2, 1, 0, -1, -2, -3$. Here $p = -0{\cdot}5999$ and $q = -0{\cdot}8016$

6.7 Explicit solutions

In general not only does similarity analysis indicate how a problem involving a differential equation may be simplified, but it also points a way to finding explicit solutions of the problem in hand. This is no less true here.

We note that for certain values of p and q, equation (6.16) can be solved explicitly [57,67].

(a) If $q = 0$ then (6.16) admits the solution

$$f \equiv U.$$

(b) If $q < 0$ and $2p + q = 0$ then (6.16) admits the solution

$$f(\xi) = \begin{cases} \xi^{1/m}\{\frac{p(m-1)}{m+1}(\Xi^{(m+1)/m} - \xi^{(m+1)/m})\}^{1/(m-1)} & \text{for } \xi < \Xi \\ 0 & \text{for } \xi \geq \Xi \end{cases}$$

for any $\Xi \in (0, \infty)$.

(c) If $q < 0$ and $p + q = 0$ then (6.16) admits the solution

$$f(\xi) = \begin{cases} \{\frac{p(m-1)}{2m}(\Xi^2 - \xi^2)\}^{1/(m-1)} & \text{for } \xi < \Xi \\ 0 & \text{for } \xi \geq \Xi \end{cases}$$

for any $\Xi \in (0, \infty)$.

(d) If $q > 0$ and $p = (m-1)q$ then (6.16) admits the solution

$$f(\xi) = \begin{cases} \{\frac{p(m-1)}{m}\Xi(\Xi - \xi)\}^{1/(m-1)} & \text{for } \xi < \Xi \\ 0 & \text{for } \xi \geq \Xi \end{cases}$$

for any $\Xi \in (0, \infty)$.

(e) If $q > 0$ and $p = 0$ then (6.16) admits the solution

$$f(\xi) = \begin{cases} \{\frac{q(m-1)^2}{2m(m+1)}(\Xi - \xi)^2\}^{1/(m-1)} & \text{for } \xi < \Xi \\ 0 & \text{for } \xi \geq \Xi \end{cases}$$

for any $\Xi \in (0, \infty)$.

The reader may observe that the combinations of parameters p and q in (a), (b), (c) and (e) above have proved to play critical roles in the previous analysis. To be specific, one or other term vanishes in one of the integral identities (6.22) - (6.24), and this is the key to obtaining the presented explicit solutions.

6.8 Relevance for the porous media equation

The explicit solutions of (6.16) on $0 < \xi < \infty$ summarized in the previous section yield the following solutions of (6.1). (See Figure 6.12.)

(a) The constant solution. Via the Transformation I, II or III with $\alpha = 0$ this yields the solution

$$u \equiv U.$$

This requires no further comment.

(b) The "dipole solution" [14]. Via the Transformation I with $\alpha = -1/m$ giving $p = 1/2m$ and $q = -1/m$ this is

$$u(x,t) = (t+\tau)^{-1/m} f(x(t+\tau)^{-1/2m}).$$

This solution has the property that $u(0,t) = 0$ for all $t \geq 0$, whilst it possesses an interface $\zeta(t)$ such that

$$u(x,t) > 0 \quad \text{for} \quad x < \zeta(t) \tag{6.47}$$

and

$$u(x,t) = 0 \quad \text{for} \quad x \geq \zeta(t). \tag{6.48}$$

The interface is given by $\zeta(t) = \Xi(t+\tau)^{1/2m}$.

(c) The "instantaneous point-source solution" [11,89,116]. Via the Transformation I with $\alpha = -1/(m+1)$ giving $p = 1/(m+1)$ and $q = -1/(m+1)$ this is

$$u(x,t) = (t+\tau)^{-1/(m+1)} f(x(t+\tau)^{-1/(m+1)}).$$

This solution has the property that $(u^m)_x(0,t) = (t+\tau)^{-1}(f^m)'(0) = 0$ for all $t \geq 0$, whilst it possesses on interface $\zeta(t)$ defined by (6.47), (6.48) of the form $\zeta(t) = \Xi(t+\tau)^{1/(m+1)}$. Subsequently, defining $u(x,t) = u(-x,t)$ for $x < 0$, $t > 0$, this function can be shown to generate a generalized solution of equation (6.1) on the whole of the domain $(-\infty,\infty) \times (0,\infty)$. Moreover this solution has the property that

$$\int_{-\infty}^{\infty} u(x,t)dx = 2\Xi^{\frac{m+1}{m-1}} \int_0^1 [\frac{(m-1)}{2m(m+1)}(1-\xi^2)]^{\frac{1}{m-1}} d\xi \text{ for all } t \geq 0.$$

Thus it represents a solution with fixed total mass. Moreover, this mass can be determined precisely using the parameter Ξ. This gives the solution its name. When $\tau = 0$ it describes a solution of fixed mass, but which at time $t = 0$ is equal to zero at all points $x \neq 0$ in $(-\infty, \infty)$. This solution is often referred to in the literature as the Barenblatt-Pattle solution [8,90], after Barenblatt [11-13] who first brought the solution to people's attention in the Soviet literature and Pattle [89] who later derived it independently. However the solution was first discovered by Zel'dovich and Kompaneec [116]. This solution has provided an essential key to present-day theory for the porous media equation [8,65,90].

(d) The "travelling-wave solution" [12,17,87]. Via the Transformation I with $\alpha = 1/(m-1)$ giving $p = 1$ and $q = 1/(m-1)$ this is

$$u(x,t) = \begin{cases} \{\frac{m-1}{m}\Xi(\Xi t - x)\}^{1/(m-1)} & \text{for } x < \Xi t \\ 0 & \text{for } x \geq \Xi t \end{cases}.$$

This solution could of course have been obtained by looking for travelling waves to start with.

(e) The "waiting-time solution" [74]. Via the Transformation II with $\alpha = -1/(m-1)$ yielding $p = 0$ and $q = 1/(m-1)$ this gives

$$u(x,t) = \begin{cases} \{\frac{(m-1)}{2m(m+1)}\frac{(\Xi - x)^2}{\tau - t}\}^{1/(m-1)} & \text{for } x < \Xi \\ 0 & \text{for } x \geq \Xi \end{cases}.$$

The remarkable aspect of this solution is that it yields an interface of the type (6.47), (6.48) which does not move in time. On the other hand, the solution itself blows up as $t \uparrow \tau$. This solution has proved an extremely valuable tool in analysing how an interface may 'wait', i.e. remain fixed, before expanding [8,90]. Hence its name.

Referring to the general similarity solutions of (6.1) which are obtainable through the Transformations I, II or III the analysis reveals the picture described in words in the following paragraphs and in Table 6.1 [57]. See Figure 6.12.

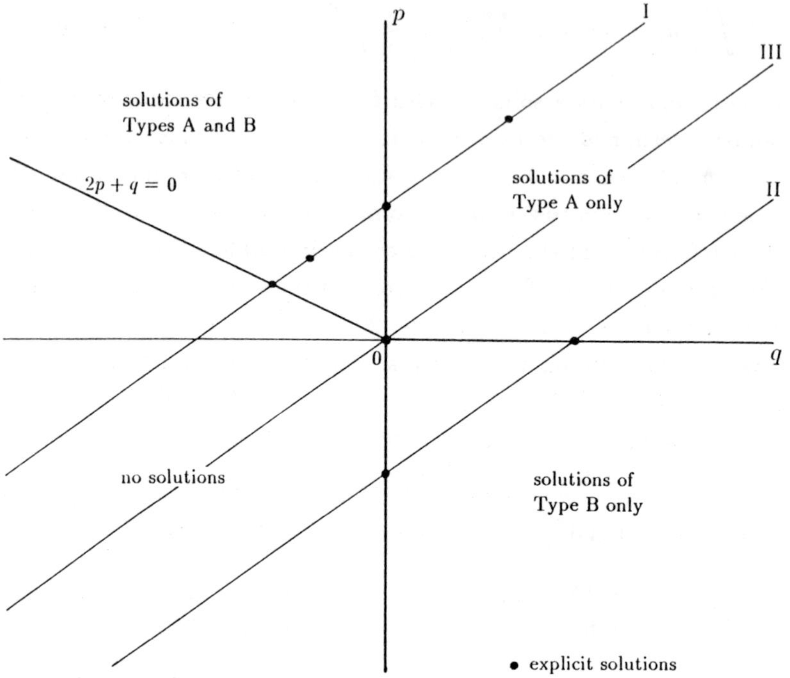

I. the line $(q,p) = (\alpha, [1+(m-1)\alpha]/2)$
II. the line $(q,p) = (-\alpha, -[1+(m-1)\alpha]/2)$
III. the line $(q,p) = (\alpha, (m-1)\alpha/2)$

Figure 6.12: The admission of solutions in terms of p and q

There is a similarity solution for Transformation I for all $\alpha \geq -1/m$. In the critical case $\alpha = -1/m$, the only similarity solutions which are admissible are those with the boundary value $u(0,t) = 0$ for $t \geq 0$. For all $\alpha \leq 0$ there is a one parameter family of solutions with compact support and a two parameter family of solutions which are positive everywhere. When $\alpha > 0$ the solutions with compact support are the only possible bounded solutions.

Bounded similarity solutions obtainable through Transformation II exist if and only if $\alpha \leq 0$. For every value $\alpha \leq 0$ there is a one parameter family of such solutions. In the critical case $\alpha = 0$ the only members are the constant solutions. For $-1/(m-1) < \alpha < 0$, the members are positive everywhere, whilst for $\alpha \geq -1/(m-1)$ the members have compact support. The critical

similarity transformation	reduced boundary conditions	parameter values for which solutions exists			solution type	sign of critical value
I. For $\tau \geq 0$		$\alpha = -1/m$	$U = 0$	$\beta > 0$	A	$(\beta^* = \infty)$
$u(x,t) = (t+\tau)^{\alpha} f(\xi)$	$u(0,t) = U(t+\tau)^{\alpha}$	$-1/m < \alpha < -1/(m+1)$	$U \geq 0$	$\beta > U^{(m+1)/2}\beta^*$	B	$\beta^* > 0$
$\xi = x(t+\tau)^{-\{1+(m-1)\alpha\}/2}$	$(u^m)_x(0,t) = \beta(t+\tau)^{\{(m+1)\alpha-1\}/2}$		$U > 0$	$\beta = U^{(m+1)/2}\beta^*$	A	
		$\alpha = -1/(m+1)$	$U \geq 0$	$\beta > U^{(m+1)/2}\beta^*$	B	$\beta^* = 0$
			$U > 0$	$\beta = U^{(m+1)/2}\beta^*$	A	
		$-1/(m+1) < \alpha \leq 0$	$U \geq 0$	$\beta > U^{(m+1)/2}\beta^*$	B	$\beta^* < 0$
			$U > 0$	$\beta = U^{(m+1)/2}\beta^*$	A	
		$\alpha > 0$	$U > 0$	$\beta = U^{(m+1)/2}\beta^*$	A	$\beta^* < 0$
II. For $\tau > t^*$		$\alpha \leq -1/(m-1)$	$U > 0$	$\beta = U^{(m+1)/2}\beta^*$	A	$\beta^* < 0$
$u(x;t) = (\tau - t)^{\alpha} f(\xi)$	$u(0,t) = U(\tau - t)^{\alpha}$	$-1/(m-1) < \alpha < 0$	$U > 0$	$\beta = U^{(m+1)/2}\beta^*$	B	$\beta^* < 0$
$\xi = x(\tau - t)^{-\{1+(m-1)\alpha\}/2}$	$(u^m)_x(0,t) = \beta(\tau - t)^{\{(m+1)\alpha-1\}/2}$	$\alpha = 0$	$U > 0$	$\beta = U^{(m+1)/2}\beta^*$	B	$\beta^* = 0$
III.						
$u(x,t) = \exp(\alpha t) f(\xi)$	$u(0,t) = U\exp(\alpha t)$	$\alpha = 0$	$U > 0$	$\beta = U^{(m+1)/2}\beta^*$	B	$\beta^* = 0$
$\xi = x\exp\{-\frac{1}{2}(m-1)\alpha t\}$	$(u^m)_x(0,t) = \beta\exp\{\frac{1}{2}(m+1)\alpha t\}$	$\alpha > 0$	$U > 0$	$\beta = U^{(m+1)/2}\beta^*$	A	$\beta^* < 0$

Table 6.1: Nontrivial (i.e. other than identically-zero) similarity solutions of (6.1) in the domain $0 < x < \infty,\ 0 < t \leq t^*$

value $\alpha = -1/(m-1)$ yields the waiting-time solution.

There is a similarly solution to be derived from Transformation III if and only if $\alpha \leq 0$. In the critical case $\alpha = 0$, these similarity solutions can only be the constant solutions. For all values $\alpha > 0$ there is a one parameter family of similarity solutions with compact support and none which are positive everywhere and bounded.

Considering all these similarity solutions, we may conclude the following about the general behaviour of solutions of the porous media equation. Solutions may admit an interface of the type (6.47), (6.48). Furthermore, such an interface cannot move backwards (there is no similarity solution where this occurs). An interface may remain stationary though. However this cannot occur indefinitely (the only similarity solution with a stationary interface blows up). In general, for large time the interface of a solution in the half plane $x > 0, t > 0$ will move at least as fast as $t^{1/m}$. Given that the lateral boundary data expands fast enough, there is no limit on the maximum asymptotic speed though (there are similarity solutions whose interfaces move exponentially for large t, and similarity solutions whose interfaces expand as an arbitrary large power of $(\tau - t)$ on domains $x > 0$, $0 < t \leq t^* < \tau$).

The above observations have long-since all be proven rigorously [8]. Moreover, it has been shown how the similarity solutions may describe the asymptotic behaviour of arbitrary solutions [55]. Notwithstanding, the similarity solutions gave the first insight into the relevant phenomena, and have proven to be invaluable tools in the attainment of the general results.

Appendix - Nomenclature

A - actual evapotranspiration [dim. LT^{-1}]
$\mathcal{A}$ - cross-sectional area [dim. L^2]
A_1 - soil classification
a - coefficient of diffusion term in differential equation
a_k - approximating function
B_2 - soil classification
b - coefficient of convection term in differential equation
b_k - approximating function
C - resistivity [dim. T]
$\mathcal{C}$ - constant
C_{11} - soil classification
C_{12} - soil classification
$\mathcal{C}^*$ - constant
c - rate of change of moisture content in soil per unit change in pressure head [dim. L^{-1}]
c_f - crop factor [dim. -]
D - thickness of aquifer [dim. L]
 - soil-moisture diffusivity [dim. L^2T^{-1}]
D_0 - reference value for soil-moisture diffusivity [dim. L^2T^{-1}]
d - thickness of root zone [dim. L]
E - surface evaporation [dim. LT^{-1}]
F - arbitrary function
f - unknown similarity solution
f_0 - representative value of unknown similarity solution function f
g - known function in right-hand side of equation
H - half-strip [dim. L $\times$ T]
h - groundwater head [dim. L]
h_i - unknown head in computational point i [dim. L]
$\hat{h}$ - approximate solution of groundwater head derived with finite element method [dim. L]
I - number of computational points [dim. -]
i - index [dim. -]

J - functional in finite element method [dim. L^4T^{-1}]

j - index [dim. -]

K - hydraulic conductivity [dim. LT^{-1}]

K_s - hydraulic conductivity at saturation [dim. LT^{-1}]

K_{xx} - component of hydraulic conductivity tensor [dim. LT^{-1}]

K_{xy} - component of hydraulic conductivity tensor [dim. LT^{-1}]

K_{xz} - component of hydraulic conductivity tensor [dim. LT^{-1}]

K_{yx} - component of hydraulic conductivity tensor [dim. LT^{-1}]

K_{yy} - component of hydraulic conductivity tensor [dim. LT^{-1}]

K_{yz} - component of hydraulic conductivity tensor [dim. LT^{-1}]

K_{zz} - component of hydraulic conductivity tensor [dim. LT^{-1}]

k - index [dim. -]

L - length [dim. L]

M - upper bound of solution of differential equation

m - exponent

N - precipitation [dim. LT^{-1}]

n - porosity [dim. -]
- exponent

P - potential evapotranspiration [dim. LT^{-1}]

$\mathcal{P}$ - set where solution is positive [dim. L]

P_c - critical potential evapotranspiration level [dim. LT^{-1}]

P_0 - open-water evaporation [dim. LT^{-1}]

p - pressure head [dim. L]
- parameter in differential equation

p_d - minimal pressure head associated with soil desiccation [dim. L]

p_r - root-soil interface pressure [dim. L]

p_s - maximal pressure head associated with soil-saturation [dim. L]

Q - exchange between saturated and unsaturated zones [dim. LT^{-1}]

Q_i - exchange between reservoirs [dim. LT^{-1}]

Q_ζ - accretion through upper boundary of aquifer [dim. LT^{-1}]

Q_η - accretion through lower boundary of aquifer [dim. LT^{-1}]

q - flux [dim. LT^{-1}]
- parameter in differential equation

$\mathbf{q}$	-	flux vector [dim. LT^{-1}]
$q^{(k)}$	-	flux generated by approximate solution of problem
q_x	-	component of flux vector [dim. LT^{-1}]
q_y	-	component of flux vector [dim. LT^{-1}]
q_z	-	vertical flux, component of flux vector [dim. LT^{-1}]
R	-	rectangle [dim. L $\times$ T]
r	-	resistance coefficient in plant uptake function [dim. L^{-2}]
r_0	-	basic value of resistance coefficient in plant uptake function [dim. L^{-2}]
S	-	storage coefficient [dim. -]
S_A	-	set of parameter values for which there is a similarity solution with compact support
S_B	-	set of parameter values for which there is a similarity solution which is positive everywhere
s	-	specific storativity [dim. L^{-1}]
T	-	transmissivity [dim. L^2T^{-1}]
T_{xx}	-	component of transmissivity tensor [dim. L^2T^{-1}]
T_{xy}	-	component of transmissivity tensor [dim. L^2T^{-1}]
T_{yx}	-	component of transmissivity tensor [dim. L^2T^{-1}]
T_{yy}	-	component of transmissivity tensor [dim. L^2T^{-1}]
t	-	time [dim. T]
U	-	initial value of solution of ordinary differential equation
u	-	unknown solution of partial differential equation
u_k	-	approximating sequence of solutions of differential equation
u_0	-	initial data function
$u_{0,k}$	-	approximating sequence for initial data function
$\overline{u}$	-	generalized supersolution of partial differential equation
$\underline{u}$	-	generalized subsolution of partial differential equation
v	-	soil-moisture particle velocity [dim. LT^{-1}]
W	-	abstraction from aquifer per unit area [dim. LT^{-1}]
W_n	-	aquifer boundary recharge [dim. L^2T^{-1}]
w	-	abstraction from ground per unit volume [dim. T^{-1}]
w_0	-	intrinsic component of plant uptake function [dim. T^{-1}]

X - similarity transformation function

x - distance, horizontal coordinate [dim. L]

Y - similarity transformation function

y - horizontal spatial coordinate [dim. L]

z - elevation, vertical coordinate [dim. L]

α - parameter

β - initial value parameter for solution of ordinary differential equation

β^* - critical value of initial value parameter for solution of ordinary differential equation

Γ - boundary of domain [dim. L]

γ - parameter

Δt - time step [dim. T]

Δx - distance step [dim. L]

Δy - distance step [dim. L]

Δz - distance step [dim. L]

δ - small positive parameter

ε - small positive parameter

ζ - interface [dim. L]

- upper elevation of aquifer [dim. L]

η - lower elevation of aquifer [dim. L]

Θ - storage in unsaturated zone [dim. L]

Θ_i - storage in reservoir [dim. L]

θ - moisture content [dim. -]

θ_d - minimal moisture content associated with desiccation [dim. -]

θ_s - moisture content at saturation [dim. -]

θ_0 - reference moisture content [dim. -]

$\tilde{\theta}$ - moisture content function [dim. -]

κ - integration constant

Λ - total soil-moisture deficit [dim. L]

Λ_r - soil-moisture deficit in the root zone [dim. L]

λ - speed of travelling wave [dim. LT^{-1}]

λ_0 - minimal speed of travelling waves [dim. LT^{-1}]
μ - specific yield [dim. -]
ν - dummy variable
Ξ - supremum of support of similarity solution
ξ - similarity variable
ξ_0 - reference value of similarity variable
ρ - root distribution function [dim. -]
σ - integration constant, parameter
τ - time variable or parameter [dim. T]
Φ - groundwater potential [dim. L]
ϕ_i - finite element basis function [dim. -]
χ - parameter
Ψ - boundary data function
Ψ_i - function in series expansion
ψ - test function
Ω - domain [dim. $L^2, L \times T$]
ω - dummy variable

$C(\Omega)$ - space of continuous functions in domain Ω
$C(\xi_1, \xi_2)$ - space of continuous functions of one variable in the open real interval (ξ_1, ξ_2)
$C^k(\Omega)$ - space of k-times continuously differentiable functions in the domain Ω
$C^k(\xi_1, \xi_2)$ - space of k-times continuously differentiable functions of one variable in the open real interval (ξ_1, ξ_2)
$C^{2,1}(\Omega)$ - space of functions of two variables which are twice continuously differentiable with respect to the first variable and once continuously differentiable with respect to the second variable in Ω
$C_0^\infty(\xi_1, \xi_2)$ - space of infinitely many times continuously differentiable functions which vanish outside of some closed sub-interval of the open real interval (ξ_1, ξ_2).

div	-	divergence operator [dim. L^{-1}]
$\frac{\partial}{\partial \nu}$	-	partial derivative
grad	-	gradient operator [dim. L^{-1}]
$W^{1,2}(\xi_1, \xi_2)$	-	Sobolev space of functions of one variable in the open real interval (ξ_1, ξ_2) which are such that the function possesses a generalized first derivative and the square of both the function and its generalized first derivative are integrable on (ξ_1, ξ_2)

References

[1] Aalto, J., "A finite element aquifer model which takes into account elevation changes of the aquifer", Nordic Hydrology 9, 173-186 (1978).

[2] Abbott, M.B., Warren, I.R., Jensen, K.H., Jonch-Clausen, T. and Havno, K., "Coupling of unsaturated and saturated zone models", Hydraulic Engineering in Water Resources Development and Management Proceedings Volume 5 Subject D Physical and Mathematical Models for Watersheds, International Association for Hydraulic Research, Cagliari, Italy, 1979, pp. 119-125.

[3] Abrahamse, A.H., Baarse, G. and Beek, E. van, "Policy Analysis of Water Management for the Netherlands, Volume XII, Model for Regional Hydrology, Agricultural Water Demands and Damages from Drought and Salinity", The Rand Corporation, Santa Monica, California, 1982.

[4] Afshar, A. and Marino, M.A., "Model for simulating soil-water content considering evapotranspiration", Journal of Hydrology 37, 309-322 (1978).

[5] Aggelides, S. and Youngs, E.G., "The dependence of the parameters in the Green and Ampt infiltration equation on the initial water content in draining and wetting states", Water Resources Research 14, 857-862 (1978).

[6] Ames, W.F., "Nonlinear Partial Differential Equations in Engineering", Academic Press, New York, 1965.

[7] Ames, W.F., "Nonlinear Partial Differential Equations in Engineering, Volume II", Academic Press, New York, 1972.

[8] Aronson, D.G., "The porous medium equation", Nonlinear Diffusion Problems (edited by A. Fasano and M. Primicerio), Lecture Notes in Mathematics 1224, Springer-Verlag, Berlin, 1986, pp. 1-46.

[9] Bailey, P.B., Shampine, L.F. and Waltman, P.E., "Nonlinear Two Point Boundary Value Problems", Academic Press, New York, 1968.

[10] Bakel, P.J.T. van, "A Systematic Approach to Improve the Planning, Design and Operation of Surface Water Management Systems: A Case Study", Instituut voor Cultuurtechniek en Waterhuishouding, Wageningen, 1986.

[11] Barenblatt, G.I., "On some unsteady motions of a liquid and a gas in a porous medium (in Russian)", Prikladnaya Matematika i Mekhanika 16, 67-78 (1952).

[12] Barenblatt, G.I., "On a class of exact solutions of the plane one-dimensional problem of unsteady filtration of a gas in a porous medium (in Russian)", Prikladnaya Matematika i Mekhanika 17, 739-742 (1953).

[13] Barenblatt, G.I., "On limiting self-similar motions in the theory of unsteady filtration of a gas in a porous medium and the theory of the boundary layer (in Russian)", Prikladnaya Matematika i Mekhanika 18, 409-414 (1954).

[14] Barenblatt, G.I. and Zel'dovich, Ya.B., "On the dipole-type solution in problems of unsteady gas filtration in the polytropic regime (in Russian)", Prikladnaya Matematika i Mekhanika 21, 718-720 (1957).

[15] Barenblatt, G.I. and Zel'dovich, Ya.B., "Intermediate asymptotics in mathematical physics", Russian Mathematical Surveys 26, 45-61 (1971). Translation of: Uspekhi Matematicheskikh Nauk 26, 115-129 (1971).

[16] Barenblatt, G.I. and Zel'dovich, Ya.B., "Self-similar solutions as intermediate asymptotics", Annual Review of Fluid Mechanics 4, 285-312 (1972).

[17] Basak, P. and Murty, V.N.N., "Nonlinear diffusion applied to groundwater contamination problems", Journal of Hydrology 35, 357-363 (1977).

[18] Bear, J., "Dynamics of Fluids in Porous Media", American Elsevier Publishing Company, New York, 1972.

[19] Bear, J., "Hydraulics of Groundwater", McGraw-Hill International Book Company, New York, 1979.

[20] Bear, J., Zaslavsky, D. and Irmay, S., "Physical Principles of Water Percolation and Seepage", United Nations Educational Scientific and Cultural Organization, Paris, 1968.

[21] Bernfeld, S.R. and Lakshmikantham, V., "An Introduction to Nonlinear Boundary Value Problems", Academic Press, New York, 1974.

[22] Bluman, G.W. and Cole, J.D., "Similarity Methods for Differential Equations", Springer-Verlag, New York, 1974.

[23] Blumenthal, K.P, Baarse, G., Beek, E. van, Koster, P.K., Langeweg, F., Baan, P.J.A., Pulles, J.W., Dijkman J.P.M. and Dorsman, W.A., "Policy Analysis for the National Water Management of the Netherlands", Government Publishing Office, The Hague, 1982.

[24] Bouwer, H., "Groundwater Hydrology", McGraw-Hill Book Company, New York, 1978.

[25] Braester, C., "Moisture variation at the soil surface and the advance of the wetting front during infiltration at constant flux", Water Resources Research 9, 687-694 (1973).

[26] Bredehoeft, J.D. and Pinder, G.F., "Digital analysis of areal flow in multiaquifer groundwater systems: a quasi three-dimensional model", Water Resources Research 6, 883-888 (1970).

[27] Bridgman, P.V., "Dimensional Analysis", Yale University Press, New Haven, 1963.

[28] Chandrasekhar, S., "Stochastic problems in physics and astronomy", Reviews of Modern Physics 15, 1-89 (1943).

[29] Chieng, S.-T., Broughton, R.S. and Foroud, N., "Drainage rates and water-table depths", Journal of the Irrigation and Drainage Division Proceedings of the American Society of Civil Engineers 104, 413-433 (1978).

[30] Childs, E.C., "An Introduction to the Physical Basis of Soil Water Phenomena", John Wiley and Sons, London, 1969.

[31] Chow, V.T., "Handbook of Applied Hydrology", McGraw-Hill Book Company, New York, 1964.

[32] Chu, S.T., "Infiltration during an unsteady rain", Water Resources Research 14, 461-466 (1978).

[33] Darcy, H., "Les Fontaines Publiques de la Ville de Dijon", Libraire des Corps Impériaux des Ponts et Chaussées et des Mines, Paris, 1856. Reprinted in: The Theory of Ground-Water Motion and Related Papers (edited by M.K. Hubbert), Hafner Publishing Company, New York, 1969.

[34] Douglas, J. and Jones, B.F., "On predictor-corrector methods for non-linear parabolic differential equations", Journal of the Society for Industrial and Applied Mathematics 11, 195-204 (1963).

[35] Eagleson, P.S., "Climate, soil, and vegetation 1: introduction to water balance dynamics", Water Resources Research 14, 705-712 (1978).

[36] Eagleson, P.S., "Climate, soil, and vegetation 2: the distribution of annual precipitation derived from observed storm sequences", Water Resources Research 14, 713-721 (1978).

[37] Eagleson, P.S., "Climate, soil, and vegetation 3: a simplified model of soil moisture movement in the liquid phase", Water Resources Research 14, 722-730 (1978).

[38] Eagleson, P.S., "Climate, soil, and vegetation 4: the expected value of annual evapotranspiration", Water Resources Research 14, 731-739 (1978).

[39] Eagleson, P.S., "Climate, soil, and vegetation 5: a derived distribution of storm surface runoff", Water Resources Research 14, 741-748 (1978).

[40] Eagleson, P.S., "Climate, soil, and vegetation 6: dynamics of annual water balance", Water Resources Research 14, 749-764 (1978).

[41] Eagleson, P.S., "Climate, soil, and vegetation 7: a derived distribution of annual water yield", Water Resources Research 14, 765-776 (1978).

[42] Feddes, R.A., "Water, Heat and Crop Growth", Veenman en Zonen, Wageningen, 1971.

[43] Feddes, R.A., Bresler, E. and Neuman, S.P., "Field test of a modified numerical model for water uptake by root systems", Water Resources Research 10, 1199-1206 (1974).

[44] Feddes, R.A., Kowalik, P., Kolińska-Malinka, K. and Zaradny, H., "Simulation of field water uptake by plants using a soil water dependent root extraction function", Journal of Hydrology 31, 13-26 (1976).

[45] Feddes, R.A., Kowalik, P.J. and Zaradny, H., "Simulation of Field Water Use and Crop Yield", Centre for Agricultural Publishing and Documentation, Wageningen, 1978.

[46] Federer, C.A., "A soil-plant-atmosphere model for transpiration and availability of soil water", Water Resources Research 15, 555-562 (1979).

[47] Freeze, R.A. and Cherry, J.A., "Groundwater", Prentice-Hall, Englewood Cliffs, New Jersey, 1979.

[48] Friedman, A., "Partial Differential Equations of Parabolic Type", Prentice-Hall, Englewood Cliffs, New Jersey, 1964.

[49] Frind, E.O. and Verge, M.J., "Three-dimensional modeling of groundwater flow systems", Water Resources Research 14, 844-856 (1978).

[50] Fuchik, S. and Kufner, A., "Nonlinear Differential Equations", Elsevier Scientific Publishing Company, Amsterdam, 1980.

[51] Genuchten, M.T. van, "A closed-form equation for predicting the hydraulic conductivity of unsaturated soils", Soil Science Society of America Journal 44, 892-898 (1980).

[52] Genuchten, M.T. van and Nielsen, D.R., "On describing and predicting the hydraulic properties of unsaturated soils", Annales Geophysicae 3, 615-628 (1985).

[53] Gilding, B.H., "Properties of solutions of an equation in the theory of infiltration", Archive for Rational Mechanics and Analysis 65, 203-225 (1977).

[54] Gilding, B.H., "A nonlinear degenerate parabolic equation", Annali della Scuola Normale Superiore di Pisa Classe di Scienze Serie IV 4, 393-432 (1977).

[55] Gilding, B.H., "Stabilization of flows through porous media", SIAM Journal on Mathematical Analysis 10, 237-246 (1979).

[56] Gilding, B.H., "On a class of similarity solutions of the porous media equation III", Journal of Mathematical Analysis and Applications 77, 381-402 (1980).

[57] Gilding, B.H., "Similarity solutions of the porous media equation", Journal of Hydrology 56, 251-263 (1982).

[58] Gilding, B.H., "The soil-moisture zone in a physically-based hydrologic model", Advances in Water Resources 6, 36-43 (1983).

[59] Gilding, B.H., "The first boundary value problem for $-u'' = \lambda u^p$", Journal of Mathematical Analysis and Applications 128, 419-442 (1987).

[60] Gilding, B.H., "The occurrence of interfaces in nonlinear diffusion-advection processes", Archive for Rational Mechanics and Analysis 100, 243-263 (1988).

[61] Gilding, B.H., "Improved theory for a nonlinear degenerate parabolic equation", Annali della Scuola Normale Superiore di Pisa Classe di Scienze Serie IV 16, 165-224 (1989).

[62] Gilding, B.H., "Localization of solutions of a nonlinear Fokker-Planck equation with Dirichlet boundary conditions", Nonlinear Analysis Theory Methods and Applications 13, 1215-1240 (1989).

[63] Gilding, B.H., "Qualitative mathematical analysis of the Richards equation", Transport in Porous Media 5, 651-666 (1991).

[64] Gilding, B.H. and Kersner, R., "Instantaneous shrinking in nonlinear diffusion-convection", Proceedings of the American Mathematical Society 109, 385-394 (1990).

[65] Gilding, B.H. and Kersner, R., "The characterization of reaction-convection-diffusion processes by travelling waves", in preparation.

[66] Gilding, B.H. and Peletier, L.A., "The Cauchy problem for an equation in the theory of infiltration", Archive for Rational Mechanics and Analysis 61, 127-140 (1976).

[67] Gilding, B.H. and Peletier, L.A., "On a class of similarity solutions of the porous media equation", Journal of Mathematical Analysis and Applications 55, 351-364 (1976).

[68] Gilding, B.H. and Peletier, L.A., "On a class of similarity solutions of the porous media equation II", Journal of Mathematical Analysis and Applications 57, 522-538 (1977).

[69] Gilding, B.H. and Wesseling, J.W., "PREDIS - a physically-based hydrologic response model", XX Congress of the International Association for Hydraulic Research Proceedings Subject D Volume V, International Association for Hydraulic Research, Moscow, 1983, pp. 26-32.

[70] Gurtin, M.E. and MacCamy, R.C., "On the diffusion of biological populations", Mathematical Biosciences 33, 35-49 (1977).

[71] Haverkamp, R., Vauclin, M., Touma, J., Wierenga, P.J. and Vachaud, G., "A comparison of numerical simulation models for one-dimensional infiltration", Soil Science Society of America Journal 41, 285-294 (1977).

[72] Hooghart, J.C. (editor), "Comparison of Models for the Unsaturated Groundwater System and the Evaporation (in Dutch)", Commissie voor Hydrologisch Onderzoek TNO Rapporten en Nota's no. 13, Commissie voor Hydrologisch Onderzoek TNO, Den Haag, 1985.

[73] Hubbert, M.K., "The theory of ground-water motion", The Journal of Geology 48, 785-944 (1940).

[74] Kalashnikov, A.S., "The occurrence of singularities in solutions of the non-steady seepage equation", U.S.S.R. Computational Mathematics and Mathematical Physics 7, 269-275 (1967). Translation of: Zhurnal Vychislitnel'noi Matematiki i Matematicheskoi Fiziki 7, 440-444 (1967).

[75] Kalashnikov, A.S., "Some problems of the qualitative theory of non-linear degenerate second-order parabolic equations", Russian Mathematical Surveys 42, 169-222 (1987). Translation of: Uspekhi Matematicheskikh Nauk 42, 135-176 (1987).

[76] Laat, P.J.M. de, "Model for Unsaturated Flow above a Shallow Water-Table, applied to a regional sub-surface flow problem", Centre for Agricultural Publishing and Documentation, Wageningen, 1980.

[77] Ladyzhenskaja, O.A., Solonnikov, V.A. and Ural'ceva, N.N., "Linear and Quasi-linear Equations of Parabolic Type", Translations of Mathematical Monographs Volume 23, American Mathematical Society, Providence, Rhode Island, 1968.

[78] Lomen, D.O. and Warrick, A.W., "Solution of the one-dimensional linear moisture flow equation with implicit water extraction functions", Soil Science Society of America Journal 40, 342-344 (1976).

[79] Lomen, D.O. and Warrick, A.W., "Time-dependent solution to the one-dimensional linearized moisture flow equation with water extraction", Journal of Hydrology 39, 59-67 (1978).

[80] Marlon-Lambert, J.M., Manoel, P.J. and Friday, R.G., "The development of a general groundwater computer modelling package", Hydrology and Water Resources Symposium Perth 1979, Institution of Engineers, Australia, 1979.

[81] Molz, F.J., "Models of water transport in the soil-plant system: a review", Water Resources Research 17, 1245-1260 (1981).

[82] Morel-Seytoux, H.J., "Derivation of equations for rainfall infiltration", Journal of Hydrology 31, 203-219 (1976).

[83] Morel-Seytoux, H.J., "Derivation of equations for variable rainfall infiltration", Water Resources Research 14, 561-568 (1978).

[84] Morton, F.I., "Estimating evapotranspiration from potential evaporation: practicality of an isonoclastic approach", Journal of Hydrology 38, 1-32 (1978).

[85] Muskat, M., "The Flow of Homogeneous Fluids through Porous Media", McGraw-Hill Book Company, New York, 1937.

[86] Neuman, S.P., Feddes, R.A. and Bresler, E., "Finite element analysis of two-dimensional flow in soils considering water uptake by roots: I. theory", Soil Science Society of America Proceedings 39, 224-230 (1975).

[87] Oleinik, O.A., Kalashnikov, A.S. and Chzhou Y.-L., "The Cauchy problem and boundary problems for equations of the type of non-stationary filtration (in Russian)", Izvestiya Akademii Nauk SSSR Seriya Matematicheskaya 22, 667-704 (1958).

[88] Ovsiannikov, L.V., "Group Analysis of Differential Equations", Academic Press, New York, 1982.

[89] Pattle, R.E., "Diffusion from an instantaneous point source with a concentration-dependent coefficient", The Quarterly Journal of Mechanics and Applied Mathematics 12, 407-409 (1959).

[90] Peletier, L.A., "The porous media equation", Applications of Nonlinear Analysis in the Physical Sciences (edited by H. Amann, N. Bazley and K. Kirchgässner), Pitman Advanced Publishing Program, Boston, 1981, pp. 229-241.

[91] Philip, J.R., "Evaporation, and moisture and heat fields in the soil", Journal of Meteorology 14, 354-366 (1957).

[92] Philip, J.R., "Theory of infiltration", Advances in Hydroscience Volume 5 (edited by V.T. Chow), Academic Press, New York, 1969, pp. 215-296.

[93] Pikul, M.F., Street, R.L. and Remson, I., "A numerical model based on coupled one-dimensional Richards and Boussinesq equations", Water Resources Research 10, 295-302 (1974).

[94] Pinder, G.F. and Frind, E.O., "Application of Galerkin's procedure to aquifer analysis", Water Resources Research 8, 108-120 (1972).

[95] Pinder, G.F. and Gray, W.G., "Finite Element Simulation in Surface and Subsurface Hydrology", Academic Press, New York, 1977.

[96] Prickett, T.A. and Lonnquist, G.G., "Selected Digital Computer Techniques for Groundwater Resource Evaluation", Illinois State Water Survey, Urbana, Illinois, 1971.

[97] Protter, M.H. and Weinberger, H.F., "Maximum Principles in Differential Equations", Prentice-Hall, Englewood Cliffs, New Jersey, 1967.

[98] Rees Vellinga, E. van (editor) "Hydrology and Water Quality of Central West-Netherlands (in Dutch)", Instituut voor Cultuurtechniek en Waterhuishouding, Wageningen, 1976.

[99] Remson, I., Hornberger, G.M. and Molz, F.J., "Numerical Methods in Subsurface Hydrology with an introduction to the finite element method", Wiley-Interscience, New York, 1971.

[100] Richards, L.A., "Capillary conduction of liquids through porous mediums", Physics 1, 318-333 (1931).

[101] Richter, J., "A simple numerical solution for the vertical flow equation of water through unsaturated soils", Soil Science 129, 138-144 (1980).

[102] Rushton, K.R. and Redshaw, S.C., "Seepage and Groundwater Flow", John Wiley and Sons, Chichester, 1979.

[103] Rushton, K.R. and Wedderburn, L.A., "Starting conditions for aquifer simulations", Ground Water 11, 37-42 (1973).

[104] Strang, G. and Fix, G.J., "An Analysis of the Finite Element Method", Prentice-Hall, Englewood Cliffs, New Jersey, 1973.

[105] Stricker, H. and Brutsaert, W., "Actual evapotranspiration over a summer period in the 'Hupsel catchment' ", Journal of Hydrology 39, 139-157 (1978).

[106] Stricker, J.N.M., "Methods of estimating evapotranspiration from meteorological data and their applicability in hydrology", Evaporation in Relation to Hydrology, Committee for Hydrological Research TNO Proceedings and Informations 28, 59-77 (1981).

[107] Stuff, R.G. and Dale, R.F., "A soil moisture budget model accounting for shallow water table influences", Soil Science Society of America Journal 42, 637-643 (1978).

[108] Swartzendruber, D., "The flow of water in unsaturated soils", Flow Through Porous Media (edited by R.J.M. de Wiest), Academic Press, New York, 1969, pp. 215-292.

[109] Todd, D.K., "Ground Water Hydrology", John Wiley and Sons, New York, 1959.

[110] Trescott, P.C., Pinder, G.F. and Larson, S.P., "Finite-Difference Model for Aquifer Simulation in Two Dimensions with results of numerical experiments", Techniques of Water-Resources Investigations of the United States Geological Survey Book 7 Chapter C1, United States Government Printing Office, Washington, 1976.

[111] Vachaud, G. and Vauclin, M. "Comments on 'A numerical model based on coupled one-dimensional Richards and Boussinesq equations' by Mary F. Pikul, Robert L. Street, and Irwin Remson", Water Resources Research 11, 506-509 (1975).

[112] Walton, W.C., "Groundwater Resource Evaluation", McGraw-Hill Kogakusha, Tokyo, 1970.

[113] Warrick, A.W., "Solution to the one-dimensional linear moisture flow equation with water extraction", Soil Science Society of America Journal 38, 573-576 (1974).

[114] Warrick, A.W., "Analytical solutions to the one-dimensional linearized moisture flow equation for arbitrary input", Soil Science 120, 79-84 (1975).

[115] Zagwijn, W.M. and Staalduinen, C.J. van (editors), "Geological Survey Maps of the Netherlands (in Dutch)", Rijks Geologische Dienst, Haarlem, 1975.

[116] Zel'dovich, Ya.B. and Kompaneec, A.S., "On the theory of propagation of heat with the heat conductivity depending upon the temperature (in Russian)", Collection Commemorating the Seventieth Birthday of Academician A.F. Ioffe, Izdat Akadmii Nauk S.S.S.R., Moscow, 1950, pp. 61-71.

[117] Zienkiewicz, O.C., "The Finite Element Method in Engineering Science", McGraw-Hill, London, 1971.

[118] Anonymous, "GROMULA computer program for modelling groundwater flow in a multi-layered system of aquifers manual", Delft Hydraulics Laboratory, Delft, 1983.

[119] Anonymous, Groundwater maps 37 east and 38 west, Dienst Grondwaterverkenning TNO, TNO, Delft, 1984.

[120] Anonymous, "Modelling the spreading of contaminants in the subsurface below a proposed silt depot with location Capelle-West (in Dutch)", Research Report R1983, Waterloopkundig Laboratorium, Delft, 1984.

Applications of the Homogenization Method to Flow and Transport in Porous Media

Ulrich Hornung
SCHI
P.O. Box 1222
W-8014 Neubiberg, Germany

1 Introduction

Porous media are a typical example for media with *microstructure*. On the scale of pores there are not only the solid matrix and the space of the pores, but also the different phases of liquids and gases filling the pores and all the interfaces between these phases. This situation has been described very often by soil physicists, soil chemists and chemical engineers. Pictures illustrating possible configurations can be found in every textbook in these fields.

Science is far from understanding all phenomena taking place on the pore scale. Let us only mention that there is more than only fluid flow, namely also interfacial forces play an important role. Furthermore, in the vicinity of the surfaces of the solid matrix (its geometry can be extremely complicated) adsorption processes, the special effects of very thin films, electric forces, and catalytic recations make a general theory difficult.

Nevertheless, for quite some time big effort has been invested into answering the question of how to formulate laws on a scale that is larger than the pore scale and to justify these laws from "first principles". I.e., to perform what is know as the *averaging* process, see, e.g., [45], [47] [48] [49], [66] and [92]. A recent survey articel is [76]. Using the words of this paper one wants to start from certain differential equations that are assumed to be valid on the *micro-scale* and - using special techniques - to transform them into equations on the *macro-scale*.

In the seventies mathematicians have developed a new method called *homogenization*. This method has two major aspects: (a) a theoretical and (b) a practial aspect. (a) From a theoretical point of view one studies partial differential equations with coefficients that are higly oscillatory, not necessarly arbitrarily, but either periodically or in a well defined sense randomly (or stochastically). Homogenization performs certain limits of the solutions of these equations and determines equations which the limits are solution of. Textbooks on this subject are [16] [13] and [78]. A more abstract theory is contained in [35] and [7]. Important contributions to the theory can be found in [10], [11], [12] [73], [74]. (b) From a practical point of view what one does is to consider on a large scale media with *microstructure*, to average out the physical and chemical processes on the small scale and to calculate *effective properties* of the media. The volume [43] contains a collection of papers devoted to applications. Especially in elasticity theory there is a large number of papers on this subject, see [86], [30], and [59].

After early attempts to set up a theory (see [65]) a first important step in applying the method of homogenization to porous media was done in 1980 in the paper [90] (see also [61]). There a mathematically rigorous proof was presented of deriving Darcy's law from the assumption that the Stokes equations hold on the pore scale. The theorem given there is applicable to steady flow of a single fluid in a periodic medium the solid matrix of which is diconnected. A formal derivation of the same result has been published in the same year in [57], see also [62]. In the meantime this result has been generalized in several ways. The paper [70] deals with nonhomogeneous boundary conditions. In [1] the geometry has been generalized to connected media, i.e., to cases where the porous matrix itself is connected in space, and not only the pore space. In [2] it was shown that certain assumptions made on the pore scale can lead to different types of equations on the large scale, such as Brinkman's law etc. The Navier-Stokes equations were taken into consideration in the paper [68]. Flow around bubbles was considered in [63]

Since the method has proved to be very powerful, a variety of applications to more complicated situations have been published. Miscible displacement problems were studied in [23] [24] [67] [69] [71]. A related problem is that of hydrodynamic dispersion. The papers [17] [18] [26] [27] [46] [83] investigate this effect in periodic media. Thermal flow was studied in [41] and [42].

The present work is both an introduction and a survey on the subject, see also [38]. In section 2.1 we motivate the procedure by giving a very simple example of a layered medium. Since everybody is familiar with this situation, it is easy to demonstrate what is behind this problem. The rest of the section 2 is devoted to the homogenization of elliptic problems with periodic coefficients. The formal procedure of asymptotic expansions and also the convergence proof using energy estimates are given.

In the section 3 media are introduced that have a spatially periodic array of obstacles. In the space left by the obstacles one studies differential equations. The first examples are elliptic equations.

The section 4 shows that for time dependent problems the method of homogenization may lead to non-local equations. The crucial point is how to scale the micro-equations. In certain cases the macro-equations may be of integro-differential type. This situation is by no means artificial. That kind of problems has been studied in the field of oil reservoir simulation for quite some time, see [14] and [4] and [36]. The models are called *double porosity models*. They apply to problems of aggregated soils and also to fractured rocks. But also the soil physicists have studied this type of models, see [15]. Interesting applications are those to *mobile and immobile water*. The problems are of great theoretical interest, see [88].

The section 5 presents a formal derivation of Darcy's law. Further, following the approach of [9] an attempt is made to derive the model for immiscible displacement in porous media. The drawback here is that certian assumptions of periodicity have to be made that are not really realistic. Finally, two relatively new problems are presented. Firstly, the problem of heterogeneous catalysis. In this case, adsoprtion of chemical species is considered and also chemicals reactions that take place on the surfaces of the porous matrix. Secondly, reactive transport through an array of cells with semipermeable membranes is described.

The last section 6 of this work gives a short outlook onto further problems and some open questions.

2 Elliptic Equations

2.1 A First Example

In the unit square $\{(x_1, x_2) : 1 < x_1, x_2 < 1\}$ in $\mathbf{R}^2$ we study a diffusion process governed by

$$\nabla \cdot (a^\varepsilon(x) \nabla u^\varepsilon(x)) = 0.$$

We make the assumption that a^ε is a function of x_2 only and is periodic with periodicity length ε, in other words we assume that

$$a^\varepsilon(x_1, x_2) = a(\frac{x_2}{\varepsilon})$$

with a fixed 1-periodic function $a : \mathbf{R} \to \mathbf{R}$. This means that we have a layered medium in which we want to calculate the effective properties. Naturally, we expect that in an averaged sense the medium can be described by a diffusion equation of the type

$$\nabla \cdot (D \nabla u^0) = 0 \tag{1}$$

where D is a tensor of the form

$$D = \begin{pmatrix} \bar{a} & 0 \\ 0 & a^* \end{pmatrix}.$$

The easiest way to find the numbers $\bar{a}$ and a^* is to study two simple boundary value problems, namely the following.

Case 1:

$$\begin{cases} u = 0, & x_1 = 0 \\ u = 1, & x_1 = 1 \\ \partial_\nu u = 0, & x_2 = 0 \text{ or } = 1 \end{cases}$$

Here one has $u^\varepsilon(x_1, x_2) = u^0(x_1, x_2)$ with $u^0(x_1, x_2) = x_1$ for all ε and hence $a^\varepsilon \nabla u^\varepsilon = a^\varepsilon \vec{e}_1$ with $\vec{e}_1 = (1, 0)$. Therefore, the total flux through the square in x_1-direction is

$$\int_0^1 a^\varepsilon \vec{e}_1 \cdot \nabla u^\varepsilon \, dx_2 = \int_0^1 a^\varepsilon(x_2) \, dx_2 = \int_0^1 a(\eta) \, d\eta.$$

This quantity equals to $\bar{a}$, since we have equation 1 independently of ε.

Case 2:

$$\begin{cases} u = 0, & x_2 = 0 \\ u = 1, & x_2 = 1 \\ \partial_\nu u = 0, & x_1 = 0 \text{ or } = 1 \end{cases}$$

Here one obtains by a simple calculation

$$u^\varepsilon(x_1, x_2) = u^0(x_1, x_2) + \varepsilon \sigma_2(\frac{x_2}{\varepsilon})$$

where $u^0(x_1, x_2) = x_2$ and

$$\sigma_2(y_2) = a^* \int_0^{y_2} \frac{d\eta}{a(\eta)} - y_2 \text{ with } a^* = \frac{1}{\int_0^1 \frac{d\eta}{a(\eta)}}.$$

One gets immediately

$$a^\varepsilon \nabla u^\varepsilon = a^* \vec{e}_2.$$

Therefore, the totals flux through the square is

$$\int_0^1 a^\varepsilon \vec{e}_2 \cdot \nabla u^\varepsilon \, dx_2 = \int_0^1 a^* dx_2 = a^*.$$

In this case we see again equation 1. In the next section we are going to see how to give a unifying theory for these two cases.

2.2 An Elliptic Boundary Value Problem

We start from the following problem. Let Ω be a bounded domain in $\mathbf{R}^n$ with smooth boundary $\partial\Omega$. Then we study the family of boundary value problems

$$\begin{cases} \nabla \cdot (a^\varepsilon(x) \nabla u^\varepsilon(x)) + f(x) = 0, & x \in \Omega, \\ u(x) = u_D(x), & x \in \partial\Omega. \end{cases} \tag{2}$$

We assume that the coefficient a^ε is rapidly oscillating, i.e. that it is of the form

$$a^\varepsilon(x) = a(\frac{x}{\varepsilon}) \tag{3}$$

for all $x \in \Omega$, where the function a is Z-periodic in $\mathbf{R}^n$ with periodicity cell $Z = \{y = (y_1, \ldots, y_n) : 0 < y_i < 1 \; for \; i = 1, \ldots, n\}$ and ε is a scale parameter. For smaller and smaller ε the coefficient a^ε oscillates more and more rapidly, and the question is natural what the solution u^ε does in the limit $\varepsilon \to 0$. We just mention that the weak form of problem 2 is

$$(a^\varepsilon \nabla u^\varepsilon, \nabla \varphi)_\Omega - (f, \varphi)_\Omega = 0$$

for all test functions φ on Ω with $\varphi | \partial\Omega = 0$. Here have used the L^2-scalar product on Ω

$$(u, v)_\Omega = \int_\Omega u(x) v(x) \, dx.$$

2.3 Formal Asymptotic Expansion

In order to derive the limit problem in a formal way one starts from the *ansatz* that the unknown function u^ε has a power series expansion with respect to ε of the form

$$u^\varepsilon(x) = u^0(x,y) + \varepsilon u^1(x,y) + \varepsilon^2 u^2(x,y) + \ldots, \tag{4}$$

where the coefficient functions $u^i(x,y)$ are Z-periodic with respect to the variable $y = \frac{x}{\varepsilon}$ (it is only in this context that superscripts denote powers of ε). The derivatives obey the law

$$\nabla = \nabla_x + \frac{1}{\varepsilon}\nabla_y,$$

where the subscripts indicate the partial derivatives with respect to x and y resp. Therefore, from equation 2 one gets immediately the formula

$$\begin{aligned}
&\varepsilon^{-2}\nabla_y \cdot (a(y)\nabla_y u^0(x,y))\\
&+\varepsilon^{-1}(\nabla_y \cdot (a(y)\nabla_y u^1(x,y)) + \nabla_y \cdot (a(y)\nabla_x u^0(x,y))\\
&\quad +a(y)\nabla_x \cdot \nabla_y u^0(x,y)))\\
&+\varepsilon^{0}(\nabla_y \cdot (a(y)\nabla_y u^2(x,y) + a(y)\nabla_x u^1(x,y))\\
&\quad +a(y)\nabla_x \cdot \nabla_y u^1(x,y) + a(y)\nabla_x \cdot \nabla_x u^0(x,y))\\
&+\varepsilon^{1}(\ldots) + \ldots + f(x) = 0.
\end{aligned} \tag{5}$$

The next step consists in comparing the coefficients of the different ε-powers in this equation. The term with ε^{-2} gives

$$\nabla_y \cdot (a(y)\nabla_y u^0(x,y)) = 0 \text{ for } y \in Z.$$

Since $u^0(x,y)$ is Z-periodic in the variable y, we obtain that $u^0(x,y) = u^0(x)$ is a function of x alone independently of y. Using this, the term with ε^{-1} in equation 5 gives

$$\nabla_y \cdot (a(y)\nabla_y u^1(x,y)) = -\nabla_y \cdot (a(y)\nabla_x u^0(x)) \text{ for } y \in Z.$$

At this point one expresses the function $u^1(x,y)$ in terms of the function $u^0(x)$. The way to do this is to first use the obvious identity

$$\nabla_x u^0(x) = \sum_{j=1}^{n} \vec{e}_j \partial_{x_j} u^0(x).$$

Therefore, we can write

$$\nabla_y \cdot (a(y)\nabla_y u^1(x,y)) = -\sum_{j=1}^{n} \partial_{y_j} a(y) \partial_{x_j} u^0(x) \text{ for } y \in Z.$$

Now for $j = 1, \ldots, n$ let $\lambda_j(y)$ be a Z-periodic solution of the *cell-problem*

$$\boxed{\nabla_y \cdot (a(y)\nabla_y \lambda_j(y)) = -\nabla_y \cdot (a(y)\vec{e}_j) \text{ for } y \in Z} \tag{6}$$

the weak form of which is

$$(a(\nabla\lambda_j + \vec{e}_j), \nabla\varphi)_Z = 0$$

for all Z-periodic test functions φ on Z.

Using these functions $\lambda_j(y)$ we can give a solution formula for $u^1(x,y)$, namely

$$u^1(x,y) = \sum_{j=1}^{n} \lambda_j(y)\partial_{x_j}u^0(x) + u^1(x),$$

where $u^1(x)$ is independent of y. Differentiating this gives immediately

$$\nabla_y u^1(x,y) = \sum_{j=1}^{n} \nabla_y \lambda_j(y)\partial_{x_j}u^0(x). \tag{7}$$

We proceed further and look at the term with ε^0 in equation 5. We get

$$\begin{aligned}&\nabla_y \cdot (a(y)\nabla_y u^2(x,y) + a(y)\nabla_x u^1(x,y))\\ &\quad + a(y)\nabla_x \cdot \nabla_y u^1(x,y) + a(y)\nabla_x \cdot \nabla_x u^0(x) + f(x) = 0 \text{ for } y \in Z.\end{aligned}$$

We integrate this identity over Z and obtain

$$\begin{aligned}&\int_Z \nabla_y \cdot (a(y)\nabla_y u^2(x,y) + a(y)\nabla_x u^1(x,y))\, dy\\ &\quad + \int_Z a(y)\nabla_x \cdot \nabla_y u^1(x,y)dy\\ &\quad + \int_Z a(y)dy\ \Delta_x u^0(x) + f(x) = 0,\end{aligned} \tag{8}$$

since the volume of Z is unity. We integrate the first integral by parts and get

$$\begin{aligned}&\int_Z \nabla_y \cdot (a(y)\nabla_y u^2(x,y) + a(y)\nabla_x u^1(x,y))\, dy\\ &\quad = \int_{\partial Z} \vec{\nu} \cdot (a(y)\nabla_y u^2(x,y) + a(y)\nabla_x u^1(x,y))\, dy.\end{aligned}$$

This boundary integral vanishes because of the Z-periodicity of the functions $u^2(x,y)$ and $u^1(x,y)$. For the second term in equation 8 we use equation 7 and get

$$\nabla_x \cdot \nabla_y u^1(x,y) = \sum_{i,j=1}^{n} \partial_{y_i}\lambda_j(y)\partial_{x_i x_j}u^0(x).$$

In this way we obtain from equation 8

$$\sum_{i,j=1}^{n} \int_Z a(y)\partial_{y_i}\lambda_j(y)\partial_{x_i x_j} u^0(x) + \int_Z a(y)dy\ \Delta_x u^0(x) + f(x) = 0.$$

It is convenient to introduce the abbreviation

$$\boxed{d_{ij} = \int_Z a(y)(\delta_{ij} + \partial_{y_i}\lambda_j(y))\ dy,}$$

with which we get the final result

$$\sum_{i,j=1}^{n} d_{ij}\partial_{x_i x_j} u^0(x) + f(x) = 0.$$

This elliptic differential equation is the *homogenized* limit of the equation in problem 2. For short we simplify the notation and write

$$\sum_{i,j=1}^{n} d_{ij}\partial_{ij} u(x) + f(x) = 0.$$

Using the tensor $D = (d_{ij})_{i,j}$ we sum up and get the following statement.

Proposition 1 *The differential operator* $\nabla \cdot (D\nabla u(x))$ *is the homogenization of the operator family* $\nabla \cdot (a^\varepsilon(x)\nabla u^\varepsilon(x))$, *i.e., the homogenization of problem 2 is*

$$\begin{cases} \nabla \cdot (D\nabla u(x)) + f(x) = 0, & x \in \Omega, \\ u(x) = u_D(x), & x \in \partial\Omega. \end{cases} \tag{9}$$

Remark: This techniques using a power series with two scales goes back to J. Keller (see, e.g. [57]). We just write down the weak form of this problem. On gets

$$(D\nabla u, \nabla\varphi)_\Omega - (f, \varphi)_\Omega = 0$$

for all test functions φ on Ω with $\varphi|\partial\Omega = 0$.

It has to be emphasized that the coefficients in the equation we have got by homogenization are in general not diagonal, or, in other words, we have derived an equation that describes an anisotropic medium. However, in general, the tensor D is symmetric and positive definite.

Proposition 2 *(a) The tensor* D *is symmetric.*

(b) If the coefficient a *in equation 3 satisfies*

$$a(y) \geq \alpha > 0 \text{ for all } y \in Z,$$

then D *is positive definite.*

Proof (a) The coefficients d_{ij} can be written as

$$d_{ij} = \int_Z a(y)(\nabla\lambda_j(y) + \vec{e}_j) \cdot \vec{e}_i \, dy.$$

The weak formulation of problem 6 is

$$\int_Z a(y)(\nabla\lambda_j(y) + \vec{e}_j) \cdot \nabla\phi(y) \, dy = 0 \text{ for all } Z\text{-periodic } \phi.$$

Therefore, one has

$$\begin{aligned} 0 &= \int_Z a(y)(\nabla\lambda_j(y) + \vec{e}_j) \cdot \nabla\lambda_i(y) \, dy \\ &= \int_Z a(y)\nabla\lambda_j(y) \cdot \nabla\lambda_i(y) \, dy + \int_Z a(y)\vec{e}_j \cdot \nabla\lambda_i(y) \, dy \end{aligned}$$

and thus

$$\begin{aligned} \int_Z a(y)\vec{e}_j \cdot \nabla\lambda_i(y) \, dy &= -\int_Z a(y)\nabla\lambda_j(y) \cdot \nabla\lambda_i(y) \, dy \\ &= -\int_Z a(y)\nabla\lambda_i(y) \cdot \nabla\lambda_j(y) \, dy = \int_Z a(y)\vec{e}_i \cdot \nabla\lambda_j(y) \, dy. \end{aligned}$$

From this we get $d_{ij} = d_{ji}$.

(b) We know from the previous calculations

$$d_{ij} = \int_Z a(y)(\nabla\lambda_j(y) + \vec{e}_j) \cdot \vec{e}_i \, dy = \int_Z a(y)(\nabla\lambda_j(y) + \vec{e}_j) \cdot (\nabla\lambda_i(y) + \vec{e}_i) \, dy.$$

Therefore, one gets for real numbers α_i

$$\alpha_i d_{ij} \alpha_j = \int_Z a(y)\nabla(\alpha_j(\lambda_j(y) + y_j)) \cdot \nabla(\alpha_i(\lambda_i(y) + y_i)) \, dy.$$

Whenever at least one of the α_i is different from zero, the quadratic form built by summing over these terms does not vanish. Q.E.D.

Remarks Though the functions λ_j are uniquely defined only up to an additive constant, the numbers d_{ij} are unique. The question which types of tensors can be obtained by homogenizing scalar equations is dealt with in [58] [64] [72] and [79].

2.4 Layered Media

Here we consider a special case of problem 2 that were already encountered in section 2.1. We assume that the function a in equation 3 depends only on y_n and not on y_i for $i \neq n$. In this case the functions λ_j can be given explicitly.

Proposition 3 *(a) If $a(y_1,\ldots,y_n) = \tilde{a}(y_n)$, then solutions of equation 6 are given by*

$$\lambda_n(y_1,\ldots,y_n) = \frac{\int_0^{y_n} \frac{d\eta}{\tilde{a}(\eta)}}{\int_0^1 \frac{d\eta}{\tilde{a}(\eta)}} - y_n$$

and $\lambda_j = 0$ for $j \neq n$.

(b) In this case the coefficients d_{ij} are given by

$$d_{ij} = \begin{cases} a^*, & i = j = n, \\ \bar{a}\delta_{ij}, & else \end{cases}$$

with $a^* = \int_0^1 \frac{d\eta}{\tilde{a}(\eta)}$ and $\bar{a} = \int_0^1 \tilde{a}(\eta)d\eta$.

Proof The formulas are easily verified by differentation and integration. Q.E.D.

Remark The physical meaning of part (b) of the proposition is that the effective *conductivity* or *diffusivity* of a layered medium is given by the *arithmetic* mean in the directions parallel to the layers but by the *geometric* mean in the direction normal to the layers. This fact is analogous to the rules that apply to electric resistances which are connected in series or parallel; see, e.g. Lurie and Cherkaev [64]; see also [23].

2.5 Energy Estimate Proof

The purpose of this chapter is to show that equation 9 holds in the interior of Ω in a strict sense. The essential part of the proof is to show estimates in certain function spaces that are satisfied by the solutions u^ε independently of ε. For this purpose we make the following assumption:

$$0 < \alpha_0 \leq a(y) \leq \alpha_1 < \infty \text{ for all } y \in Z.$$

First we introduce some convenient notations. The usual scalar product of $H = L^2(\Omega)$ and its norm are denoted by

$$(.,.) \text{ and } \|.\|.$$

Whenever the domain of integration is different from Ω, this will be indicated by an appropriate subscript, e.g.,

$$(\phi,\psi)_Z = \int_Z \phi(y)\psi(y)dy.$$

We use the function spaces

$$V = W^{1,2}(\Omega) = H^1(\Omega) \text{ and } V_0 = W_0^{1,2}(\Omega) = H_0^1(\Omega),$$

and the space of test functions $\mathcal{D} = C_0^\infty(\Omega)$. The weak form of problem 2 is

$$(a^\varepsilon \nabla u^\varepsilon, \nabla\phi) = (f,\phi) \text{ for all } \phi \in V_0. \tag{10}$$

Lemma 1 *The norms $\|u^\varepsilon\|_V$ are uniformly bounded independently of ε.*

Proof We assume that the boundary data have been extended to a function $u_D \in V$. Then we get from equation 10

$$(a^\varepsilon \nabla u^\varepsilon, \nabla(u^\varepsilon - u_D)) = (f, u^\varepsilon - u_D)$$

and therefore

$$\begin{aligned} \alpha_0 \|\nabla u^\varepsilon\|^2 &\le (a^\varepsilon \nabla u^\varepsilon, \nabla u^\varepsilon) \\ &\le |(a^\varepsilon \nabla u^\varepsilon, \nabla u_D)| + |(f, u^\varepsilon - u_D)| \\ &\le |(a^\varepsilon \nabla u^\varepsilon, \nabla u_D)| + \|f\|(\|u^\varepsilon\| + \|u_D\|) \\ &\le \alpha_1 |(\nabla u^\varepsilon, \nabla u_D)| + C_1 \|\nabla u^\varepsilon\| + C_2 \\ &\le C_3 \|\nabla u^\varepsilon\| + C_2 \end{aligned}$$

and thus

$$\|\nabla u^\varepsilon\| \le C.$$

Q.E.D.

We need a basic fact about weak convergence.

Lemma 2 *Let $g \in L^2(Z)$ be periodically extended to all $\mathbf{R}^n$ and $g^\varepsilon(x) = g(\frac{x}{\varepsilon})$. Then $g^\varepsilon \rightharpoonup \hat{g}$ weakly in H, where*

$$\hat{g} = \int_Z g(y)\, dy.$$

Proof: Without loss of generality we can assume $\hat{g} = 0$. Let $\phi \in C^\infty(\bar{\Omega})$; then we can write

$$\int_\Omega g^\varepsilon(x)\phi(x) dx = \sum_{k\in K} \int_{\varepsilon Z^k} g^\varepsilon(x)\phi(x)\, dx$$

with a index set K the number $|K|$ of elements of which satisfies $|K|\varepsilon^n \le C$. Let a step function $\hat{\phi}$ be defined by

$$\hat{\phi}(x) = \int_{\varepsilon Z^k} \phi(\tilde{x})\, d\tilde{x}, \text{ if } x \in \varepsilon Z^k;$$

then we can write

$$\int_\Omega g^\varepsilon(x)\phi(x)\, dx = \sum_{k\in K} \int_{\varepsilon Z^k} g^\varepsilon(x)(\phi(x) - \hat{\phi}(x))\, dx + \sum_{k\in K} \int_{\varepsilon Z^k} g^\varepsilon(x)\, dx\, \hat{\phi}(x).$$

The latter sum vanishes, since $\hat{g} = 0$; and since ϕ is uniformly continuous, we obtain

$$|\int_\Omega g^\varepsilon(x)\phi(x)\,dx| \leq \varepsilon C \sum_{k\in K} \int_{\varepsilon Z^k} |g^\varepsilon(x)|\;dx$$
$$\leq \varepsilon C \sum_{k\in K} \varepsilon^n \int_Z |g(y)|\,dy \leq \varepsilon C.$$

Q.E.D.

In the proof of the convergence result the following functions play an important role. We define $\lambda_j^\varepsilon(x) = \varepsilon\lambda_j(\frac{x}{\varepsilon})$, which satisfy

$$(a^\varepsilon(\nabla\lambda_j^\varepsilon + \vec{e}_j), \nabla\phi) = 0 \text{ for all } \phi \in V_0. \tag{11}$$

Lemma 3 *For the functions λ_j^ε the convergence*

$$\lambda_j^\varepsilon \to 0$$

holds strongly in H, and the convergence

$$a^\varepsilon(\nabla\lambda_j^\varepsilon + \vec{e}_j)\cdot\vec{e}_i \rightharpoonup d_{ij}$$

holds weakly in H.

Proof We have

$$\int_\Omega |\lambda_j^\varepsilon(x)|^2\,dx = \sum_{k\in K} \int_{\varepsilon Z^k} |\lambda_j^\varepsilon(x)|^2\,dx$$
$$\leq \varepsilon^{n+2} \sum_{k\in K} \int_Z |\lambda_j(y)|^2\,dy \to 0.$$

Therefore, we get the strong convergence as stated. From the definition of the d_{ij} we have

$$\int_Y a(y)(\delta_{ij} + \partial_i\lambda_j(y))\,dy = \;d_{ij};$$

now lemma 2 applies directly for the weak convergence. Q.E.D.

Theorem 1 *Let u^ε be the solution of problem 2 and u the solution of problem 9. Then*

$$u^\varepsilon \to u$$

converges strongly in H.

Proof Since problem 9 has a unique solution u, it suffices to show that each sequence taken from $\{u^\varepsilon : \varepsilon > 0\}$ has a subsequence that converges to u. For simplicity of notation we drop subscripts of ε indicating subsequences. We may assume that $u^\varepsilon \rightharpoonup \eta$ weakly in V and hence $u^\varepsilon \to \eta$ strongly in H; furthermore, we have $a^\varepsilon \partial_i u^\varepsilon \rightharpoonup \xi_i$ weakly in H, all this at least for a subsequence. Let $\phi \in \mathcal{D}$ be arbitrarily chosen; then we get from equation 10 in the limit

$$\sum_i (\xi_i, \partial_i \phi) = (f, \phi). \tag{12}$$

Using $\lambda_i^\varepsilon \phi$ as a test function in equation 10 gives

$$(a^\varepsilon \nabla u^\varepsilon, \phi \nabla \lambda_i^\varepsilon + \lambda_i^\varepsilon \nabla \phi) = (f, \lambda_i^\varepsilon \phi),$$

and using $u^\varepsilon \phi$ as a test function in equation 11 gives

$$(a^\varepsilon (\nabla \lambda_i^\varepsilon + \vec{e}_i), \phi \nabla u^\varepsilon + u^\varepsilon \nabla \phi) = 0.$$

Subtracting the last two equations yields

$$(a^\varepsilon \partial_i u^\varepsilon, \phi) + (a^\varepsilon (\nabla \lambda_i^\varepsilon + \vec{e}_i), u^\varepsilon \nabla \phi) - (a^\varepsilon \nabla u^\varepsilon, \lambda_i^\varepsilon \nabla \phi) + (f, \lambda_i^\varepsilon \phi) = 0.$$

We get for the first of these terms

$$(a^\varepsilon \partial_i u^\varepsilon, \phi) \to (\xi_i, \phi);$$

for the second term we know

$$\sum_j (a^\varepsilon (\partial_j \lambda_i^\varepsilon + \delta_{ij}), u^\varepsilon \partial_j \phi) \to \sum_j (d_{ji}, \eta \partial_j \phi)$$

from lemma 3, and the third and fourth term tend to zero. Therefore we have in the limit

$$(\xi_i, \phi) + \sum_j (d_{ji}, \eta \partial_j \phi) = 0$$

or after integration by parts

$$(\xi_i, \phi) = \sum_j (d_{ji} \partial_j \eta, \phi),$$

and since ϕ was arbitrary

$$(\xi_i, \partial_i \phi) = \sum_j (d_{ji} \partial_j \eta, \partial_i \phi).$$

Together with equation 12 and proposition 2 this gives

$$\sum_{i,j} (d_{ij} \partial_j \eta, \partial_i \phi) = (f, \phi),$$

which is the weak form the differential equation 9, since $\mathcal{D}$ is dense in V_0; i.e., we have shown $\eta = u$. Q.E.D.

Remark: There are also estimates for $u^\varepsilon - u$ available, see, e.g., [13] p. 111.

2.6 Non-Scalar Coefficients

Here we treat anisotropic media, i.e., we allow that the coefficients of the original problem are non-scalar. We study the family of boundary value problems

$$\begin{cases} \nabla \cdot (A^\varepsilon \nabla u^\varepsilon(x)) + f(x) = 0, & x \in \Omega, \\ u(x) = u_D(x), & x \in \partial\Omega. \end{cases} \tag{13}$$

Analogously to section 2 we assume that A^ε is a (not necessarily symmetric) tensor of the form

$$A^\varepsilon(x) = A(\frac{x}{\varepsilon})$$

where $A(y)$ is a tensor-valued function on Z for which $|A(y)|$ ($|.|$ being any norm) is bounded and $A(y)$ is positive definite, both uniformly in $y \in Z$. We go through the same formal asymtotic analysis as in section 2.3. For $j = 1, \ldots, n$ we define *cell-problems* for functions $\lambda_j(x, y)$ that solve

$$\nabla_y \cdot (A(y) \nabla_y \lambda_j(y)) = -\nabla_y \cdot (A(y)\vec{e}_j) \text{ for } y \in Z, \tag{14}$$

the weak form of which is

$$(A(\nabla\lambda_j + \vec{e}_j), \nabla\phi)_Z = 0 \text{ for all } Z\text{-periodic } \phi. \tag{15}$$

Using these functions we now get the coefficients from

$$d_{ij} = (A(\nabla\lambda_j + \vec{e}_j), \vec{e}_i)_Z.$$

In this way one gets in the same formal way as before a homogenization result.

Proposition 4 *Under the assumptions of this section the homogenization of problem 2 is 9 with the tensor $D = (d_{ij})_{i,j}$.*

This statement can also be made rigorous.

Theorem 2 *Under the assumptions of this section let u^ε be the solution of problem 13 and u the solution of problem 9. Then*

$$u^\varepsilon \to u$$

converges strongly in H.

We have to modify the proof of theorem 1 in the following way. Instead of the functions λ_j we now use $\lambda_j^*(y)$ which are defined as solutions of the *adjoint cell-problem*

$$\nabla_y \cdot (A^*(y) \nabla_y \lambda_j^*(y)) = -\nabla_y \cdot (A^*(y)\vec{e}_j) \text{ for } y \in Z, \tag{16}$$

where A^* is the transpose of A. Then the functions $\lambda_j^{*\varepsilon}(x) = \varepsilon\lambda_j^*(\frac{x}{\varepsilon})$ satisfy

$$(A^{*\varepsilon}(\nabla\lambda_j^{*\varepsilon} + \vec{e}_j), \nabla\phi) = 0 \text{ for all } \phi \in V_0 \tag{17}$$

with $A^{*\varepsilon}(x) = A^*(\frac{x}{\varepsilon})$. In this way one gets analogously to the scalar case

$$\sum_{i,j}(d_{ji}^*\partial_j\eta, \partial_i\phi) = (f, \phi),$$

where the coefficients are defined by

$$d_{ji}^* = (A^*(\nabla_y\lambda_i^* + \vec{e}_i), \vec{e}_j)_Z.$$

The proof of the theorem will then be complete after one derives the following proposition.

Proposition 5 *The homogenized tensor of the adjoint coefficient tensor is identical to the adjoint of the homogenized tensor, i.e. one has*

$$d_{ji}^* = d_{ij}.$$

Especially, if the coefficient tensor A is symmetric, so is its homogenization D.

Proof The part

$$(A\vec{e}_j, \vec{e}_i)_Z = (A^*\vec{e}_i, \vec{e}_j)_Z$$

is trivial. From equations 15 and 17 we get (compare the proof of proposition 2)

$$\begin{aligned}(A\nabla\lambda_j, \vec{e}_i)_Z &= (A^*\vec{e}_i, \nabla\lambda_j)_Z = -(A^*\nabla\lambda_i^*, \nabla\lambda_j)_Z \\ &= -(A\nabla\lambda_j, \nabla\lambda_i^*)_Z = (A\vec{e}_j, \nabla\lambda_i^*)_Z = (A^*\nabla\lambda_i^*, \vec{e}_j)_Z.\end{aligned}$$

Now the result follows immediately from the definition of the coefficients d_{ji}^* and d_{ij}. Q.E.D.

Similarly to proposition 2 one can show the positive definiteness of D in this case.

2.7 Non-Uniformly Oscillating Coefficients

In this section we are going to generalize the preceeding results to the following situation. We study the family of boundary value problems

$$\begin{cases} \nabla\cdot(A^\varepsilon(x)\nabla u^\varepsilon(x)) + f(x) = 0, & x \in \Omega, \\ u(x) = u_D(x), & x \in \partial\Omega. \end{cases} \tag{18}$$

Now we assume that $A^\varepsilon(x)$ is a (not necessarily symmetric) tensor which is of the form

$$A^\varepsilon(x) = A(x, \frac{x}{\varepsilon})$$

where $A(x,y)$ depends smoothly on its first argument x and is Z-periodic with respect to y. Further we make the assumption that $|A(x,y)|$ ($|.|$ being any norm) is bounded and $A(x,y)$ is positive definite, both uniformly in $x \in \Omega$. We go through the same formal asymtotic analysis as in section 2.3. The difference is that now we have to solve a cell-problem for each point $x \in \Omega$, i.e., for $j = 1, \ldots, n$ we determine functions $\lambda_j(x,y)$ that solve the *cell-problem*

$$\nabla_y \cdot (A(x,y)\nabla_y \lambda_j(x,y)) = -\nabla_y \cdot (A(x,y)\vec{e}_j) \text{ for } y \in Z, \tag{19}$$

the weak form of which is

$$(A(x,.)(\nabla\lambda_j(x,.) + \vec{e}_j), \nabla\phi(.))_Z = 0 \text{ for all } Z\text{-periodic } \phi. \tag{20}$$

Using these functions we now get the coefficients from

$$d_{ij}(x) = (A(x,.)(\nabla\lambda_j(x,.) + \vec{e}_j), \vec{e}_i)_Z.$$

As before one gets a homogenization result.

Proposition 6 *Under the assumptions of this section the homogenization of problem 2 is 9 with the tensor function $D(x) = (d_{ij}(x))_{i,j}$.*

This statement can also be made rigorous.

Theorem 3 *Under the assumptions of this section let u^ε be the solution of problem 13 and u the solution of problem 9. Then*

$$u^\varepsilon \longrightarrow u$$

converges strongly in H.

The general case is different in that one no longer gets the relation 17. The proof in this case can be found in the book [16].

2.8 Non-Uniformly Periodic Media

Using the results of section 2.6 it is easy to generalize the homogenization process to problems in which the underlying geometry is not strictly periodic. Of course, we do not make the attempt to treat absolutely arbitrary problems. Therefore, we restrict ourselves in that we consider problems for which the geometry is a diffeomorphic image of a periodic structure. We make the following assumption. Let a family of diffeomorphisms $\psi^\varepsilon : \mathbf{R}^n \to \mathbf{R}^n$ with $\varepsilon > 0$ be defined such that

$$\Psi(x) = \lim_{\varepsilon \to 0} \varepsilon (\psi^\varepsilon)^{-1}(x) \text{ for } x \in \Omega$$

is a diffeomorphisms from Ω onto a domain $\tilde{\Omega} = \Psi(\Omega)$; we make the additional assumption that the mapping $(\psi^\varepsilon)^{-1} - \frac{1}{\varepsilon}\Psi$ tends to 0 uniformly on Ω. A natural example is $\psi^\varepsilon(y) = \Psi^{-1}(\varepsilon y)$ with a given diffeomorphism Ψ; but also any $\psi^\varepsilon(y) = \Psi^{-1}(\varepsilon y) + o(\varepsilon^2)$ would do. Now we study a boundary value problem of the form 2 with a coefficient

$$a^\varepsilon(x) = a((\psi^\varepsilon)^{-1}(x)) \text{ for all } x \in \Omega$$

where a is a continuous Z-periodic function $a : \mathbf{R}^n \to \mathbf{R}$.

Remark The previous situation of section 2 is a special case of this more general setting if one uses $\psi^\varepsilon(y) = \varepsilon y$ for $y \in \mathbf{R}^n$. Further, it is only for simplicity of notations that here we consider a differential equation with a scalar coefficient.

The idea is to transform the problem onto the domain $\tilde{\Omega}$ using the mapping Ψ. To this end we need the tensor functions $B(z) = (b_{ij}(z))_{i,j}$ with

$$b_{ij}(z) = \sum_k \frac{\partial z_i}{\partial x_k}(z) \frac{\partial z_j}{\partial x_k}(z)$$

and

$$\tilde{A}^\varepsilon(z) = a^\varepsilon(\Psi^{-1}(z)) \, |\frac{\partial x}{\partial z}(z)| \, B(z)$$

where $|\frac{\partial x}{\partial z}(z)|$ is the functional determinant of the mapping $\Psi^{-1} : \tilde{\Omega} \to \Omega$. Further, we need

$$\tilde{f}(z) = f(\Psi^{-1}(z)) \, |\frac{\partial x}{\partial z}(z)| \text{ and } \tilde{u}_\Psi(z) = u_\Psi(\Psi^{-1}(z)).$$

Lemma 4 *Under the assumtions of this section problem 2 is equivalent to*

$$\begin{cases} \nabla_z \cdot (\tilde{A}^\varepsilon(z) \nabla_z \tilde{u}^\varepsilon(z)) + \tilde{f}(z) = 0, & z \in \tilde{\Omega}, \\ \tilde{u}^\varepsilon(z) = \tilde{u}_\Psi(z), & z \in \partial\tilde{\Omega}. \end{cases}$$

Proof We start from the weak form of 2, namely equation 10. According to the chain rule we have

$$\nabla_x = \nabla_z \cdot \frac{\partial z}{\partial x}$$

where $\frac{\partial z}{\partial x}$ is the Jacobian of the mapping Ψ. Therefore, we get

$$\begin{aligned} (a^\varepsilon \nabla_x u^\varepsilon, \nabla_x \phi) &= \int_\Omega a^\varepsilon(x) \nabla_x u^\varepsilon(x) \cdot \nabla_x \phi(x) \, dx \\ &= \int_{\tilde{\Omega}} a^\varepsilon(\Psi^{-1}(z)) B(z) \nabla_z u^\varepsilon(z) \cdot \nabla_z \phi(z) \, |\frac{\partial x}{\partial z}(z)| \, dz \\ &= (\tilde{A}^\varepsilon \nabla_z \tilde{u}^\varepsilon, \nabla_z \tilde{\phi})_{\tilde{\Omega}}. \end{aligned}$$

Similarly we have

$$(f,\phi) = (|\frac{\partial x}{\partial z}|\, f, \tilde{\phi})_{\tilde{\Omega}} = (\tilde{f}, \tilde{\phi})_{\tilde{\Omega}}.$$

Therefore, equation 10 has been transformed to

$$(\tilde{A}^{\varepsilon}\nabla_z \tilde{u}^{\varepsilon}, \nabla_z \tilde{\phi})_{\tilde{\Omega}} = (\tilde{f}, \tilde{\phi})_{\tilde{\Omega}}.$$

This is the weak form of the differential equation stated in the lemma, and one gets the conclusion. Q.E.D.

Instead of homogenizing problem 2 directly we now treat the boundary value problem of lemma 4.

Lemma 5 *The family of tensor functions $\tilde{A}^{\varepsilon}$ satisfies the assumptions of section 2.6, namely with the tensor function*

$$\tilde{A}(z,y) = a(y)\ |\frac{\partial x}{\partial z}(z)|\ B(z),$$

defined on $\tilde{\Omega} \times \mathbf{R}^n$, one has

$$\int_{\tilde{\Omega}} |\tilde{A}^{\varepsilon}(z) - \tilde{A}(z, \frac{z}{\varepsilon})|\ dz \rightarrow 0$$

for $\varepsilon \rightarrow 0$.

Proof One gets for the integral in question

$$\int_{\tilde{\Omega}} |a(\frac{1}{\varepsilon}\varepsilon(\psi^{\varepsilon})^{-1}(\Psi^{-1}(z))) - a(\frac{z}{\varepsilon})|\ |\frac{\partial x}{\partial z}|\ |B(z)|\ dz$$

which tends to zero due to the uniform convergence of $(\psi^{\varepsilon})^{-1}\Psi^{-1} - \frac{1}{\varepsilon}id \rightarrow 0$ and the continuity of the function a. Q.E.D.

In order to use these results and to homogenize the original problem we need the following notations. Let $\tilde{\Psi}(z) = (\tilde{s}_{ij}(z))_{i,j}$ be the homogenization of the tensor family $\tilde{A}^{\varepsilon}(z)$ in the sense of section 2.6 and let $\Psi(x) = (d_{ij}(x))_{i,j}$ have the coefficients

$$d_{ij}(x) = \sum_{k,l} \tilde{s}_{ij}(\Psi(x))\ \frac{\partial x_k}{\partial z_j}(x)\ \frac{\partial x_l}{\partial z_i}(x)\ |\frac{\partial z}{\partial x}(x)|.$$

Proposition 7 *Under the assumtions of this section the homogenization of problem 2 is problem 9.*

Proof In the same way as in lemma 4 one proves that the differential equation

$$\nabla \cdot (\tilde{\Psi}\nabla\tilde{u}) = \tilde{f} \text{ on } \tilde{\Omega}$$

transforms back onto Ω such that one gets the boundary value problem 9. Q.E.D.

2.9 Nonlinear Problems

We are not going to treat the details of the mathematical analysis for nonlinear problems. The basic approach for the formal derivation of the homogenized problems is to linearize nonlinear functions whenever they occur, e.g. to write

$$f(u^\varepsilon(x)) = f(u^0(x,y) + \varepsilon u^1(x,y) + \ldots) = f(u^0(x,y)) + \varepsilon u^1(x,y) f'(u^0(x,y)) + \ldots.$$

For rigorous convergence proofs the estimates needed require careful treatment. Examples are given in [16] section 16 and in [13] section 5.1. A nonlinear problem of special interest for flow through porous media is the *dam problem.* The question of homogenization in this case is treated in [33] and [34] [81].

3 Media with Obstacles

In this section we assume that the medium in which we study diffusion processes has a periodic arrangement of obstacles. The formal description of this geometry goes along the following lines.

3.1 Solid Obstacles

We assume that in the standard periodicity cell Z there is a standard obstacle $X \subset\subset Z$ with smooth boundary Γ. The remainder is denoted by $Y = Z \setminus X$. We assume that this standard geometry is repeated periodically all over $\mathbf{R}^n$. The geometric structure within the fixed domain Ω is obtained by intersecting the ε-multiple of this periodic geometry with Ω. The result is denoted by

$$\Omega^\varepsilon = \Omega \cap (\varepsilon Y) \text{ and } \Gamma^\varepsilon = \Omega \cap (\varepsilon\Gamma).$$

In order to avoid too many technical difficulties we assume that $\partial\Omega \cap \varepsilon\Gamma = \emptyset$. In this situation we want to study a diffusion process that takes place only in the domain left by the obstacles, namely in Ω^ε. A simple model is the following boundary value problem.

$$\begin{cases} \nabla \cdot (a^\varepsilon(x)\nabla u^\varepsilon(x)) + f(x) = 0, & x \in \Omega^\varepsilon, \\ \vec{\nu} \cdot a^\varepsilon(x)\nabla u^\varepsilon(x) = 0, & x \in \Gamma^\varepsilon, \\ u(x) = u_D, & x \in \partial\Omega. \end{cases} \tag{21}$$

For simplicity, we assume that

$$a^\varepsilon(x) = a(\frac{x}{\varepsilon})$$

with a Z-periodic function $a : Y \to \mathbf{R}$. The zero flux condition on Γ^ε is used to model the obstacles as impermeable. Problem 21 can be looked at as a generalization of the situation from section 2. In order to do this one has to introduce the characteristic function of Y, namely

$$\chi(x) = \begin{cases} 1, & x \in Y, \\ 0, & x \in X, \end{cases}$$

and to define $\chi^\varepsilon(x) = \chi(\frac{x}{\varepsilon})$. Then one gets as the weak form of 21 the equation

$$(\chi^\varepsilon a^\varepsilon \nabla u^\varepsilon, \nabla\phi) = (f, \phi) \text{ for all } \phi \in V_0, \tag{22}$$

which has the same formal structure as we had in equation 10. The well-posedness of the ε-problem follows from standard analysis. Therefore, the same analysis applies as in section 2, the difference being that now the *cell-problems* are, written in strong form, to determine Z-periodic functions $\lambda_j(y)$ which satisfy

$$\boxed{\begin{array}{ll} \nabla_y \cdot (a(y)\nabla\lambda_j(y)) = -\nabla_y \cdot (a(y)\vec{e}_j), & y \in Y, \\ \vec{\nu} \cdot \nabla_y\lambda_j(y) = -\vec{\nu} \cdot \vec{e}_j, & y \in \Gamma, \end{array}} \tag{23}$$

and in weak form

$$(\chi a \nabla(\lambda_j + \vec{e}_j), \nabla\phi)_Z = 0 \text{ for all } Z\text{-periodic } \phi.$$

Remark In the case that a is constant on Y, the differential equation for λ_j simplifies to

$$\Delta\lambda_j(y) = 0 \text{ for } y \in Y.$$

In any case, the tensor $D = (d_{ij})_{i,j}$ is defined by

$$d_{ij} = (a(\nabla\lambda_j + \vec{e}_j), \vec{e}_i)_Y$$

analogously to section 2. In this way one gets the following result.

Proposition 8 *The homogenization of problem 21 is problem 9.*

All the rest remains almost litteraly the same as before. For the convergence proof there is a detail to be treated carefully, namely the fact that now the functions u^ε are defined only on Ω^ε, i.e., on ε-dependent domains. Before one can go through the same proof as before, one has to extend these functions into all Ω. This is done by means of the following lemma.

Lemma 6 *(a) If a function $u \in H^{1,2}(Y)$ is given on Y there is an extension $\tilde{u}$ into X, and thus onto all Z, such that*

$$\|\tilde{u}\|_{H^{1,2}(Z)} \leq C\|u\|_{H^{1,2}(Y)}$$

(b) A function $u^\varepsilon \in H^{1,2}(\Omega^\varepsilon)$ can be extended to a function $\tilde{u}^\varepsilon$ into all Ω such that

$$\|\tilde{u}^\varepsilon\|^2_{H^{1,2}(\Omega)} \leq C\|u^\varepsilon\|^2_{H^{1,2}(\Omega^\varepsilon)}$$

Proof: See also [32]. (a) According to our general assumption the data of the problem are sufficiently smooth. Therefore, we can find a neighborhood U of the interface Γ such that the field of normal vectors $\vec{\nu}$ on Γ allows to construct a parametrization of U. Then we extend the function u from Y into all U by reflection at the interface Γ, and further into all X in any smooth manner. This procedure gives us a function u^* in all Z. We denote the average of u^* over Y by

$$m = \frac{1}{|Y|}\int_Y u^* \, dy.$$

Now we assume that ψ is a smooth function on Z with compact support in the interior of X such that $1 - \psi$ is identically 1 on U. Then we define the extension by

$$\tilde{u} = (1 - \psi)(u^* - m) + m.$$

Obviously $\tilde{u}$ is identical to u on Y. Its gradient is

$$\nabla\tilde{u} = (u^* - m)\nabla(1-\psi) + (1-\psi)\nabla u^*$$

Therefore we get

$$\|\tilde{u}\|^2_{H^{1,2}(Z)} = \|u\|^2_{H^{1,2}(Y)} + \|\tilde{u}\|^2_{H^{1,2}(X)}$$

with the estimate

$$\|\tilde{u}\|^2_{H^{1,2}(X)} = \int_X (|\ \tilde{u}\ |^2 + |\ \nabla\tilde{u}\ |^2)\ dy.$$

The inequality

$$\int_X |\ \tilde{u}\ |^2\ dy \leq C\int_Y |\ u\ |^2\ dy$$

is obvious. On the other hand we have

$$\int_X |\ \nabla\tilde{u}\ |^2\ dy \leq \|\nabla(1-\psi)\|_{L^\infty}\int_X |\ u^* - m\ |^2\ dy + \|1-\psi\|_{L^\infty}\int_U |\ \nabla u^*\ |^2\ dy.$$

From this we get the conclusion of the first part of the lemma.

(b) This is a simple consequence of (a) obtained by scaling and summation. Q.E.D.

Using this extension lemma one gets the following result that is essential for the convergence proof.

Lemma 7 *There is an extension $\tilde{u}^\varepsilon$ of the function u^ε into all Ω such that*

$$\|\tilde{u}^\varepsilon\|_V \leq C$$

independently of ε.

Proof In the same way as in section 2 one first shows that the norms $\|u^\varepsilon\|_{H^1(\Omega^\varepsilon)}$ are uniformly bounded. Then one applies the previous lemma and gets the result. Q.E.D.

Now we are ready to derive the convergence theorem.

Theorem 4 *Let u^ε be the solution of problem 21 and u the solution of problem 8, then*

$$u^\varepsilon \to u$$

converges strongly in H.

Proof The proof is litteraly the same as for theorem 1. Q.E.D.

What makes this problem relatively simple is the fact that we have imposed homogeneous Neumann boundary conditions on the surfaces Γ^ε of the obstacles. The nonhomogeneous case is treated in [31].

3.2 Thin Obstacles

The situation becomes different from what we have in the previous section if the obstacles are no longer "full" n-dimensional sets, but have lower dimension, say are of the form of a $(n-1)$-dimensional hyper-surface. One assumes that the obstacle $\Gamma \subset\subset Z$ is the diffeomorphic image of a $(n-1)$-dimensional bounded domain. Then one studies the same boundary value problem as in problem 21 where one now has to consider the zero flux condition on both "sides" of Γ^ε. The mathematical analysis of this situation is given in [7].

3.3 Non-Uniformly Periodic Obstacles

It is natural to combine the techniques of section 2.8 with those of section 3.1. We chose a family of diffeomorphisms ψ^ε as in 2.8. Further, we assume that there is a standard obstacle Y with boundary Γ given and repeated periodically in $\mathbf{R}^n$. Now the geometry within the domain Ω is given by

$$\Omega^\varepsilon = \Omega \cap \psi^\varepsilon(Y) \text{ and } \Gamma^\varepsilon = \Omega \cap \psi^\varepsilon(\Gamma).$$

The boundary value problem that we are interested in is the same as problem 21, where in this case the coefficient function is

$$a^\varepsilon(x) = a((\psi^\varepsilon)^{-1}(x)) \text{ for all } x \in \Omega^\varepsilon,$$

and a is a periodic function $a : Y \to \mathbf{R}$. In exactly the same way as in section 2.8 one can transform this problem onto $\tilde{\Omega}$. Thus we get the following result.

Proposition 9 *Under the assumtions of this section the homogenization of problem 21 is problem 9.*

4 Parabolic Equations

In the foregoing analysis we have seen that the differential equations may change their character in the process of homogenization. E.g., an equation with a scalar coefficient can become anisotropic in the sense that its coefficient is changed into a tensor. When we go over to parabolic equations and treat time dependent problems we will encounter other effects with change of type. Of course, if we considered parabolic equations in the same way as we studied elliptic equations, namely just by introducing a term with a time derivarive, in principle nothing new would come up. A term of the form $\partial_t u$ could formally be dealt with as another inhomogeneity, i.e. similarly to a source term f. New aspects come into play, if we allow more complex models on the micro-scale.

4.1 Fractured Media

Here we use a configuration on the micro-scale which is different from what we had in section 3.1 in that we now allow flow to take place in the interior of the aggregates. We assume that in the standard periodicity cell Z there is a standard aggregate $X \subset\subset Z$ with smooth boundary Γ the remainder of which is denoted by $Y = Z \setminus X$. As before, we assume that this standard geometry is repeated periodically all over $\mathbf{R}^n$. The geometric structure within the fixed domain Ω is obtained by intersecting the ε-multiple of this periodic geometry with Ω. The result is

$$\Pi^\varepsilon = \Omega \cap (\varepsilon X)\ ,\ \Omega^\varepsilon = \Omega \cap (\varepsilon Y) \text{ and } \Gamma^\varepsilon = \Omega \cap (\varepsilon \Gamma).$$

First we study a diffusion process in which two scales for the diffusivity occur. We start from the following initial boundary value problem.

$$\begin{cases} \partial_t v^\varepsilon(t,x) = \nabla \cdot (a^\varepsilon(x)\nabla v^\varepsilon(t,x)), & t>0,\ x\in\Omega^\varepsilon, \\ \partial_t w^\varepsilon(t,x) = \varepsilon^2\nabla\cdot(b^\varepsilon(x)\nabla w^\varepsilon(t,x)), & t>0,\ x\in\Pi^\varepsilon, \\ a^\varepsilon(x)\vec{\nu}\cdot\nabla v^\varepsilon(t,x) = \varepsilon^2 b^\varepsilon(x)\vec{\nu}\cdot\nabla w^\varepsilon(t,x), & t>0,\ x\in\Gamma^\varepsilon, \\ \varepsilon d^\varepsilon(x) b^\varepsilon(x)\vec{\nu}\cdot\nabla w^\varepsilon(t,x) = v^\varepsilon(t,x) - c w^\varepsilon(t,x), & t>0,\ x\in\Gamma^\varepsilon, \\ v^\varepsilon(t,x) = v_D(t,x), & t>0,\ x\in\partial\Omega, \\ v^\varepsilon(t,x) = v_I(x), & t=0,\ x\in\Omega, \\ w^\varepsilon(t,x) = w_I(x), & t=0,\ x\in\Omega. \end{cases} \tag{24}$$

We make the assumption that the coefficients a_k^ε and d^ε are oscillating, i.e., that we have

$$a_k^\varepsilon(x) = a_k(\frac{x}{\varepsilon}) \text{ and } d^\varepsilon(x) = d(\frac{x}{\varepsilon}),$$

where the functions a_k and d are Z-periodic. The parameter d has been introduced in order to allow different types of models, namely for $d = 0$ one gets

$$v^\varepsilon(x) = c w^\varepsilon(x)$$

on the interface Γ^ε, whereas for $d > 0$ one has a boundary condition of *Robin* type. In any case we assume $c > 0$. The formal asymptotics is not exactly the same as in section 2 since we have to take care of the factors ε and ε^2 which appear in the model. The starting point is as usual the *ansatz* that the unknowns can be written as a power series in the form

$$v^\varepsilon(t,x) = v^0(t,x,y) + \varepsilon v^1(t,x,y) + \varepsilon^2 v^2(t,x,y)\dots$$

and

$$w^\varepsilon(t,x) = w^0(t,x,y) + \varepsilon w^1(t,x,y) + \varepsilon^2 w^2(t,x,y)\dots$$

with the usual assumption of Z-periodicity of the coefficient functions. For simplicity of notation we drop the variable t in the following. The differential equation for v^ε in Ω^ε gives

$$\begin{aligned}\varepsilon^0\partial_t v^0(x,y) + \varepsilon^1\dots &= \varepsilon^{-2}\nabla_y\cdot(a(y)\nabla_y v^0(x,y))\\ &+\varepsilon^{-1}(\nabla_y\cdot(a(y)\nabla_y v^1(x,y)) + \nabla_y\cdot(a(y)\nabla_x v^0(x,y))\\ &\quad +a(y)\nabla_x\cdot\nabla_y v^0(x,y)))\\ &+\varepsilon^0(\nabla_y\cdot(a(y)\nabla_y v^2(x,y) + a(y)\nabla_x v^1(x,y))\\ &\quad +a(y)\nabla_x\cdot\nabla_y v^1(x,y) + a(y)\nabla_x\cdot\nabla_x v^0(x,y))\\ &+\varepsilon^1(\dots)+\dots,\ y\in Y.\end{aligned} \tag{25}$$

The differential equation for w^ε in Π^ε gives

$$\varepsilon^0\partial_t w^0(x,y) + \varepsilon^1\dots = \varepsilon^0\nabla_y\cdot(b(y)\nabla_y w^0(x,y)) + \varepsilon^1\dots,\ y\in X \tag{26}$$

The two conditions on the interface Γ^ε yield

$$\begin{aligned}&\varepsilon^{-1}a(y)\vec{\nu}\cdot\nabla_y v^0(x)\\ &\quad+\varepsilon^0 a(y)\vec{\nu}\cdot(\nabla_y v^1(x) + \nabla_x v^0(x))\\ &\quad+\varepsilon^1(a(y)\vec{\nu}\cdot(\nabla_y v^2(x) + \nabla_x v^1(x)))\\ &\quad+\varepsilon^2\dots = \varepsilon^1 b(y)\vec{\nu}\cdot\nabla_y w^0(x,y) + \varepsilon^2\dots,\ y\in\Gamma\end{aligned} \tag{27}$$

and

$$\begin{aligned}&\varepsilon^0 d(y)b(y)\vec{\nu}\cdot\nabla_y w^0(x,y) + \varepsilon^1\dots\\ &\quad= \varepsilon^0(v^0(x,y) - cw^0(x,y)) + \varepsilon^1\dots,\ y\in\Gamma.\end{aligned} \tag{28}$$

Now we have to compare the coefficients of the different ε-powers in these equations. The term with ε^{-2} in equation 25 gives

$$\nabla_y\cdot(a(y)\nabla_y v^0(x,y)) = 0 \text{ for } y\in Y.$$

Since $v^0(x,y)$ is Z-periodic in the variable y, we obtain that $v^0(x,y) = v^0(x)$ is a function of x alone independently of y. Using this, the term with ε^{-1} in equation 25 gives

$$\nabla_y \cdot (a(y)\nabla_y v^1(x,y)) = -\nabla_y \cdot (a(y)\nabla_x v^0(x)) \text{ for } y \in Y.$$

One expresses the function $v^1(x,y)$ in terms of the function $v^0(x)$ in the same way as in section 2.5. Using the functions $\lambda_j(y)$ from 23 with a instead of a we get

$$v^1(x,y) = \sum_{j=1}^{n} \lambda_j(y)\partial_{x_j} v^0(x) + v^1(x),$$

where $v^1(x)$ is independent of y. We proceed further and look at the term with ε^0 in equation 25. We get

$$\begin{aligned}\partial_t v^0(x,y) &= \nabla_y \cdot (a(y)\nabla_y v^2(x,y) + a(y)\nabla_x v^1(x,y)) \\ &+ a(y)\nabla_x \cdot \nabla_y v^1(x,y) + a(y)\nabla_x \cdot \nabla_x v^0(x) \text{ for } y \in Y.\end{aligned}$$

We integrate this identity over Y and obtain

$$\begin{aligned}&|Y|\partial_t v^0(x) \\ &\quad = \int_Y \nabla_y \cdot (a(y)\nabla_y v^2(x,y) + a(y)\nabla_x v^1(x,y))\, dy \\ &\quad + \int_Y a(y)\nabla_x \cdot \nabla_y v^1(x,y) dy + \int_Y a(y) dy\ \Delta_x v^0(x),\end{aligned} \tag{29}$$

We integrate the first integral on the right side by parts and get

$$\begin{aligned}&\int_Y \nabla_y \cdot (a(y)\nabla_y v^2(x,y) + a(y)\nabla_x v^1(x,y))\, dy \\ &\quad = \int_{\partial Y} a(y)\vec{\nu} \cdot (\nabla_y v^2(x,y) + \nabla_x v^1(x,y))\, d\Gamma(y).\end{aligned}$$

We have $\partial Y = \partial Z + \Gamma$; the boundary integral over ∂Z vanishes because of the Z-periodicity of the functions $v^2(x,y)$ and $v^1(x,y)$. For the integral over Γ we have a look at the term ε^1 in 27 and get

$$\int_\Gamma a(y)\vec{\nu} \cdot (\nabla_y v^2(x,y) + \nabla_x v^1(x,y))\, d\Gamma(y) = \int_\Gamma b(y)\vec{\nu} \cdot \nabla_y w^0(x,y)\, d\Gamma(y).$$

For the second term in equation 29 we get in the well known way

$$\nabla_x \cdot \nabla_y v^1(x,y) = \sum_{i,j=1}^{n} \partial_{y_i} \sigma(y) \partial_{x_i x_j} v^0(x).$$

Therefore, we obtain from equation 29

$$|Y|\partial_t v^0(x,y) - \int_\Gamma b(y)\vec{\nu}\cdot\nabla_y w^0(x,y)\, d\Gamma(y)$$
$$= \sum_{i,j=1}^{n}\int_Y a(y)\partial_{y_i}\sigma(y)dy\ \partial_{x_i x}v^0(x) + \int_Y a(y)dy\ \Delta_x v^0(x).$$

The term with ε^0 in equation 26 yields

$$\partial_t w^0(x,y) = \nabla_y\cdot(b(y)\nabla_y w^0(x,y)),\ y\in X,$$

and the term with ε^1 in equation 28 gives

$$d(y)b(y)\vec{\nu}\cdot\nabla_y w^0(x,y) = v^0(x) - cw^0(x,y), y\in\Gamma.$$

In this way we have shown the following result. For clarification we write the time variable t again.

Proposition 10 *The homogenization of the initial boundary value problem 24 is*

$$\begin{cases} |Y|\partial_t v(t,x) + S(t,x) = \nabla\cdot(D\nabla v(t,x)), & t>0,\ x\in\Omega,\\ S(t,x) = -\int_\Gamma b(y)\vec{\nu}\cdot\nabla_y w(t,x,y)\ d\Gamma(y), & t>0,\ x\in\Omega,\\ \partial_t w(t,x,y) = \nabla_y\cdot(b(y)\nabla_y w(t,x,y)), & t>0,\ x\in\Omega,\ y\in X,\\ d(y)b(y)\vec{\nu}\cdot\nabla_y w(t,x,y) = v(t,x) - cw(t,x,y), & t>0,\ x\in\Omega,\ y\in\Gamma,\\ v(t,x) = v_D(t,x), & t>0,\ x\in\partial\Omega,\\ v(t,x) = v_I(x), & t=0,\ x\in\Omega,\\ w(t,x) = w_I(x), & t=0,\ x\in\Omega. \end{cases} \tag{30}$$

The homogenized problem can be rewritten in several different ways. We use the average

$$\bar{w}(t,x) = \frac{1}{|X|}\int_X w(t,x,y)dy.$$

Proposition 11 *The source/sink term $S(t,x)$ can be represented as*

$$S(t,x) = \int_X \nabla_y\cdot(b(y)\nabla_y w(t,x,y))\ dy = \int_X \partial_t w(t,x,y)\ dy = |X|\partial_t\bar{w}(t,x).$$

If the function $r:[0,\infty)\times X\to\mathbf{R}$ solves the initial boundary value problem

$$\begin{cases} \partial_t r(t,y) = \nabla_y\cdot(b(y)\nabla_y r(t,y)), & t>0,\ y\in X,\\ d(y)b(y)\vec{\nu}\cdot\nabla_y r(t,y) = -cr(t,y), & t>0,\ y\in\Gamma,\\ r(t,y) = 1, & t=0,\ y\in X \end{cases} \tag{31}$$

then one has

$$w(t,x,y) = (w_I(x) - \frac{1}{c}v_I(x))r(t,y)$$
$$+\frac{1}{c}(v(t,x) - \int_0^t r(t-s,y)\partial_t v(s,x)\, ds);$$

and with

$$\vec{\rho}(t) = \int_X r(t,y)\, dy$$

one can write

$$S(t,x) = -(w_I(x) - \frac{1}{c}v_I(x))\frac{d}{dt}\rho(t) - \frac{1}{c}\int_0^t \frac{d}{dt}\vec{\rho}(t-s)\partial_t v(s,x)\, ds.$$

Proof: The representation of S as a volume integral is obtained using integration by parts, where now $\vec{\nu}$ is the inner normal on ∂X. Let $\tilde{w}$ be the expression given in the proposition for w. Then we get

$$\partial_t\tilde{w}(t,x,y) = (w_I(x) - \frac{1}{c}v_I(x))\partial_t r(t,y)$$
$$+\frac{1}{c}(\partial_t v(t,x) - r(0,y)\partial_t v(t,x) - \int_0^t \partial_t r(t-s,y)\partial_t v(s,x)\, ds)$$

This is equal to

$$\nabla_y \cdot (b(y)\nabla_y w(t,x,y)) = (w_I(x) - \frac{1}{c}v_I(x))\nabla_y \cdot (b\nabla_y r(t,y))$$
$$-\frac{1}{c}\int_0^t \nabla_y \cdot (b\nabla_y r(t-s,y))\partial_t v(s,x)\, ds.$$

Therefore $\tilde{w}$ satifies the differential equation required. Moreover, we have

$$d(y)b(y)\vec{\nu}\cdot\nabla_y w(t,x,y) =$$
$$= (w_I(x) - \frac{1}{c}v_I(x))d(y)b(y)\vec{\nu}\cdot\nabla_y r(t,y)$$
$$-\frac{1}{c}\int_0^t d(y)b(y)\vec{\nu}\cdot\nabla_y r(t-s,y)\partial_t v(s,x)\, ds.$$

This equals to

$$v(t,x) - cw(t,x,y) = v(t,x) - (cw_I(x) - v_I(x))r(t,y)$$
$$-v(t,x) + \int_0^t r(t-s,y)\partial_t v(s,x)\, ds.$$

Therefore, $\tilde{w}$ also satifies the proper boundary condition. The initial condition $\tilde{w}(0,y) = w_I(x)$ is trivial; therefore, we get $\tilde{w} = w$. The final representation of S is just a matter of differentiation with respect to t and integration over X. Q.E.D.

Proposition 12 *If X is a ball of radius R and $b = a$ is a constant, then the kernel $\vec{\rho}$ is given by*

$$\vec{\rho}(t) = |X|\frac{6}{\pi^2}\sum_{k=1}^{\infty}\frac{1}{k^2}e^{-\pi^2 k^2 at}$$

Proof: The cell function r in this case is

$$r(t,y) = \frac{2}{\pi}\sum_{k=1}^{\infty}\frac{(-1)^{k-1}}{k}e^{-\pi^2 k^2 at}\sin(\pi k\frac{|y|}{R}).$$

The kernel $\vec{\rho}$ is obtained by integration over X. Q.E.D.

Corollary 1 *If the initial data are zero, one gets the problem*

$$\begin{cases} |Y|\partial_t v(x) + S(x) = \nabla\cdot(D\nabla v(x)), & t>0,\ x\in\Omega, \\ S(x) = -\frac{1}{c}\int_0^t \frac{d}{dt}\vec{\rho}(t-s)\partial_t v(s,x)\,ds, & t>0,\ x\in\Omega, \\ v(t,x) = v_D(t,x), & t>0,\ x\in\partial\Omega, \\ v(t,x) = 0, & t=0,\ x\in\Omega, \\ w(t,x) = 0, & t=0,\ x\in\Omega. \end{cases} \tag{32}$$

A problem of this type with $d = 0$ was dealt with in [91]; a convergence proof for a problem with a lower order term was given in [5], see also chapter 2 of [6]. In the next section we give another convergence proof. The well-posedness of the initial boundary value problem of corollary 1 was studied independently in [56].

4.2 Convergence Proof

The existence and uniqueness of the solutions of the ε-problems is standard and not treated here in detail. We introduce the follwoing shorthand notations:

$$Q = [0,T]\times\Omega,\ U = Q\times X,\ Q^\varepsilon = [0,T]\times\Omega^\varepsilon,\ R^\varepsilon = [0,T]\times\Pi^\varepsilon.$$

In order to simplify the calculations, we treat only the following special case. We assume $c = 1$, $d = 0$, $v_D = 0$, and a, b to be constant. In this special case the weak form of problem 24 is

$$(\partial_t v^\varepsilon,\varphi)_{\Omega^\varepsilon} + (\partial_t w^\varepsilon,\psi)_{\Pi^\varepsilon} + a(\nabla v^\varepsilon,\nabla\varphi)_{\Omega^\varepsilon} + \varepsilon^2 b(\nabla w^\varepsilon,\nabla\psi)_{\Pi^\varepsilon} = 0 \tag{33}$$

for all test functions $\varphi\in H^1(\Omega^\varepsilon)$ with $\varphi|_{\partial\Omega} = 0$, $\psi\in H^1(\Pi^\varepsilon)$ and $\varphi|_{\Gamma^\varepsilon} = \psi|_{\Gamma^\varepsilon}$ (in other words, we use functions from $H_0^1(\Omega)$ which we denote by φ on Ω^ε and by ψ on Π^ε). First of all we give estimates of the solutions v^ε and w^ε that are independent of ε.

Proposition 13 *For the solution of the micro-model problem 24 the estimates*

$$\|v^\varepsilon\|_{L^2(0,T;H^1(\Omega^\varepsilon))} \le C,\ \ \varepsilon\|\nabla w^\varepsilon\|_{L^2(0,T;L^2(\Pi^\varepsilon))} \le C,$$
$$\|\partial_t v^\varepsilon\|_{L^\infty(0,T;L^2(\Omega^\varepsilon))} \le C,\ \ \|\partial_t w^\varepsilon\|_{L^\infty(0,T;L^2(\Pi^\varepsilon))} \le C,$$

and

$$\|\nabla\partial_t v^\varepsilon\|_{L^2(0,T;L^2(\Omega^\varepsilon))} \le C,\ \ \varepsilon\|\nabla\partial_t w^\varepsilon\|_{L^2(0,T;L^2(\Pi^\varepsilon))} \le C$$

hold independently of ε.

Proof: Plugging $\varphi = v^\varepsilon$ and $\psi = w^\varepsilon$ into equation 33 we get

$$(\partial_t v^\varepsilon, v^\varepsilon)_{\Omega^\varepsilon} + (\partial_t w^\varepsilon, w^\varepsilon)_{\Pi^\varepsilon} + a(\nabla v^\varepsilon, \nabla v^\varepsilon)_{\Omega^\varepsilon} + \varepsilon^2 b(\nabla w^\varepsilon, \nabla w^\varepsilon)_{\Pi^\varepsilon} = 0.$$

We integrate this over $[0,t]$ and obtain

$$\frac{1}{2}(\|v^\varepsilon(t)\|_{\Omega^\varepsilon} + \|w^\varepsilon(t)\|_{\Pi^\varepsilon})$$
$$+\int_0^t \left(a(\nabla v^\varepsilon, \nabla v^\varepsilon)_{\Omega^\varepsilon} + \varepsilon^2 b(\nabla w^\varepsilon, \nabla w^\varepsilon)_{\Pi^\varepsilon}\right) d\tau = \frac{1}{2}(\|v_I\|_{\Omega^\varepsilon} + \|w_I\|_{\Pi^\varepsilon}).$$

Hence we get the first two estimates. Now we differentiate equation 33 and get

$$(\partial_{tt} v^\varepsilon, \varphi)_{\Omega^\varepsilon} + (\partial_{tt} w^\varepsilon, \psi)_{\Pi^\varepsilon} + a(\partial_t \nabla v^\varepsilon, \nabla\varphi)_{\Omega^\varepsilon} + \varepsilon^2 b(\partial_t \nabla w^\varepsilon, \nabla\psi)_{\Pi^\varepsilon} = 0.$$

Here we use the test functions $\varphi = \partial_t v^\varepsilon$ and $\psi = \partial_t w^\varepsilon$ and get by integration over time

$$\tfrac{1}{2}(\|\partial_t v^\varepsilon(t)\|_{\Omega^\varepsilon} + \|\partial_t w^\varepsilon(t)\|_{\Pi^\varepsilon})$$
$$+\int_0^t \left(a(\partial_t\nabla v^\varepsilon, \partial_t\nabla v^\varepsilon)_{\Omega^\varepsilon} + \varepsilon^2 b(\partial_t\nabla w^\varepsilon, \partial_t\nabla w^\varepsilon)_{\Pi^\varepsilon}\right) d\tau = \frac{1}{2}(\|\hat{v}_I\|_{\Omega^\varepsilon} + \|\hat{w}_I\|_{\Pi^\varepsilon}).$$

where $\hat{v}_I = \nabla\cdot a(\nabla v_I)$ and $\hat{w}_I = \varepsilon^2\nabla\cdot b(\nabla w_I)$. In this way we get all the remaining estimates. Q.E.D.

From now on we identify v^ε with its extension according to lemma 6; let $\vec{\xi}^\varepsilon = \nabla v^\varepsilon$ (extended by $\varepsilon^2\nabla w^\varepsilon$ in the interior of Π^ε). Note that the extension lemma from [32] pp. 593-597 implies that $\nabla\cdot\vec{\xi}^\varepsilon$ is uniformly bounded in $L^2(Q)$. We we also identify w^ε with its H^1-extension defined by $w^\varepsilon = v^\varepsilon$ on Ω^ε. Then we get

Proposition 14 *There is a subsequence such that*

$$\begin{array}{ll} v^\varepsilon \rightharpoonup v^* & \textit{weakly in } L^2(0,T;H^1(\Omega)), \\ v^\varepsilon \rightharpoonup v^* & \textit{weak* in } L^\infty(Q), \\ \partial_t v^\varepsilon \rightharpoonup \partial_t v^* & \textit{weakly in } L^2(Q), \\ v^\varepsilon \to v^* & \textit{strongly in } L^2(0,T;L^2(\Omega)), \\ \vec{\xi}^\varepsilon \rightharpoonup \vec{\xi}^* & \textit{weakly in } (L^2(Q))^n \end{array}$$

with some v^* *and* $\vec{\xi}^*$, *such that* $\nabla\cdot\vec{\xi}^*$ *is bounded in* $L^2(Q)$.

Proof: This follows from proposition 13. Q.E.D.

In order to prove the convergence result, we use the notion of *two-scale convergence* which was introduced in [75] and developed further in [3]. The idea behind this concept was used in [5].

Definition 1 *The sequence $\{w^\varepsilon\} \subset L^2(Q)$ is said to two-scale converge to a limit $w \in L^2(Q \times Z)$ iff for any $\sigma \in C^\infty(Q; C^\infty_{per}(Z))$ ("per" denotes Z-periodicity) one has*

$$\lim_{\varepsilon\to 0} \int_Q w^\varepsilon(t,x)\sigma(t,x,\frac{x}{\varepsilon})\; dx\; dt = \int_Q \int_Z w(t,x,y)\sigma(t,x,y)\; dy\; dx\; dt.$$

Lemma 8 *From each bounded sequence in $L^2(Q)$ one can extract a subsequence which two-scale converges to a limit $w \in L^2(Q \times Z)$.*

Proof: See [75]. Q.E.D.

Lemma 9 *Let w^ε and $\varepsilon\nabla w^\varepsilon$ be bounded sequences in $L^2(Q)$. Then there exists a function $w \in L^2(Q; H^1_{per}(Z))$ and a subsequence such that both w^ε and $\varepsilon\nabla w^\varepsilon$ two-scale converge to w and $\nabla_y w$, resp.*

Proof: See [3] and [75]. Q.E.D.

Proposition 15 *There is a subsequence such that*

$$w^\varepsilon \to w^*$$

$$\partial_t w^\varepsilon \to \partial_t w^*$$

$$\varepsilon\nabla w^\varepsilon \to \nabla_y w^*$$

in the two-scale sense with some $w^* \in L^2(Q; H^1_{per}(Z)) \cap H^1(0,T; L^2(\Omega \times Z))$.

Proof: This follows from the estimates of proposition 13 and lemma 9. Q.E.D.

Proposition 16 *The function w^* defined in proposition 15 satisfies*

$$w^*|_{t=0} = w_I \;\; a.e. \;\; on \;\; \Omega \times X.$$

Proof: Let $\varphi \in C_0^\infty(\Omega)$ and $\omega \in C_0^\infty(0,T)$ with $\omega(T) = 0$. Furthermore, let $\sigma \in C_0^\infty(X)$. We extend σ by zero to $Z \setminus X$ and Z-periodically to $\mathbf{R}^n$. We define σ^ε by $\sigma^\varepsilon(x) = \sigma(\frac{x}{\varepsilon})$. Then we have

$$(\partial_t w^\varepsilon, \varphi\sigma^\varepsilon\omega)_{R^\epsilon} = -(w_I, \varphi\sigma^\varepsilon)_{\Pi^\epsilon}\, \omega(0) - (w^\varepsilon, \varphi\sigma^\varepsilon\partial_t\omega)_{R^\epsilon}.$$

Passing to the limit yields

$$\int_U \partial_t w^*\sigma\; dy\; \varphi\; dx\; \omega\; dt = -\int_\Omega \int_X w_{jI}\sigma\; dy\; \varphi\; dx - \int_U w^*\sigma\; dy\; \varphi\; dx\; \partial_t\omega\; dt$$

and thus the equality to be proved. Q.E.D.

Proposition 17 *For the subsequences in propositions 14 and 15 the relations*

$$v^* = w^*$$

hold on $Q \times \Gamma$.

Proof: Let $\vec{\rho} \in C^\infty(X)$, extend it Z-periodically and define $\vec{\rho}^\varepsilon(x) = \vec{\rho}(\frac{x}{\varepsilon})$. Then with $\psi \in C_0^\infty(\Omega)$ and $\omega \in C_0^\infty(0,T)$ we get

$$\begin{aligned} 0 &= \int_0^T \varepsilon(v^\varepsilon - w^\varepsilon, \vec{\nu} \cdot \vec{\rho}^\varepsilon \psi\omega)_{\Gamma^\varepsilon}\, dt = \int_0^T \varepsilon(\nabla \cdot (\vec{\rho}^\varepsilon\, (v^\varepsilon - w^\varepsilon)\psi), \omega)_{\Pi^\varepsilon}\, dt \\ &= \int_0^T \varepsilon(\nabla \cdot \vec{\rho}^\varepsilon\, (v^\varepsilon - w^\varepsilon), \psi)_{\Pi^\varepsilon}\, \omega\, dt + \int_0^T \varepsilon(\vec{\rho}^\varepsilon \cdot \nabla(v^\varepsilon - w^\varepsilon), \psi)_{\Pi^\varepsilon}\, \omega\, dt + O(\varepsilon) \end{aligned}$$

Now we determine the limits of these two terms separately. We have

$$\lim_{\varepsilon\to 0} \int_0^T \varepsilon(\nabla \cdot \vec{\rho}^\varepsilon\, (v^\varepsilon - w^\varepsilon), \psi\chi_{\Pi^\varepsilon})_\Omega\, \omega\, dt = \int_U \nabla_y \cdot \vec{\rho}\, (v^* - w^*)\, dy\, \psi\, dx\, \omega\, dt$$

and

$$\lim_{\varepsilon\to 0} \int_0^T \varepsilon(\vec{\rho}^\varepsilon \cdot \nabla(v^\varepsilon - w^\varepsilon), \psi\chi_{\Pi^\varepsilon})_\Omega\, \omega\, dt = \int_U \vec{\rho} \cdot \nabla_y(v^* - w^*)\, dy\, \psi\, dx\, \omega\, dt.$$

Therefore, we conclude

$$\int_U \nabla_y \cdot (\vec{\rho}\, (v^* - w^*))\, dy\, \psi\, dx\, \omega\, dt = 0$$

and hence

$$\int_\Gamma \vec{\nu} \cdot \vec{\rho}\, (v^* - w^*)\, d\Gamma(y) = 0.$$

Since $\vec{\rho}$ was arbitrary, we have $v^* = w^*$ on Γ. Q.E.D.

Proposition 18 *The function* w^* *from proposition 15 satisfies*

$$(\partial_t w^*, \sigma)_U + b(\nabla_y w^*, \nabla_y \sigma)_U = 0 \quad \forall \sigma \in L^2(Q; H^1(X)).$$

Proof: We choose $\sigma \in C_0^\infty(Q; C_{\text{per}}^\infty(Z))$ such that $\sigma = 0$ on Y. We define $\sigma^\varepsilon(t,x) = \sigma(t, x, \frac{x}{\varepsilon})$. Then we get by integration over $(0,T)$

$$(\partial_t w^\varepsilon, \sigma^\varepsilon)_{R^\varepsilon} + \varepsilon^2 b(\nabla w^\varepsilon, \nabla \sigma^\varepsilon)_{R^\varepsilon} = (\partial_t w^\varepsilon, \sigma^\varepsilon)_Q + \varepsilon^2 b(\nabla w^\varepsilon, \nabla \sigma^\varepsilon)_Q = 0.$$

Since

$$(\partial_t w^\varepsilon, \sigma^\varepsilon)_Q \to (\partial_t w^*, \sigma)_U$$

and

$$\varepsilon^2(\nabla w^\varepsilon, \nabla \sigma^\varepsilon)_Q \to (\nabla_y w^*, \nabla_y \sigma)_U,$$

we get the result. Q.E.D.

Proposition 19 *The function $\vec{\xi^*}$ from proposition 14 satisfies*

$$\vec{\xi^*} = D\nabla v^*. \tag{34}$$

Proof: We introduce the functions $\mu_j(y) = \lambda_j(y) + y_j$ which we identify with their extensions according to lemma 6 (a). Let $\mu_j^\varepsilon(x) = \mu_j(\frac{x}{\varepsilon})$, $\vec{\eta}_j = \nabla_y \mu_j$, and $\vec{\eta}_j^\varepsilon(x) = \vec{\eta}_j(\frac{x}{\varepsilon})$. It is known (see, e.g., [32] or [54]) that

$$\mu_j^\varepsilon \rightharpoonup \mu_j^* \text{ weakly in } H^1(\Omega)$$

$$\nabla \mu_j^\varepsilon \rightharpoonup \vec{e}_j \text{ weakly in } (L^2(\Omega))^n$$

$$\vec{\eta}_j^\varepsilon \rightharpoonup D\vec{e}_j \text{ weakly in } (L^2(\Omega))^n$$

for some μ_j^* which satisfies $\nabla \mu_j^* = \vec{e}_j$. The extension lemma from [32] implies $\nabla \cdot \vec{\eta}_j^\varepsilon = 0$ in Ω. Let $\varphi \in C_0^\infty(\Omega)$ and $\omega \in C_0^\infty(0,T)$. Then we get

$$(\vec{\eta}_j^\varepsilon, \nabla v^\varepsilon \varphi\omega)_Q + (\vec{\eta}_j^\varepsilon, v^\varepsilon \nabla\varphi\omega)_Q = 0.$$

Passing to the limit $\varepsilon \to 0$ and by using the "div-curl"-lemma from [73] we get

$$(\vec{\eta}_j^\varepsilon, \nabla v^\varepsilon \varphi\omega)_Q = (\nabla \mu_j^\varepsilon, \vec{\xi^\varepsilon} \varphi\omega)_Q \to (\vec{e}_j, \vec{\xi^*} \varphi\omega)_Q$$

and

$$(\vec{\eta}_j^\varepsilon, v^\varepsilon \nabla\varphi\omega)_Q \to (D\vec{e}_j, v^* \nabla\varphi\omega)_Q.$$

Therefore, we have

$$(\xi^*, \vec{e}_j \varphi\omega)_Q = (D\nabla v^*, \vec{e}_j \varphi\omega)_Q.$$

Since φ, and ω were arbitrary, we get the result. Q.E.D.

Proposition 20 *For the the limit of the subsequence in proposition 14 the relation*

$$(|Y|\partial_t v^* + S, \varphi)_Q + (D\nabla v^*, \nabla\varphi))_Q = 0 \quad \forall \varphi \in L^2(0,T; H^1(\Omega)) \textit{ with } \varphi|_{\partial\Omega} = 0 \tag{35}$$

holds.

Proof: Let $\varphi \in C_0^\infty(Q)$. Then we use φ for both test functions φ and ψ in the equation 33 and get by integration over $(0,T)$

$$(\partial_t v^\varepsilon, \varphi)_{Q^\varepsilon} + (\partial_t w^\varepsilon, \varphi)_{R^\varepsilon} + a(\nabla v^\varepsilon, \nabla\varphi)_{Q^\varepsilon} + \varepsilon^2 b(\nabla w^\varepsilon, \nabla\varphi)_{R^\varepsilon} = 0$$

Now propositions 14, 15, and 18 imply by taking the limits

$$|Y|(\partial_t v^*, \varphi)_Q + (\partial_t w^*, \varphi)_U + (D\nabla v^*, \nabla\varphi)_Q = 0$$

Q.E.D.

Theorem 5 *The limit functions $\vec{v}^*$ and $\vec{w}^*$ solve the macro-model problem 30. And the extensions of the sequences v^ε and w^ε satisfy*

$$\begin{array}{ll} v^\varepsilon \rightharpoonup v & \textit{weakly in } L^2(0,T;H^1(\Omega)) \\ \partial_t v^\varepsilon \rightharpoonup \partial_t v & \textit{weakly in } L^2(Q) \\ v^\varepsilon \to v & \textit{strongly in } C([0,T],L^2(\Omega)) \\ w^\varepsilon \rightharpoonup w & \textit{in the two-scale sense} \\ \partial_t w^\varepsilon \rightharpoonup \partial_t w & \textit{in the two-scale sense} \\ \varepsilon\nabla_x w^\varepsilon \rightharpoonup \nabla_y w & \textit{in the two-scale sense} \end{array}$$

Proof: We have only to collect the results from propositions 16, 17 18, and 20. The convergence of the whole sequences follows since the limit functions are uniquely determined, see [56]. Q.E.D.

4.3 Miscible Displacement Influenced by Mobile and Immobile Water

The model that we are going to study in this section is slightly different from what we had in the preceeding section. Now we allow also convection of a dissolved substances. We use the same geometry as before and start from the following family of problems.

$$\left\{ \begin{array}{ll} \vec{u}^\varepsilon(x) = -k^\varepsilon(x)\nabla p^\varepsilon(x), & x \in \Omega^\varepsilon, \\ \nabla\cdot\vec{u}^\varepsilon(x) = 0, & x \in \Omega^\varepsilon, \\ \vec{\nu}\cdot\vec{u}^\varepsilon(x) = 0, & x \in \Gamma^\varepsilon, \\ \theta\partial_t v^\varepsilon(t,x) & \\ \quad = \nabla\cdot(a^\varepsilon(x)\nabla v^\varepsilon(t,x)) - \vec{u}^\varepsilon(x)\cdot\nabla v^\varepsilon(t,x), & t>0,\ x\in\Omega^\varepsilon, \\ \vartheta\partial_t w^\varepsilon(t,x) = \varepsilon^2\nabla\cdot(b^\varepsilon(x)\nabla w^\varepsilon(t,x)), & t>0,\ x\in\Pi^\varepsilon, \\ a^\varepsilon(x)\vec{\nu}\cdot\nabla v^\varepsilon(t,x) = \varepsilon^2 b(x)^\varepsilon\vec{\nu}\cdot\nabla w^\varepsilon(t,x), & t>0,\ x\in\Gamma^\varepsilon, \\ v^\varepsilon(t,x) = w^\varepsilon(t,x), & t>0,\ x\in\Gamma^\varepsilon, \\ v^\varepsilon(t,x) = v_D(t,x), & t>0,\ x\in\partial\Omega, \\ v^\varepsilon(t,x) = v_I(x), & t=0,\ x\in\Omega^\varepsilon, \\ w^\varepsilon(t,x) = w_I(x), & t=0,\ x\in\Pi^\varepsilon. \end{array}\right. \tag{36}$$

The physical interpretation is that a porous medium consists of two parts: In Ω^ε steady flow of water takes place according to Darcy's law. Here a solute is transported and it undergoes diffusion and convection. In the second part of the medium Π^ε the water is assumed to be at rest. The solute transport is only influenced by diffusion here. The important aspect is that the difussivity is orders of magnitude smaller in Π^ε than in Ω^ε. The words used by soil physicists in this context are *mobile water* in Ω^ε and *immobile water* in Π^ε. This type of problems was studied in [50] and [52]. The homogenized limit problem of the system 36 is the following.

$$\begin{cases} \vec{u}(x) = -K\nabla p(x), & x \in \Omega, \\ \nabla \cdot \vec{u}(x) = 0, & x \in \Omega, \\ |Y|\theta\partial_t v(t,x) + S(t,x) & \\ \ = \nabla \cdot (D\nabla v(t,x)) - \vec{u}(x) \cdot \nabla v(t,x), & t > 0,\ x \in \Omega, \\ S(t,x) = \int_X \partial_t w(t,x,y)\, dy, & t > 0,\ x \in \Omega, \\ \vartheta\partial_t w(t,x,y) = \nabla_y \cdot (b(x)\nabla_y w(t,x,y)), & t > 0,\ x \in \Omega,\ y \in X \\ v(t,x) = w(t,x,y), & t > 0,\ x \in \Omega,\ y \in \Gamma, \\ v(t,x) = v_D(t,x), & t > 0,\ x \in \partial\Omega, \\ v(t,x) = v_I(x), & t = 0,\ x \in \Omega, \\ w(t,x) = w_I(x), & t = 0,\ x \in \Omega. \end{cases} \tag{37}$$

Here the relation between k^ε and K is the same as between a^ε and D. The convergence proof is almost the same as in section 4.2. The papers [50] and [52] give also numerical results for one-dimensional infiltration experiments that lead to *break-through curves*. There comparisons are also given with this system and the standard approach used in soil science.

4.4 Flow and Transport in Unsaturated Fractured Media

In the paper [6] the problem of immiscible displacement of two liquids, such as water and oil, in a fractured porous medium is studied. There both the micro-system is given and also the macro-system which can be obtained by a formal homogenization procedure. We do not repeat the rather complicated system of equations. Systems of that kind have already been used in oil reservoir simulation for several years, see [36].

In a certain sense the problem of unsaturated flow is a limit case of the immiscible displacement problem. One comes to this conclusion by considering the two fluids water and air. If one makes the assumption that the conductivity of the air is large enough such that the air pressure is everywhere always at atmospheric pressure, one gets rid of the equation describing the air flow. Here we shortly give such a system. At the same time we couple the water flow with the transport of a solute. After going through all the formal calculations one is lead to the following system.

For the water flow one has

$$\begin{cases} |Y|\partial_t\theta + S = -\nabla_x \cdot \vec{Q} & , x \in \Omega \\ \vec{Q} = -K(\nabla_x\Psi + \vec{e}) & , x \in \Omega \\ S = \int_\Gamma \vec{\nu} \cdot \vec{q}\, d\Gamma(y) & , x \in \Omega \\ \partial_t\vartheta = -\nabla_y \cdot \vec{q} & , x \in \Omega,\ y \in Y \\ \vec{q} = -k\nabla_y\psi & , x \in \Omega,\ y \in Y \\ \Psi = \psi & , x \in \Omega,\ y \in \Gamma \end{cases}$$

For the concentration of the solute one has

$$\begin{array}{ll} |Y|\partial_t(\theta v) + R = \nabla_x \cdot (D\nabla_x v - v\vec{Q}) & ,x \in \Omega \\ R = \int_\Gamma \vec{\nu} \cdot \vec{r}\, d\Gamma(y) & ,x \in \Omega \\ \partial_t(\vartheta w) = -\nabla_y \cdot \vec{r} & ,x \in \Omega,\ y \in Y \\ \vec{r} = -d\nabla_y w + w\vec{q} & ,x \in \Omega,\ y \in Y \\ v = w & ,x \in \Omega,\ y \in \Gamma \end{array}$$

A system of this type was described in more detail in [51].

5 Problems on the Pore Scale

5.1 Derivation of Darcy's Law

We start from the following problem on the pore scale. The geometry is the same as in section 3

$$\begin{cases} \varepsilon^2 \Delta \vec{u}^\varepsilon(x) = \nabla p^\varepsilon(x), & x \in \Omega^\varepsilon, \\ \nabla \cdot \vec{u}^\varepsilon(x) = 0, & x \in \Omega^\varepsilon, \\ \vec{u}^\varepsilon(x) = 0, & x \in \Gamma^\varepsilon. \end{cases} \tag{38}$$

The reason that one chooses the ε^2 to enter in this way into the problem is that we want to scale the velocity field $\vec{u}^\varepsilon$ such that it has a weak limit $\vec{u}$. Here we give the formal asymptotics of this problem. We are also going to sketch the proof; for the details we refer to the book [86] and the paper [90]. As usual we start from the assumption that there is a power series of the unknown functions of the form

$$\vec{u}^\varepsilon(x) = \vec{u}^0(x,y) + \varepsilon \vec{u}^1(x,y) + \varepsilon^2 \vec{u}^2(x,y) \ldots$$

and

$$p^\varepsilon(x) = p^0(x,y) + \varepsilon p^1(x,y) + \varepsilon^2 p^2(x,y) \ldots$$

with Z-periodicity of the coefficient functions. From this one gets the following equations

$$\begin{aligned} &\varepsilon^0 \Delta_y \vec{u}^0(x,y) + \varepsilon^1(\ldots) + \ldots \\ &\quad = \varepsilon^{-1} \nabla_y p^0(x,y) + \varepsilon^0(\nabla_y p^1(x,y) + \nabla_x p^0(x,y)) \\ &\quad\quad + \varepsilon^1(\ldots) + \ldots, \; y \in Y, \end{aligned} \tag{39}$$

$$\begin{aligned} &\varepsilon^{-1} \nabla_y \cdot \vec{u}^0(x,y) + \varepsilon^0(\nabla_y \cdot \vec{u}^1(x,y) + \nabla_x \cdot \vec{u}^0(x,y)) \\ &\quad\quad + \varepsilon^1(\ldots) + \ldots = 0, \; y \in Y, \end{aligned} \tag{40}$$

$$\varepsilon^0 \vec{u}^0(x,y) + \varepsilon^1 \vec{u}^1(x,y) + \ldots = 0, \; y \in \Gamma. \tag{41}$$

Comparing the coefficients goes along the following lines. The ε^{-1} term in equation 39 gives

$$\nabla_y p^0(x,y) = 0, \; y \in Y,$$

hence $p^0(x,y) = p^0(x)$ independently of y. The ε^0 term in equation 39 gives

$$\Delta_y \vec{u}^0(x,y) = \nabla_y p^1(x,y) + \nabla_x p^0(x), \; y \in Y$$

and the ε^{-1} term in equation 40 yields

$$\nabla \cdot \vec{u}^0(x,y) = 0, \; y \in Y.$$

In the usual way one writes

$$\nabla_x p^0(x) = \sum_j \vec{e}_j \partial_j p^0(x).$$

The *cell-problems* are now to find Z-periodic vector fields $\vec{\kappa}_j(y)$ with components $\kappa_{ji}(y)$ that solve the Stokes problems

$$\boxed{\begin{array}{ll} \Delta_y \vec{\kappa}_j(y) = \nabla_y \pi_j(y) - \vec{e}_j, & y \in Y, \\ \nabla \cdot \vec{\kappa}_j(y) = 0, & y \in Y, \\ \vec{\kappa}_j(y) = 0, & y \in \Gamma, \end{array}} \tag{42}$$

where the functions $\pi_j(y)$ are the corresponding Z-periodic pressure fields. Using these cell-functions we get easily

$$\vec{u}^0(x,y) = -\sum_j \vec{\kappa}_j(y) \partial_j p^0(x).$$

One defines the averaged vector field

$$\bar{u}(x) = \int_Y \vec{u}^0(x,y) dy$$

and gets for its i-th component

$$\bar{u}_i(x) = -\sum_j k_{ij} \partial_{x_j} p^0(x)$$

with

$$\boxed{k_{ij} = \int_Y \kappa_{ji}(y) dy.}$$

If one introduces the tensor $K = (k_{ij})_{i,j}$, one gets for short

$$\bar{u}(x) = -K \nabla p^0(x)$$

which is *Darcy's law*. It remains to show that the velocity field $\bar{u}$ is divergence free. The term with ε^0 in equation 40 gives

$$\nabla_y \cdot u^1(x,y) + \nabla_x \cdot u^0(x,y) = 0, \; y \in Y$$

Integrating this over Y gives

$$\begin{aligned} \nabla_x \cdot \bar{u}(x) &= \int_Y \nabla \cdot \vec{u}^0(x,y) dy = -\int_Y \nabla_y \cdot \vec{u}^1(x,y) dy \\ &= -\int_{\partial Y} \vec{\nu} \cdot \vec{u}^1(x,y) d\Gamma(y) = -\int_\Gamma \vec{\nu} \cdot \vec{u}^1(x,y) d\Gamma(y) - \int_{\partial Z} \vec{\nu} \cdot \vec{u}^1(x,y) d\Gamma(y). \end{aligned}$$

The boundary integral over Γ is zero due to the term with ε^1 in equation 41, and the boundary integral over ∂Z is zero due to the Z-periodicity of $\vec{u}^1(x,y)$ with respect to y. Therefore, we have shown

Proposition 21 *The homogenization of the Stokes problem 38 is*

$$\bar{u}(x) = -K\nabla p(x), \ \nabla \cdot \bar{u} = 0 \tag{43}$$

One gets similarly to section 2 the symmetry and positive definiteness of the tensor K.

Lemma 10 *The tensor K is symmetric and positive definite.*

Proof The weak form of the cell problems for the vector fields $\vec{\kappa}_j$ is

$$((\vec{\kappa}_j, \vec{\phi}))_Y = (\vec{e}_j, \vec{\phi})_Y$$

for all Z-periodic vector fields $\vec{\phi}$ with $\nabla \cdot \vec{\phi} = 0$ in Y and $\vec{\phi} = 0$ on Γ, where we have used the bilinear form

$$((\vec{\phi}, \vec{\psi}))_Y = \sum_{l,k} (\partial_l \phi_k, \partial_l \psi_k)_Y.$$

Thus one gets with $\vec{\kappa}_i$ as a test function

$$((\vec{\kappa}_j, \vec{\kappa}_i))_Y = (\vec{e}_j, \vec{\kappa}_i)_Y = k_{ji}.$$

With indices j and i interchanged one gets

$$((\vec{\kappa}_i, \vec{\kappa}_j))_Y = (\vec{e}_i, \vec{\kappa}_j)_Y = k_{ij}.$$

Since the left hand sides are equal, so are the right hand sides. Therefore, K is symmetric. If real numbers $\alpha_1, \ldots, \alpha_n$ are given, then one has

$$\alpha_i k_{ij} \alpha_j = \alpha_i (\vec{e}_i, \vec{\kappa}_j)_Y \alpha_j = ((\alpha_i \vec{\kappa}_i, \alpha_j \vec{\kappa}_j))_Y.$$

The sum over these terms is positive, whenever one of the α_i is different from zero. Q.E.D.

Theorem 6 *Let $\vec{u}^\epsilon$ be extended by zero to $\Omega \setminus \Omega^\epsilon$. Then there exists an extension $\tilde{p}^\epsilon$ of the pressure p^ϵ such that*

$$\begin{aligned} \vec{u}^\epsilon &\rightharpoonup \vec{u} \quad \text{weakly in } (L^2(\Omega))^n \\ \tilde{p}^\epsilon &\rightarrow p \quad \text{strongly in } L_0^2(\Omega) \end{aligned}$$

with $\vec{u} \in (L^2(\Omega))^n$ and $p \in L_0^2(\Omega)$ which solve the system 43.

Proof: This result goes back to the paper [90]. Generalizations can be found in [1] and [70]. An explicit formula for the extension of the pressure is given in [63]. Q.E.D.

Let us note that the result of the previous theorem remains also valid for the case of Navier-Stokes equations with the convergence of the pressure taking place in some appropriate functional spaces; for more details see [68].

5.2 Immiscible Displacement Problems

The result of the previous section applies to flow of a single fluid in a nondeformable porous medium. For two immiscible fluids the situation is more complicated, since one has to consider the interface between the two phases. In general this seems to be a very diffucult question. If one makes the assumption that the geometry of the interface varies only slightly from cell to cell, one can perform a perturbation analysis similar to the single fluid case. In this chapter we give an outline of an idea that can be found in [9]. It should be pointed out that the underlying assumption is that the flow problem with a capillary interface has a unique solution and depends continously on the data. The validity of this assumption may be very limited. Therefore, the result of this section has only limited applicability.

We consider the following problem on the micro-scale. We look for a velocity field $\vec{u}^\varepsilon$ defined on Ω^ε (that has no jump across the interface Γ^ε_*) the restrictions of which to Ω^ε_i we denote by $\vec{u}^\varepsilon_i$, and a pressure field p^ε (that may have a jump across Γ^ε_*) such that

$$\begin{cases} \varepsilon^2\kappa_i\Delta\vec{u}^\varepsilon_i = \nabla p^\varepsilon_i, & x\in\Omega^\varepsilon_i,\ i=w,n \\ \nabla\cdot\vec{u}^\varepsilon_i = 0, & x\in\Omega^\varepsilon_i,\ i=w,n \\ \vec{u}^\varepsilon_w = 0, & x\in\Gamma^\varepsilon_s \\ \vec{u}^\varepsilon_i\cdot\vec{\nu} = 0 & x\in\Gamma^\varepsilon_*,\ i=w,n \\ [T^\varepsilon]\cdot\vec{\nu} = \varepsilon\sigma 2H^\varepsilon\vec{\nu}, & x\in\Gamma^\varepsilon_* \end{cases} \tag{44}$$

Here we have

$$T^\varepsilon_i = -p^\varepsilon_i I + \varepsilon^2\kappa_i 2S^\varepsilon_i \text{ for } i=w,n,$$

I is the identity, S^ε is the stress tensor with components

$$s^\varepsilon_{jk} = \frac{1}{2}(\partial_j v^\varepsilon_k + \partial_k v^\varepsilon_j),$$

and $[\cdots]$ denotes the jump across Γ^ε_*. Further, H^ε is the mean curvature of the surfave Γ^ε_*.

In the sense of distributions one gets from 44

$$\nabla\cdot T^\varepsilon = \varepsilon\sigma 2H^\varepsilon\vec{\nu}\delta_{\Gamma^\varepsilon_*}$$

in Ω^ε. In the special case that the fluid is at rest, one has $\nabla p^\varepsilon_i = 0$ and $[-p^\varepsilon] = \varepsilon\sigma 2H^\varepsilon$ on Γ^ε_*.

We introduce the spaces

$$\mathcal{W}^\varepsilon = \{\vec{u} : \vec{u}\in\mathbf{W}^{1,2}(\Omega^\varepsilon) \text{ with } \vec{u}=0 \text{ on } \Gamma^\varepsilon\},$$

$$\mathcal{W}^\varepsilon_0 = \{\vec{u} : \vec{u}\in\mathcal{W} \text{ with } \vec{u}=0 \text{ on } \partial\Omega\}$$

and

$$\mathcal{V}^\varepsilon = \{\vec{u} : \vec{u}\in\mathcal{W} \text{ with } \nabla\cdot\vec{u}=0 \text{ in } \Omega^\varepsilon\}$$

of vector fields on Ω^ε.

Furthermore, we introduce the sesquilinear form

$$E(\vec{u},\varphi) = E_w(\vec{u}_w,\varphi) + E_n(\vec{u}_n,\varphi)$$

with

$$E_i(\vec{u},\varphi) = \varepsilon^2\kappa_i\frac{1}{2}\sum_{j,k}\int_{\Omega_i}(\partial_k v_j + \partial_j v_k)(\partial_k\phi_j + \partial_j\phi_k)\, d\Omega.$$

The number $E(\vec{u},\vec{u})$ is the energy dissipating within the domain Ω^ε.

Using integration by parts one gets easily the following formula.

Lemma 11 *If $\vec{u}^\varepsilon \in \mathcal{V}^\varepsilon$ and $\varphi \in \mathcal{W}^\varepsilon$, then one has*

$$\begin{aligned}&(-\varepsilon^2\kappa\Delta\vec{u}_i^\varepsilon + \nabla p^\varepsilon,\varphi)_{\Omega_i^\varepsilon}\\ &\quad = E_i(\vec{u}_i^\varepsilon,\varphi) - (p_i^\varepsilon,\nabla\cdot\varphi)_{\Omega_i^\varepsilon} \pm \int_{\Gamma_*^\varepsilon}\vec{\nu}\cdot T_i^\varepsilon\cdot\varphi\, d\Gamma,\end{aligned}$$

where the " + " and the " − " apply for $i = w$ and $i = n$, respectively.

From this we get the following weak form of the micro-problem.

Proposition 22 *A vector field $\vec{u}^\varepsilon$ solves the micro problem 44 iff $\vec{u}^\varepsilon \in \mathcal{W}^\varepsilon$ with $\vec{u}_w^\varepsilon\cdot\vec{\nu} = \vec{u}_n^\varepsilon\cdot\vec{\nu} = 0$ on Γ_*^ε such that there exists a pressure field $p^\varepsilon \in L^2(\Omega^\varepsilon)$ with*

$$\begin{cases} E(\vec{u}^\varepsilon,\varphi) - (p^\varepsilon,\nabla\cdot\varphi)_{\Omega^\varepsilon} = \varepsilon\sigma(2H^\varepsilon\vec{\nu},\varphi)_{\Gamma_*^\varepsilon} & \forall\varphi\in\mathcal{W}_0^\varepsilon\\ (\nabla\cdot\vec{u}^\varepsilon,\psi)_{\Omega^\varepsilon} = 0 & \forall\psi\in L^2(\Omega^\varepsilon)\end{cases}$$

The second condition ensures that $\vec{u}^\varepsilon$ is divergence free. We just mention that this problem can be reformulated as a minimization problem.

For the aymptotic expansion we make the following assumptions.

$$\vec{u}^\varepsilon(x) = \varepsilon^0\vec{u}^0(x,y) + \varepsilon^1\vec{u}^1(x,y) + \varepsilon^2\ldots$$

$$p^\varepsilon(x) = \varepsilon^0 p^0(x,y) + \varepsilon^1 p^1(x,y) + \varepsilon^2\ldots$$

$$\Gamma_*^\varepsilon(x,y) = \varepsilon^1\Gamma_*^0(x,y) + \varepsilon^2\xi(x,y) + \varepsilon^3\ldots$$

with $y = \frac{x}{\varepsilon}$ and all coefficients in the series being periodic functions of the fast variable y. By $\Gamma_*^0(x)$ we mean a surface in y-space; whereas $\xi(x,y)$ denotes a vector valued function on $\Gamma_*^0(x)$ that is parallel to the normal vector $\vec{\nu}^0(x,y)$ on $\Gamma_*^0(x)$. In other words, we assume that the free boundary problem 44 is a perturbation of a zero order problem. The perturbation is thought of as being of order ε.

First we identify the zero order problem. We get the following expansions.

$$\begin{aligned}&\varepsilon^0\kappa_i\Delta_y\vec{u}_i^0(x,y) + \varepsilon^1\ldots\\ &\quad = \varepsilon^{-1}\nabla_y p_i^0(x,y) + \varepsilon^0(\nabla_y p_i^1(x,y) + \nabla_x p_i^0(x,y)) + \varepsilon^1\ldots \quad y\in Y_i(x) \qquad (45)\end{aligned}$$

$$\varepsilon^{-1}\nabla_y \cdot \vec{u}_i^0(x,y) + \varepsilon^0(\nabla_y \cdot \vec{u}_i^1(x,y) + \nabla_x \cdot \vec{u}_i^0(x,y)) + \varepsilon^1 \ldots = 0,\ y \in Y_i(x) \quad (46)$$

$$\varepsilon^0\vec{u}_w^0(x,y) + \varepsilon^1\vec{u}_w^1(x,y) + \varepsilon^2 \ldots = 0,\ y \in \Gamma_s \quad (47)$$

When calculating the perturbations of the conditions on the interface $\Gamma_*^0(x)$ we have to take into account the perturbation of this interface also. We get

$$(\varepsilon^0\vec{u}_i^0(x,y) + \varepsilon^1 \ldots) \cdot \vec{\nu} = 0,\ y \in \Gamma_*(x) \quad (48)$$

The expansion of the stress tensor is

$$T^\varepsilon(x) = T^0(x,y) + \varepsilon T^1(x,y) + \varepsilon^2 \ldots$$

with

$$T_{jk}^0 = -p^0(x,y)\delta_{jk}$$

and

$$T_{jk}^1 = -(p^1(x,y) + \nabla_y p^0(x,y) \cdot \xi(x,y))\delta_{jk} + \kappa(\nabla_{y_j}\vec{u}_k^0(x,y) + \nabla_{y_k}\vec{u}_j^0(x,y))$$

and for the curvature we have

$$\varepsilon\sigma 2H^\varepsilon(x) = \varepsilon^0\sigma 2H^0(x,y) + \varepsilon^1\sigma 2H^1(x,y) + \varepsilon^2 \ldots$$

Therefore, from the ε^0 term of the jump condition on the interface we get

$$[-p^0(x,y)] = \sigma 2H^0(x,y),\ y \in \Gamma_*(x) \quad (49)$$

From the ε^{-1} term in equation 45 we get

$$\nabla_y p_i^0(x,y) = 0,\ y \in Y_i(x)$$

which yields

$$p_i^0(x,y) = p_i^0(y)$$

From 49 we get

$$[-p^0(x)] = \sigma 2H^0(x,y),\ y \in \Gamma_*^0(x)$$

and therefore

$$[-p^0(x)] = \sigma 2H^0(x),\ y \in \Gamma_*^0(x) \quad (50)$$

i.e., the interface $\Gamma_*^0(x)$ has a mean curvature which is independent of the fast variable y. Hence the zero order problem is to find a periodic surface $\Gamma_*^0(x)$ having mean curvature $H^0(x)$ which is given by equation 50.

Now come to the first order problem. From the ε^0 term in 45, the ε^{-1} term in 46, the ε^0 terms in 47 and 48 we get

$$\begin{cases} \kappa_i \Delta_y \vec{u}_i^0(x,y) = \nabla_y p_i^1(x,y) + \nabla_x p_i^0(x), & y \in Y_i(x) \\ \nabla_y \cdot \vec{u}_i^0(x,y) = 0, & y \in Y_i(x) \\ \vec{u}_w^0(x,y) = 0, & y \in \Gamma_s \\ \vec{u}_i^0(x,y) \cdot \vec{\nu} = 0, & y \in \Gamma_*^0(x) \end{cases} \tag{51}$$

It has to be kept in mind that the velocity field $\vec{u}^0$ is continuous across the interface $\Gamma_*^0(x)$, whereas the pressure field $p^1(x,.)$ is allowed to have a jump. The average

$$\bar{v}_i(x) = \int_{Y_i} \vec{u}_i^0(x,y)\, dy$$

then satisfies

$$\bar{v}_i^0(x) = -\sum_j \int_{Y_i} \vec{u}_i^j(x,y)\, dy\; \partial_j p^0(x) \tag{52}$$

which is the generalized *Darcy law.*

5.3 Elastic Porous Media

In soil mechanics porous media are considered to be deformable. Very often one makes the assumtion that the soil behaves like an elastic body. Models for this phenomenon go back to *Biot* [19] [20] [21]. A formal derivation of this type of models was given in [28] for one-phase flow. The problem of two-phase immiscible flow was studied in [8] and [9], see also [37].

5.4 Chromatography

In chemical engineering an important effect is known which is called chromatography. Here certain chemical species are dissolved in a fluid that is flowing through a porous medium. These are not only transported by convection and diffusion but they also penetrate into the porous matrix. If one assumed that the species under consideration may diffuse within the interior of the solid matrix one comes to models that were studied theoretically in [91] and practically in [87]. A nonlinear problem of this type is investigated in [29].

5.5 Semipermeable Membranes

In biology a situation is of interest in which chemical species are flowing within the space between cells. The cells are surrounded by membranes that are partially permeable. If one assumes that the membranes are semipermeable, special nonlinear interface conditions have to be formulated. One case is to use the set $M \subset \mathbf{R}^2$ defined by

$$(s,q) \in M \text{ iff } s \leq 0 \text{ and } q \geq 0 \text{ and } sq = 0.$$

The condition to be postulated on the interfaces Γ^ε is a relation between the difference s of the outer and inner contration v and w, resp., and the normal flux q.

The micro-problem is the following.

$$\begin{cases} \varepsilon^2 \Delta \vec{u}^\varepsilon(x) = \nabla p^\varepsilon(x), & x \in \Omega^\varepsilon \\ \nabla \cdot \vec{u}^\varepsilon(x) = 0, & x \in \Omega^\varepsilon \\ \vec{u}(x) = \vec{u}_D, & x \in \Gamma_D \\ \vec{\nu} \cdot \vec{u}^\varepsilon(x) = 0, & x \in \Gamma_N \\ \vec{u}^\varepsilon(x) = 0, & x \in \Gamma^\varepsilon \\ \partial_t v^\varepsilon(t,x) = a\Delta v^\varepsilon(t,x) - \vec{u}^\varepsilon(x) \cdot \nabla v^\varepsilon(t,x) + f(\vec{v}^\varepsilon(t,x)) & ,t>0,\ x \in \Omega^\varepsilon \\ \partial_t w^\varepsilon(t,x) = \varepsilon^2 c \Delta w^\varepsilon(t,x) + g(\vec{w}^\varepsilon(t,x)) & ,t>0,\ x \in \Pi^\varepsilon \\ a\vec{\nu}^\varepsilon \cdot \nabla v^\varepsilon(t,x) = \varepsilon^2 c \vec{\nu}^\varepsilon \cdot \nabla w^\varepsilon(t,x) & ,t>0,\ x \in \Gamma^\varepsilon \\ (v^\varepsilon(t,x) - w^\varepsilon(t,x), -\varepsilon^2 c \vec{\nu}^\varepsilon \cdot \nabla w^\varepsilon(t,x)) \in M & ,t>0,\ x \in \Gamma^\varepsilon \\ v^\varepsilon(t,x) = v_D(t,x) & ,t>0,\ x \in \Gamma_D \\ \vec{\nu} \cdot \nabla v^\varepsilon(t,x) = 0 & ,t>0,\ x \in \Gamma_N \\ v^\varepsilon(t,x) = v_I(x) & ,t=0,\ x \in \Omega^\varepsilon \\ w^\varepsilon(t,x) = w_I(x) & ,t=0,\ x \in \Pi^\varepsilon \end{cases} \tag{53}$$

In the paper [55] the folllowing convergence result is made precise and also proved.

Proposition 23 *The homogenized problem of the system 53 The flow:*

$$\begin{cases} \vec{u}(x) = -K\nabla p(x), & x \in \Omega \\ \nabla \cdot \vec{u}(x) = 0, & x \in \Omega \\ \vec{u}(x) = u_D(x), & x \in \Gamma_D \\ \vec{\nu} \cdot \vec{u}(x) = 0, & x \in \Gamma_N \end{cases}$$

The global problem:

$$\begin{cases} |Y|\partial_t v(t,x) + S(t,x) & \\ \quad = d\nabla \cdot (D\nabla v(t,x)) - \vec{u}(x) \cdot \nabla v(t,x) + |Y| f(\vec{v}(t,x)) & ,t>0,\ x \in \Omega \\ v(t,x) = v_D(t,x) & ,t>0,\ x \in \Gamma_D \\ \vec{\nu} \cdot \nabla v(t,x) = 0 & ,t>0,\ x \in \Gamma_N \\ v(t,x) = v_I(x) & ,t=0,\ x \in \Omega \end{cases}$$

where the sink term

$$S(t,x) = -\int_\Gamma c\vec{\nu} \cdot \nabla_y w(t,x,y)\, d\Gamma(y),\ t>0,\ x \in \Omega$$

is defined in terms of the local problems:

$$\begin{cases} \partial_t w(t,x,y) = c\Delta_y w(t,x,y) + g(\vec{w}(t,x,y)) & ,t>0,\ x\in\Omega,\ y\in X \\ (v(t,x)-w(t,x,y), -c\vec{\nu}\cdot\nabla_y w(t,x,y)) \in M & ,t>0,\ x\in\Omega,\ y\in\Gamma \\ w(t,x,y) = w_I(x) & ,t=0,\ x\in\Omega,\ y\in X. \end{cases}$$

A problem that is related to this type of problems was treated in [44].

5.6 Heterogeneous Catalysis

In the paper [53] a model is described that takes into account diffusion, convection, adsorption, and reaction of chemicals in porous media. The basic assumptions are very similar to section 5.4. The important difference is that the chemical substances are not allowed to penetrate into the solid matrix but that they are adsorbed to its surface Γ^ε. The concentration of the solute in the fluid is denoted by v^ε (for simplicity the model is written only for one substance; the generalization to a whole set of chemical species is obvious) and the concentration of the same substance adsorbed to the surfaces is denoted by w^ε. I.e., v^ε is a concentration per volume, whereas w^ε is a concentration per surface. In the following model it is also allowed that the chemical substance diffuses on the surface Γ^ε. This gives an extra term with the Laplace-Beltrami operator Δ^ε on Γ^ε. The adsorption rate is f^ε. And the reaction rate is a^ε. For simplicity the governing equations are taken linear.

The problem is to look for functions $\vec{u}^\varepsilon, v^\varepsilon$, and w^ε that satisfy the following system of equations:

$$\begin{cases} \varepsilon^2\Delta\vec{u}^\varepsilon(x) = \nabla p^\varepsilon(x), & x\in\Omega^\varepsilon \\ \nabla\cdot\vec{u}^\varepsilon(x) = 0, & x\in\Omega^\varepsilon \\ \vec{u}(x) = \vec{u}_D, & x\in\Gamma_D \\ \vec{\nu}\cdot\vec{u}^\varepsilon(x) = 0, & x\in\Gamma_N \\ \vec{u}^\varepsilon(x) = 0, & x\in\Gamma^\varepsilon \\ \partial_t v^\varepsilon(t,x) = d\Delta v^\varepsilon(t,x) - \vec{u}^\varepsilon(x)\cdot\nabla v^\varepsilon(t,x), & t>0,\ x\in\Omega^\varepsilon \\ -d\vec{\nu}^\varepsilon\cdot\nabla v^\varepsilon(t,x) = \varepsilon f^\varepsilon(t,x), & t>0,\ x\in\Gamma^\varepsilon \\ \partial_t w^\varepsilon(t,x) - \varepsilon^2 E\Delta^\varepsilon w^\varepsilon(t,x) + a^\varepsilon(x)w^\varepsilon(t,x) = f^\varepsilon(t,x), & t>0,\ x\in\Gamma^\varepsilon \\ v^\varepsilon(t,x) = v_D(t,x), & t>0,\ x\in\Gamma_D \\ \vec{\nu}\cdot\nabla v^\varepsilon(t,x) = 0, & t>0,\ x\in\Gamma_N \\ v^\varepsilon(t,x) = v_I(x), & t=0,\ x\in\Omega^\varepsilon \\ w^\varepsilon(t,x) = w_I(x), & t=0,\ x\in\Gamma^\varepsilon \end{cases} \tag{54}$$

where we have used the abbreviation

$$f^\varepsilon(t,x) = f(t,x,\frac{x}{\varepsilon}) \text{ with } f(t,x,y) = c(y)v(t,x) - b(y)w(t,x,y).$$

The limit problem for the fluid flow is obtained in the same way as in section 5.1. The diffusion term is homogenized in a way that is similar to section 4.1. In the limit the Laplace-Beltrami operator Δ^Γ on the standard surface Γ comes into play.

Proposition 24 *The homogenization of the initial boundary value problem 54 is to find functions $\vec{u}$, v, and w that satisfy the following system of equations:*

$$\begin{cases} \vec{u}(x) = -K\nabla p(x), & x \in \Omega \\ \nabla \cdot \vec{u}(x) = 0, & x \in \Omega \\ \vec{u}(x) = u_D(x), & x \in \Gamma_D \\ \vec{\nu} \cdot \vec{u}(x) = 0, & x \in \Gamma_N \\ |Y|\partial_t v(t,x) + \int_\Gamma f(t,x,y)d\Gamma(y) & \\ \quad = d\nabla \cdot (D\nabla v(t,x)) - \vec{u}(x)\nabla v(t,x), & t > 0, x \in \Omega \\ v(t,x) = v_D(t,x), & t > 0, x \in \Gamma_D \\ \vec{\nu} \cdot \nabla v(t,x) = 0, & t > 0,\ x \in \Gamma_N \\ \partial_t w(t,x,y) - E\Delta_y^\Gamma w(t,x,y) + a(y)w(t,x,y) & \\ \quad = f(t,x,y), & t > 0,\ x \in \Omega, y \in \Gamma \\ v(t,x) = v_I(x), & t = 0,\ x \in \Omega \\ w(t,x,y) = w_I(x), & t = 0,\ x \in \Omega, y \in \Gamma. \end{cases} \tag{55}$$

Proof: The convergence of the homogenization process is proved in the paper [54]. Q.E.D.

Numerical simulations for this type of model and also for nonlinear versions of it can be found in [53].

6 Further Aspects

In this article we do not have the space to treat all aspects of the theory of homogenization applied to flow and transport through porous media. One interesting problem is a more careful analysis of the kind of convergence of the ε-problems to the homogenized problems which is called the question of *correctors*. Theorems in this direction can be found in the book [16], see also [77].

The homogenization of immiscible displacement problems on a larger scale are treated in the papers [23] [24] [25].

A problem that seems not to have got its final answer is the formulation of the appropriate transmission conditions for coupled *surface-subsurface* flow. In this situation one has at the same time to consider *open channel* flow and also flow through a porous medium. Papers on this subject are [42] and [60].

Another question is related to random media. There are two major approaches. First, one can study porous media the permeability if whis is random. In this case one uses random fields both for the coefficients and for the solutions of the differential equations. Under certain assumptions on can prove asymptotic results. Usually one considers the limit problem of the correlation length tending to zero. This is done in [16], see also [80]. Secondly, there is also the - more fundamental - problem of media with a random pore structure. In a similar way as for the periodic case one wants to derive Darcy's law in the limit. This is done in [84], see also [39] [82].

Finally, let us mention the aspect of spatial variability. Most authors use an approach that is based on assumptions of some kind of stochastic properties of porous media on a large scale. It is not quite clear if the method of homogenization can give new insights in this respect, see [85].

References

[1] ALLAIRE G. *Homogenization of the Stokes flow in a connected porous medium* Asympt. Anal. **2** (1989) 203-222

[2] ALLAIRE G. *Homogenization of the Navier-Stokes equations in open sets with tiny holes* Arch. Rat. Mech. Anal, submitted

[3] ALLAIRE G. *Homogénéisation et convergence à deux échelles. Application à un problème de convection diffusion* C. R. Acad. Sci. Paris **312** Ser. I (1991) 581-586

[4] ARBOGAST T. *The double porosity model for single phase flow in naturally fractured reservoirs* M. F. Wheeler (Ed.) "Numerical Simulation in Oil Recovery" The IMA Volumes in Mathematics and Its Applications **11** Springer, Berlin (1988) 23-45

[5] ARBOGAST T., DOUGLAS J., HORNUNG U. *Derivation of the double porosity model of single phase flow via homogenization theory* SIAM J. Appl. Math. **21** (1990) 823-836

[6] ARBOGAST T., DOUGLAS J., HORNUNG U. *Modeling of naturally fractured reservoirs by formal homogenization techniques* "Frontiers in Pure and Applied Mathematics" North-Holland, Amsterdam (1991)

[7] ATTOUCH H. *Variational Convergence for Functions and Operators* Pitman, Boston (1984)

[8] AURIAULT J. -L. *Nonsaturated deformable porous media: quasistatics* TiPM **2** (1987) 45-64

[9] AURIAULT J. -L., LEBAIGUE O., BONNET G. *Dynamics of two immiscible fluids flowing through deformable porous media* TiPM **4** (1989) 105-128

[10] AVELLANEDA M. *Iterated homogenization, differential effective medium theory and applications* Comm. Pure Appl. Math. **40** (1987) 527-554

[11] AVELLANEDA M., FANG-HUA LIN *Compactness methods in the theory of homogenization* Comm. Pure Appl. Maths. **40** (1987) 803-847

[12] AVELLANEDA M., FANG-HUA LIN *Compactness methods in the theory of homogenization II: Equations in nondivergence form* Comm. Pure Appl. Math. **42** (1989) 139-172

[13] BAKHVALOV N., PANASENKO G. *Homogenization: Averaging Processes in Periodic Media* Kluwer, Dordrecht (1989)

[14] BARENBLATT G. I., ZHELTOV I. P., KOCHINA I. N. *Basic concepts in the theory of seepage of homogeneous liquids in fissured rocks (strata)* J. Appl. Math. Mech. **24** (1960) 1286-1303

[15] BARKER J. A. *Block-geometry functions characterizing transport in densely fissured media* J. Hydrology **77** (1985) 263-279

[16] BENSOUSSAN A., LIONS J. L., G. PAPANICOLAOU *Asymptotic Analysis for Periodic Structures* North-Holland, Amsterdam (1978)

[17] BHATTACHARYA R. N., GUPTA V. K. *A theoretical explanation of solute dispersion in saturated porous media at the Darcy scale* Wat. Res. Res. 10 (1983) 938-944

[18] BHATTACHARYA R. N., V. K. GUPTA, H. F. WALKER *Asymptotics of solute dispersion in periodic porous media* SIAM J. Appl. Math. **49** (1989) 86-98

[19] BIOT M. A. *Theory of propagation of elastic waves in a fluid-saturated porous solid. I. Low frequence range II. Higher frequency range* J. Acoust. Soc. Amer. **28** (1956) 168-178, 179-191

[20] BIOT M. A. *Theory of deformation of a porous viscoelasctic anisotropic solid* Appl. Phys. **27** (1956) 459-467

[21] BIOT M. A. *Mechanics of deformation and acoustic propagation in porous media* J. Appl. Phys. **33** (1862) 1482-1498

[22] BOSSE M. -P., SHOWALTER R. E. *Homogenization of the layered medium equation* Applic. Anal. **32** (1989) 183-202

[23] BOURGEAT A. P. *Nonlinear homogenization of two-phase flow equations* J. H. Lightbourne, S. M. Rankin (Eds.) "Physical Mathematics and Nonlinear Partial Differential Equations" Dekker (1985) 207-212

[24] BOURGEAT A. P. *Homogenization of two-phase flow equations* Proceedings Symposia Pure Mathem. **45** (1986) 157-163

[25] BOURGEAT A. P., MIKELIĆ A. *Homogenization of the two-phase immiscible flow in one dimensional porous medium* Equipe d'Analyse Numerique Lyon Saint-Etienne, preprint **116** (1991)

[26] BRENNER H. *Dispersion resulting from flow through spatially periodic porous media* Philos. Trans. R. Soc. Lond., A 297 (1980) 81-133

[27] BRENNER H., ADLER P. M. *Dispersion resulting from flow through spatially periodic porous media II: surface and intraparticle transport* Philos. Trans. R. Soc. Lond., A 307 (1982) 149-200

[28] BURRIDGE R., KELLER J. *Poroelasticity equations derived from microstructure* J. Acoust. Soc. Amer. **70** (1981) 1140-1146

[29] CANON É., JÄGER W. *Homogenization for nonlinear adsorption-diffusion processes in porous media* to appear

[30] CAPRIZ G. *Continua with Microstructure* Springer (1989)

[31] CIORANESCU D., DONATO P. *Homogenization du probleme de Neumann non homogene dans des ouverts perfores* Asympt. Anal. **1** (1988) 115-138

[32] CIORANESCU D., SAINT JEAN PAULIN J. *Homogenization in open sets with holes* J. Math. Anal. Appl. **71** (1979) 590-607

[33] CODEGONE M., RODRIGUES J.-F. *On the homogenization of the rectangular dam problem* Rend. Sem. Mat. Univers. Politecn. Torino **39** (1981) 125-136

[34] CODEGONE M., RODRIGUES J.-F. *Homogenization of the dam problem with layer-structure* Fasano A., Primicerio M. (eds.): Free Boundary Problems. Theory and Applications, Pitman, Boston (1983) 98-104

[35] DE GIORGI E., SPAGNOLO S. *Sulla convergenza degli integrali dell'energia per operatori ellittici del secondo ordine* Boll. Un. Mat. It. **8** (1973) 391-411

[36] DOUGLAS J., ARBOGAST T. *Dual-porosity models for flow in naturally fractured reservoirs* Cushman J. (Ed.) "Dynamics of Fluids in Hierarchical Porous Media" Academic Press (1990) Chapter VII, 177-221

[37] ENE H. I. *The use of the homogenization method to describe the viscoelastic behaviour of a porous saturated medium* Rock and Soil Rheology, Lecture Notes in Earth Sciences, Springer-Verlag (1987) 33-41

[38] ENE H. I. *Application of the homogenization method to transpoprt in porous media* Dynamics of Fluids in Hierarchical Porous Media, Academic Press (1990) Chapter VIII (223-241)

[39] ENE H. I. *Estimations du tenseur de perméabilité* C. R. Acad. Sci. Paris **312 II** (1991) 1269-1272

[40] ENE H. I., SANCHEZ-PALENCIA E. *Equations et phénomenes de surface pour l'écoulement dans un modèle de milieu poreux* J. Mécan. **14** (1975) 73-108

[41] ENE H. I., POLIŠEVSKI D. *Thermal Flow in Porous Media* D. Reidel, Dordrecht (1987) 194 pages

[42] ENE H. I., SANCHEZ-PALENCIA E. *On thermal equation for flow in porous media* Int. J. Eng. Sci. 20 (1982) 623-630

[43] ERICKSEN J. L., KINDERLEHRER D., KOHN R., LIONS J.-L. *Homogenization and Effective Moduli of Materials and Media* Springer, Berlin, 1986

[44] FRIEDMAN A., KNABNER P. *A Transport Model with Micro- and Macro-Structure* J. Differ. Equ., to appear

[45] GRAY W. G. *A derivation of the equations for multi-phase transport* Chem. Engr. Sci. 30 (1975) 229-233

[46] GUPTA V. K., BHATTACHARYA R. N. *A new derivation of the Taylor-Aris theory of solute dispersion in a capillary* Wat. Res. Res. 19 (1983) 945-951

[47] HASSANIZADEH M., GRAY W. G. *General conservation equations for multi-phase systems: 1. averaging procedure* Advances in Water Resources 2 (1979) 131-144

[48] HASSANIZADEH M., GRAY W. G. *General conservation equations for multi-phase systems: 2. mass, momenta, energy, and entropy equations* Adv. Wat. Res 2 (1979) 191-203

[49] HASSANIZADEH M., GRAY W. G. *General conservation equations for multi-phase systems: 3. constitutive theory for porous media flow* Adv. Wat. Res. 3 (1980) 25-40

[50] HORNUNG U. *Miscible Displacement in Porous Media Influenced by Mobile and Immobile Water* Technical Report, Department of Mathematics, Arizona State University, Tempe, **102** (1987)

[51] HORNUNG U. *Homogenization of Miscible Displacement in Unsaturated Aggregated Soils* G. Dal Maso, G. F. Dell'Antonio (Eds.) "Composite Media and Homogenization Theory", Progress in Nonlinear Differential Equations and Their Applications, Birkhäuser, Boston (1991) 143-153

[52] HORNUNG U. *Compartment and micro-structure models for miscible displacement in porous media* Rocky Mountain J. Math. **21** (1991)

[53] HORNUNG U., JÄGER W. *A model for chemical reactions in porous media* J. Warnatz, W. Jäger (Eds.) "Complex Chemical Reaction Systems. Mathematical Modeling and Simulation" Chemical Physics **47** Springer, Berlin (1987) 318-334

[54] HORNUNG U., JÄGER W. *Diffusion, convection, adsorption, and reaction of chemicals in porous media* J. Diff. Equat. **92** (1991) 199-225

[55] HORNUNG U., JÄGER W., MIKELIĆ A. *Reactive transport through an array of cells with semipermeable membranes* in preparation

[56] HORNUNG U., SHOWALTER R. *Diffusion models for fractured media* J. Math. Anal. Applics. **147** (1990) 69-80

[57] KELLER J. B. *Darcy's law for flow in porous media and the two-space method* R. L. Sternberg (Ed.) "Nonlinear Partial Differential Equations in Engineering and Applied Sciences" Dekker (1980) 429-443

[58] KOHN R. V., MILTON G. W. *On bounding the effective conductivity of anisitropic composites* Ericksen J. L., Kinderlehrer D., Kohn R., Lions J.-L. (Eds.) "Homogenization and Effective Moduli of Materials and Media" Springer, Berlin (1986) 97-125

[59] KUNIN I. A. *Elastic Media with Microstructure. 2: Three-Dimensional Models* Springer, Berlin (1983)

[60] LEVY, SANCHEZ-PALENCIA *On boundary conditions for fluid flow in porous media* Inter. J. Engin. Sci. **13** (1975) 923-940

[61] LIONS J.-L. *Some Methods in the Mathematical Analysis of Systems and Their Control* Gordon and Breach, New York (1981)

[62] LIONS J.-L. *Homogenization in perforated materials and in porous media* Continuum Models of Discrete Systems 4, Proc. Int. Conf., Stockholm 1981, (1981) 81-99

[63] LIPTON R., AVELLANEDA A. *A Darcy law for slow viscous flow past a stationary array of bubbles* Proc. Royal Soc. Edinburgh **114A** (1990) 71-79

[64] LURIE K. A., CHERKAEV A. V. *Exact estimates of the conductivity of a binary mixture of isotropic materials* Proc. Royal Soc. Edinburgh **104A** (1986) 21-38

[65] MARCHENKO V. A., KHROUSLOV E. J. *Boundary problems in domains with finely granulated boundaries* (Russian) Naukova Dumka, Kiev (1974)

[66] MARLE C. M. *On macroscopic equations governing multiphase flow with diffusion and chemical reactions in porous media* Int. J. Eng. Sci. 20 (1982) 643-662

[67] MIKELIĆ A. *A convergence theorem for homogenization of two-phase miscible flow through fractured reservoirs with uniform fracture distribution* Applic. Anal. **33** (1989) 203-214

[68] MIKELIĆ A. *Homogenization of nonstationary Navier-Stokes equations in a domain with a grained boundary* Ann. Mat. Pura Appl., to appear

[69] MIKELIĆ A. *Mathematical theory of stationary miscible filtration* J. Diff. Equat., to appear

[70] MIKELIĆ A., AGANOVIĆ I. *Homogenization in a porous medium under a nonhomogeneous boundary condition* Boll. Un. Mat. Ital. (A) **1** (1987) 171-180

[71] MIKELIĆ A., AGANOVIĆ I. *Homogenization of stationary flow of miscible fluids in a domain with a grained boundary* SIAM J. Math. Anal. **19** (1988) 287-294

[72] MILTON G. M. *The field equation recursion method* G. Dal Maso, G. F. Dell'Antonio (Eds.) "Composite Media and Homogenization Theory", Progress in Nonlinear Differential Equations and Their Applications, Birkhäuser, Boston (1991) 223-245

[73] MURAT F. *Compacité par compensation* Ann. Scuola Norm. Sup. Pisa, Ser. 4 **5** (1978) 489-507

[74] MURAT F., TARTAR L. *Calcul des variations et homogenisation* Cours de l'Ecole d'Eté d'Analyse Numérique CEA-EDF-INRIA (1983)

[75] NGUETSENG G. *A general convergence result for a functional related to the theory of homogenization* SIAM J. Math. Anal. **20** (1989) 608-629

[76] NITSCHE L. C., BRENNER H. *Eulerian kinematics of flow through spatially periodic models of porous media* Archive Rat. Mech. Anal. **107** (1989) 225-292

[77] OELSCHLÄGER K. *Homogenization of diffusion processes in a divergence-free random field* Ann. Probab. **16** (1988) 1084-1126

[78] OLEINIK O. A., KOZLOV S.M., ZHIKOV *Homogenization of Differential Operators* North-Holland, Amsterdam (1991) to appear

[79] OLEINIK O. A., SHAMAEV A. S., YOSIFIAN G. A. *The Neumann problem for second order elliptic equations with rapidly oscillating periodic coefficients in a perforated domain* Colombini F., Marino A., Modica L. Spagnolo S. (Eds.) "Pertial Differential Equations and the Calulus of Variations Vol. II" Birkhäuser, Boston (1989) 879-904

[80] PAPANICOLAOU G. C., VARADHAN S. R. S. *Boundary value problems with rapidly oscillating random coefficients* Colloquia Mathematica Societatis János Bolyai, North-Holland, Amsterdam (1982) 835-873

[81] RODRIGUES J.-F. *Some remarks on the homogenization of the dam problem* Manuscr. Math. **46** (1984) 65-82

[82] RUBINSTEIN J., KELLER J. B. *Lower bounds on permeability* Phys. Fluids **30** (1987) 2919-2921

[83] RUBINSTEIN J., MAURI R. *Dispersion and convection in periodic porous media* SIAM J. Appl. Math. **46** (1986) 1018-1023

[84] RUBINSTEIN J., TORQUATO S. *Flow in random porous media: Mathematical formulation, variational principles, and rigorous bounds* J. Fluid Mech. **206** (1989) 25-46

[85] SÀEZ A. E., OTERO C. J., RUSINEK I. *The effective homogeneous behavior of heterogeneous porous media* TiPM **4** (1989) 213-238

[86] SANCHEZ-PALENCIA E. *Non-Homogeneous Media and Vibration Theory* Springer Lecture Notes in Physics **129** (1980) 398 pp.

[87] SATTEL-SCHWIND K. *Untersuchung über Diffusionsvorgänge bei der Gelpermeations-Chromatographie von Poly-p-Methylstyrol* Dissertation, Fachbereich Chemie, Universität Heidelberg (1988)

[88] SHOWALTER R. *Diffusion models with microstructure* Transp. Porous Media **6** (1991), to appear

[89] SIEBEL K. *Diffusion in dispersen Medien. Homogenisierung* Diplomarbeit, Fakultät für Mathematik, Universität Heidelberg (1988)

[90] TARTAR L. *Incompressible fluid flow in a porous medium - convergence of the homogenization process* E. Sanchez-Palencia (Ed.) "Non-Homogeneous Media and Vibration Theory" Lecture Notes in Physics **127** Springer, Berlin (1980) 368-377

[91] VOGT C. *A homogenization theorem leading to a Volterra integro - differential equation for permeation chromatography* Preprint **155** SFB 123 (1982)

[92] WHITAKER S. *Flow in porous media II: the governing equations for immiscible, two-phase flow* TiPM **1** (1986) 105-125

Acknowledgement The author is very grateful to Willi Jäger (Heidelberg) for many encouraging and stimulating discussions on homogenization and its applications.

Finite-Element-Approximation of Solute Transport in Porous Media with General Adsorption Processes

Peter Knabner
Universität Augsburg
Institut für Mathematik
Universitätsstraße 8
D-W-8900 Augsburg
Germany

1 Introduction

In this paper we study the numerical approximation of solute transport through porous media with adsorption. We restrict ourselves to one-species-models, but we allow for multiple adsorption sites, which may be in equilibrium or in non-equilibrium. This leads to a linear partial differential equation coupled with a family of algebraic equations (equilibrium) and ordinary differential equations (non-equilibrium) (see (2.1)). We consider only one-step methods in time, i. e. involving only one preceeding time level and a finite element discretization in space. There are various ways to deal with the nonlinearity in the problem including explicit linearization, i. e. evaluation at the old time level, predictor-corrector methods, implicit and also implicit-explicit methods.
In chapter 2 we describe a unified approach and deduce in detail the discrete set of nonlinear equations to be solved. For their solution we study the two main approaches, Newton-type and fixed point type methods. The problem gets easier, if all nonlinearities are of explicit type, i. e. described by an adsorption isotherm (see (2.24)). In chapter 3 we concentrate on this situation, whereas chapter 4 deals with the general case. As an equilibrium description is only an approximation for a fast reaction, it is reasonable to use a *kinetic approximation*, i. e. to substitute an equilibrium description by a kinetic description with large rate parameters. Such a method can only work if the discrete numerical problem and the method to solve it are stable uniformly in the rate parameter. Therefore we put some stress on the classification with respect to this question, which leads to different answers for the different solution procedures. In chapter 5 it is demonstrated that the extension to an aggregated medium where the aggregation induces physical non-equilibrium does not lead to any new aspects. In chapter 6 some numerical experiments are described.

Although this paper is mathematically completely rigorous in notation, assertions and arguments, an attempt has been made to make it accessible also to the theoretically inclined scientist. For this reason the theorem–proof structure has been avoided. Instead the main assertions are marked by $\diamondsuit$ and the corresponding assumptions by •. Smoothness assumptions on nonlinearities are not stated explicitly, but follow implicitly from the use of corresponding derivatives. Chapter 2 partially is of expository character, containing various comments on the finite-element-method, order of convergence estimates, data structures and solution of sets of linear equations, which are well known to numerical analysts.

The models to be studied have been derived e.g. in [8]. For the reader's convenience we repeat the necessary essentials:
If diffusion/dispersion and convection are considered to be the mechanisms for solute transport, then conservation of mass yields for $c = c(\underline{x}, t)$ [mol/m^3], the concentration of adsorbate in solution, the linear *diffusion–convection* (or advection) *equation*

$$\frac{\partial}{\partial t}(\Theta c)(\underline{x}, t) - \nabla \cdot (\Theta \underline{\underline{D}} \nabla c - \underline{q} c)(\underline{x}, t) = G(\underline{x}, t). \tag{1.1}$$

Hereby $t[s]$, $\underline{x} = (x_1, \ldots, x_N)$ [m], Θ [-], $\underline{\underline{D}}$ [m^2/s] and $\underline{q}$ [m/s] denote time, space, the water content, the sum of molecular diffusion and mechanical dispersion and the water flux, respectively. G [mol/m^3/s] is a source/sink to be specified later on. N is the space dimension of the model.

The scalar-, vector- and matrix-valued functions Θ, $\underline{q}$, $\underline{\underline{D}}$ in $\underline{x}$, t are determined by the water flow and assumed to be given functions such that

$$\frac{\partial}{\partial t}\Theta = -\nabla \cdot \underline{q},$$

i. e. conservation of volume holds true for the water flow and $\underline{\underline{D}}(\underline{x}, t)$ is symmetric and positive definite uniformly in $\underline{x}, t$ and $\Theta(\underline{x}, t) \geq \Theta_0 > 0$ for some Θ_0 and all $\underline{x}, t$. Therefore (1.1) can be written in the form

$$\Theta \frac{\partial}{\partial t} c - \nabla \cdot (\Theta \underline{\underline{D}} \nabla c) + \underline{q} \nabla c = G, \tag{1.2}$$

showing that (1.1) is a uniformly parabolic differential equation. We consider (1.1) in an N-dimensional bounded domain Ω and the time interval $[0,T]$ and set

$$Q_T := \Omega \times (0,T]. \tag{1.3a}$$

We use the abbreviation

$$(Lc)(\underline{x},t) = \nabla \cdot (\Theta \underline{\underline{D}} \nabla c - \underline{q} c) \quad \text{for} \quad (\underline{x},t) \in Q_T, \tag{1.3b}$$

i. e. the basic equation (1.1) is

$$\frac{\partial}{\partial t}(\Theta c) - Lc = G \quad \text{in} \quad Q_T. \tag{1.4}$$

With respect to the boundary conditions we assume that $\partial\Omega$, the boundary of Ω, is subdivided into S_1, S_2 and S_3, such that S_2 is an inflow and S_3 an outflow or noflow boundary. Therefore

$$\begin{aligned} \underline{q} \cdot \underline{\hat{n}} &< 0 \quad \text{on} \quad S_2, \\ \underline{q} \cdot \underline{\hat{n}} &\geq 0 \quad \text{on} \quad S_3, \end{aligned} \tag{1.5}$$

where $\underline{\hat{n}}$ denotes the outward normal vector at $\partial\Omega$. The boundary and initial conditions now read:

$$\begin{aligned} c(\underline{x},t) &= c_f(\underline{x},t) \quad \text{for} \quad \underline{x} \in S_1, \\ ((\Theta \underline{\underline{D}} \nabla c - \underline{q} c) \cdot \underline{\hat{n}})(\underline{x},t) &= F(\underline{x},t) \quad \text{for} \quad \underline{x} \in S_2, \\ \Theta \underline{\underline{D}} \nabla c \cdot \underline{\hat{n}} &= 0 \quad \text{for} \quad \underline{x} \in S_3 \quad \text{and} \quad t \in (0,T], \\ c(\underline{x},0) &= c_0(\underline{x}), \quad \underline{x} \in \Omega. \end{aligned} \tag{1.6}$$

Hereby c_f, F and c_0 are given non-negative functions. If adsorption or more general a heterogeneous reaction with the surface of the porous skeleton takes place, this acts as a sink for c, i. e.

$$G = -\varrho \frac{\partial}{\partial t} s + \widetilde{G}, \tag{1.7}$$

where $s = s(\underline{x}, t)$ [mol/kg] is the *adsorbed concentration* and $\varrho = \varrho(\underline{x})$ [kg/m^3] is the known bulk density. For the sake of notational simplicity we neglect further sources by setting

$$\widetilde{G} := 0. \tag{1.8}$$

If the reaction is considered to be *in equilibrium*, then the equilibrium equation has to be added:

$$f(c(\underline{x}, t), s(\underline{x}, t)) = 0 \tag{1.9}$$

or in explicit form

$$s(\underline{x}, t) = \varphi(c(\underline{x}, t)), \tag{1.10}$$

where φ is the *adsorption isotherm.*

If the reaction is *not in equilibrium*, its kinetics has to be described by

$$\frac{\partial}{\partial t} s(\underline{x}, t) = k f(c(\underline{x}, t), s(\underline{x}, t)), \tag{1.11a}$$

$$s(\underline{x}, 0) = s_0(\underline{x}). \tag{1.11b}$$

(1.9) is formally included in (1.11a) by setting the rate parameter $k = \infty$.

We unify and extend these concepts by allowing for the *heterogeneity of the adsorbent surface* which we describe by a set Λ. Examples for Λ are $\Lambda = \{1, \ldots, \ell\}$ corresponding to a subdivision of a representative grain surface into ℓ chemically different collections of adsorption sites or a nondiscrete set being a description of a representative, i. e. rescaled, grain surface, e.g. the unit sphere. Then s also is a function of λ:

$$s = s(\lambda, \underline{x}, t),$$

and in (1.11) k and f depend on λ. $k_\lambda = \infty$ is allowed. In (1.4), (1.7) s has to be substituted by the discrete or continuous average

$$\int_\Lambda s(\lambda, \underline{x}, t)\, d\lambda.$$

This leads to the equations (2.1), whose numerical approximation is considered in this paper.

We conclude with a table of contents:

2 Derivation of the discrete equations

We consider the following model for transport and adsorption in porous media of one substance:

$$\frac{\partial}{\partial t}(\Theta c)(\underline{x},t) + \varrho(\underline{x})\frac{\partial}{\partial t}\int_\Lambda s(\lambda,\underline{x},t)\,d\lambda - (Lc)(\underline{x},t) = 0, \tag{2.1a}$$

$$\frac{\partial}{\partial t}s(\lambda,\underline{x},t) = k_\lambda f(\lambda,\underline{x},t,c(\underline{x},t),s(\lambda,\underline{x},t)) \tag{2.1b}$$

for $(\underline{x},t) \in Q_T, \lambda \in \Lambda$,

$$s(\lambda,\underline{x},0) = s_0(\lambda,\underline{x}) \quad \text{for} \quad \underline{x} \in \Omega \tag{2.1c}$$

supplemented by the initial and boundary conditions (1.6).

Also two-species-systems fulfilling conservation properties like binary ion exchange can be transformed to a system of the form (2.1) according to [8]. In this case f depends on $\underline{x}$ and t by the solution of a boundary-value problem for the linear diffusion-convection equation (1.2) with $G = 0$. Therefore in the algorithms to be developed for (2.1) a discretization of this equation has to be included.

In (2.1b) we allow for rate parameters $k_\lambda < \infty$ and $k_\lambda = \infty$, i. e. for adsorption sites in non-equilibrium, where (2.1b) describes the kinetics of the reaction, and for adsorption sites in equilibrium, where (2.1b) is to be interpreted as the equilibrium relation

$$0 = f(\lambda,\underline{x},t,c(\underline{x},t),s(\lambda,\underline{x},t)). \tag{2.2}$$

Often the *adsorption isotherm for f* is known explicitly, i. e. a function φ such that

$$f(\lambda,\underline{x},t,c,s) = 0 \iff s = \varphi(\lambda,\underline{x},t,c). \tag{2.3}$$

In the equilibrium case s_0 is usually given by

$$s_0(\lambda, \underline{x}, t) := \varphi(\lambda, \underline{x}, t, c_0(\underline{x}, t)).$$

Then it is equivalent to use the isotherm to describe the equilibrium relation instead of using a rate function which has been obtained e.g. from a molecular description of the adsorption mechanism. If only the isotherm is known from batch experiments, this is the only possibility to describe the equilibrium. In these cases one sets

$$f(\lambda, \underline{x}, t, c, s) = \varphi(\lambda, \underline{x}, t, c) - s. \tag{2.4}$$

If the kinetics of the reaction has to be described and only the isotherm is known, then it is popular to use also (2.4) in (2.1b) with some rate parameter $k_\lambda < \infty$. We will call f *to be of explicit type*, if it is in the form (2.4).

In the numerical approximation of (2.1) with (1.6) the following steps have to be accomplished:

I. Discretization of Λ
II. Discretization of time
III. Discretization of space
IV. (Iterative) solution of the discrete set of (nonlinear) equations

If Λ is a finite set, nothing has to be done with respect to I. Otherwise we select a finite partition of Λ and approximate s_0 and f in λ for fixed $\underline{x}$, t (and c, s) by a constant on each subset of the partition. Then also s is constant in λ on each such subset for fixed $\underline{x}$, t according to (2.1), if solutions of (2.1), (1.6) are unique (consider [7] for this question). This means that we have reached a problem of the form (2.1) for a finite set Λ, say

$$\Lambda = \{1, \ldots, \ell\}, \tag{2.5}$$

where the average reduces to a weighted sum:

$$\fint_{\Lambda} s(\lambda)\,d\lambda = \sum_{\lambda=1}^{\ell} \alpha(\lambda) s(\lambda), \tag{2.6}$$

with $\alpha(\lambda) > 0$ and $\sum_{\lambda=1}^{\ell} \alpha(\lambda) = 1$.

In II we use *finite differences*: Let

$$t^0 := 0, \quad t^{m+1} := t^m + \Delta t\,, \quad m = 0, 1, \ldots, \tag{2.7}$$

where $\Delta t = \Delta t^{\,m+1}$ is determined by a step size control discussed later on. A superscript m refers to the time level $t = t^m$, i. e.

$$\begin{aligned} &\Theta^m := \Theta(.,t^m), \quad \underline{\underline{D}}^m := \underline{\underline{D}}(.,t^m), \quad \underline{q}^m := \underline{q}(.,t^m), \quad c_f^m := c_f(.,t^m), \\ &F^m := F(.,t^m), \quad L^m c := \nabla\cdot(\Theta^m \underline{\underline{D}}^m \nabla c - \underline{q}^m c). \end{aligned} \tag{2.8}$$

Then we approximate (2.1), (1.6) by

$$\begin{aligned} &\frac{\Theta^{m+1}c^{m+1} - \Theta^m c^m}{\Delta t} + \sum_{\lambda=1}^{\ell} \varrho_\lambda \frac{s_\lambda^{m+1} - s_\lambda^m}{\Delta t} - \\ &\qquad - r_1 L^{m+1} c^{m+1} - \overline{r_1} L^m c^m = 0 \quad \text{in} \quad \Omega, \end{aligned} \tag{2.9a}$$

$$\begin{aligned} \frac{1}{k_\lambda}\frac{s_\lambda^{m+1} - s_\lambda^m}{\Delta t} &= f(\lambda, ., t^m + r_3^\lambda\,\Delta t\,, r_2^\lambda c^{m+1} + \overline{r_2^\lambda} c^m, r_3^\lambda s_\lambda^{m+1} + \overline{r_3^\lambda} s_\lambda^m) \\ &\qquad \text{in} \quad \Omega, \quad \lambda = 1, \ldots, \ell, \end{aligned} \tag{2.9b}$$

$$\begin{aligned} c^{m+1} &= c_f^{m+1} \quad \text{in} \quad S_1, \\ (\Theta^{m+1}\underline{\underline{D}}^{m+1}\nabla c^{m+1} - \underline{q}^{m+1}c^{m+1})\cdot\underline{\hat{n}} &= F^{m+1} \quad \text{in} \quad S_2, \\ \Theta^{m+1}\underline{\underline{D}}^{m+1}\nabla c^{m+1}\cdot\underline{\hat{n}} &= 0 \quad \text{in} \quad S_3, \end{aligned} \tag{2.9c}$$

$$c^0 = c_0, \quad s_\lambda^0 = s_0(\lambda, .) \quad \text{in} \quad \Omega, \quad \lambda = 1, \ldots, \ell. \tag{2.10}$$

The system (2.9) determines c^{m+1}, s_λ^{m+1}, the approximations of $c(., t^{m+1})$ and $s(\lambda, ., t^{m+1})$ from the known c^m, s_λ^m. For $m = 0$ (2.10) is used to start this time stepping.

Here r_1, r_2^λ, $r_3^\lambda \in [0, 1]$ are implicity parameters, whose selection will be discussed later on. We use the abbreviation

$$\overline{r} = 1 - r. \tag{2.11}$$

Note that the implicity parameters are allowed to be different at different places. Finally we set

$$\varrho_\lambda := \alpha(\lambda)\varrho \quad \text{for} \qquad \lambda = 1, \ldots, \ell, \tag{2.12}$$

and

$$\frac{1}{k_\lambda} := 0 \quad \text{for} \quad k_\lambda = \infty \tag{2.13}$$

i. e. for an equilibrium situation.

The formulas (2.9a), (2.9c) represent a *linear elliptic boundary value problem* for c^{m+1} (depending on s_λ^{m+1}), whereas (2.9b) is a *nonlinear algebraic equation* for $s_\lambda^{m+1}(\underline{x})$ (depending on $c^{m+1}(\underline{x})$) for each $\underline{x} \in \overline{\Omega}$.

In III. we apply the *Galerkin approach* with *linear finite elements*. To this end we note the weak formulation of (2.9a), obtained by multiplication with appropriate test functions η and integration by parts:

$$\begin{aligned}\int_\Omega \frac{\Theta^{m+1}c^{m+1}-\Theta^m c^m}{\Delta t}\,\eta\, d\underline{x} + \int_\Omega \sum_{\lambda=1}^{\ell} \varrho_\lambda \frac{s_\lambda^{m+1}-s_\lambda^m}{\Delta t}\eta\, d\underline{x} + \\ + \int_\Omega r_1(\Theta^{m+1}\underline{\underline{D}}^{m+1}\nabla c^{m+1} - \underline{q}^{m+1}c^{m+1})\cdot\nabla\eta\, d\underline{x} + \\ + \int_\Omega \overline{r}_1(\Theta^m\underline{\underline{D}}^m\nabla c^m - \underline{q}^m c^m)\cdot\nabla\eta\, d\underline{x} - \\ - \int_{S_2}(r_1F^{m+1}+\overline{r}_1F^m)\eta\, d\underline{\sigma} + \\ + \int_{S_3} r_1\underline{q}^{m+1}\cdot\underline{\hat{n}}\, c^{m+1}\eta + \overline{r}_1\underline{q}^m\cdot\underline{\hat{n}}\, c^m\eta\, d\underline{\sigma} = 0\end{aligned} \tag{2.14}$$

for all $\eta \in H_D^1(\Omega)$ and $d\underline{\sigma}$ is the surface measure on S.

Hereby, $H^1(\Omega)$ is the space of functions η on Ω such that η and $(\partial/\partial x_i)\eta$, $i = 1,\ldots,N$ (in the sense of distributions) are square-integrable on Ω and $H_D^1(\Omega)$ is the subspace of functions η, for which $\eta = 0$ on S_1 (in the sense of trace).

In the *Galerkin approach* one selects a finite-dimensioned subspace $S_h \subset H^1(\Omega)$. An approximation of c^{m+1} in S_h has to fulfill (2.14) for all testfunctions $\eta \in S_h \cap \{\eta \mid \eta = 0 \quad \text{on} \quad S_1\}$ and in addition the Dirichlet boundary condition in (2.9c).

The space S_h is constructed by *linear finite elements*. For simplicity we restrict ourselves to the space dimensions $N = 1$ and $N = 2$. Ω is subdivided into intervals $\tau \in \mathcal{T}_h$ or in triangles $\tau \in \mathcal{T}_h$, such that no vertex of a triangle is in the interior of another side (cf. e.g. Figure 1). We assume for this end that for $N = 1$ Ω is an interval, i. e. $\Omega = (0, L)$, for some $L > 0$, and for $N = 2$ that Ω is a polygon. Let P_j, $j = 1,\ldots,n$, be the vertices of the finite elements τ of the *triangulation* $\mathcal{T}_h$. The P_j are called *nodes*. Suppose that a numbering has been chosen such that

$j = \overline{n}+1,\dots,n$ correspond exactly to the nodes lying on S_1. The discretization parameter h is the maximal length of a side of a $\tau \in \mathcal{T}_h$, i. e. h decreases as the triangulation is refined.

We define S_h to consist of those continuous functions η on $\overline{\Omega}$ which are affine-linear on each $\tau \in \mathcal{T}_h$. S_h is a subspace of $H^1(\Omega)$ and a function $\eta \in S_h$ is determined by its *nodal values* $\eta(P_j)$, $j = 1,\dots,n$, i. e. n is the dimension of S_h. In particular a basis of S_h is defined by

$$\eta_i(P_j) := \left\{\begin{matrix} 1, & i = j \\ 0, & i \neq j \end{matrix}\right\} \quad \text{for} \quad i,j = 1,\dots,n. \tag{2.15}$$

Therefore the test space $S_h \cap \{\eta \mid \eta = 0 \quad \text{on} \quad S_1\}$ is the span of η_i, $i = 1,\dots,\overline{n}$. We demand in addition that the approximation of s_λ^{m+1} belongs to S_h for each $\lambda = 1,\dots,\ell$ and keep the old notation, i. e.

$$c^m = \sum_{i=1}^{n} c_i^m \eta_i, \quad s_\lambda^m = \sum_{i=1}^{n} s_{\lambda,i}{}^m \eta_i \tag{2.16a}$$

for $m = 0,1,\dots;\quad \lambda = 1,\dots,\ell$, where

$$c_i^m = c^m(P_i), \quad s_{\lambda,i}{}^m = s_\lambda^m(P_i) \tag{2.16b}$$

are the nodal values of the approximations. Therefore the determination of c^{m+1} and s_λ^{m+1} is equivalent to the determination of the vectors $(c_i^{m+1})_{i=1,\dots,n}$ and $(s_{\lambda,i}{}^{m+1})_{i=1,\dots,n}$, which we again denote by c^{m+1} and s_λ^{m+1}.

In this way we reach a linear set of equations for the unknown nodal values:

$$\begin{aligned} (\widetilde{M}^{m+1} + r_1\,\Delta t\,\widetilde{K}^{m+1})c^{m+1} + \sum_{\lambda=1}^{\ell} \widetilde{M_\lambda} s_\lambda^{m+1} = \\ = (\widetilde{M}^m - \overline{r_1}\,\Delta t\,\widetilde{K}^m)c^m + \sum_{\lambda=1}^{\ell} \widetilde{M_\lambda} s_\lambda^m + \widetilde{b}^{m+1}. \end{aligned} \tag{2.17}$$

Hereby for $i = 1, \ldots, \overline{n}$, $j = 1, \ldots, n$:

$$\begin{aligned}(\widetilde{M}^m)_{ij} &:= \int_\Omega \Theta^m \eta_j \eta_i \, d\underline{x}, \\ (\widetilde{M_\lambda})_{ij} &:= \int_\Omega \varrho_\lambda \eta_j \eta_i \, d\underline{x} \quad \lambda = 1, \ldots, \ell,\end{aligned} \tag{2.18a}$$

$$\begin{aligned}(\widetilde{K}^m)_{ij} &:= \int_\Omega \Theta^m \underline{\underline{D}}^m \nabla \eta_j \cdot \nabla \eta_i - \underline{q}^m \cdot \nabla \eta_i \, \eta_j \, d\underline{x} + \\ &\quad + \int_{S_3} \underline{q}^m \cdot \underline{\hat{n}} \, \eta_j \eta_i \, d\underline{\sigma}, \\ \widetilde{b}_i^m &:= \Delta t \int_{S_2} (r_1 F^{m+1} + \overline{r_1} F^m) \eta_i \, d\underline{\sigma},\end{aligned} \tag{2.18b}$$

The equations (2.17) have to be supplemented by an approximation of the Dirichlet boundary condition on S_1, which is given by

$$c_i^{m+1} = c_f^{m+1}(P_i), \quad i = \overline{n} + 1, \ldots, n. \tag{2.19}$$

In general the integrals in (2.18) cannot be evaluated exactly, but have to be approximated by a quadrature rule. This should be done in such a way that the order of convergence in Δt and h, which is achieved by (2.17), (2.18) and the approximation of the algebraic equations, is not deteriorated. For linear initial boundary problems as (1.2), (1.6) the requirements in accuracy for the quadrature formulas are low (compare [13]).

Denote the piecewise linear interpolant of a function f by $I(f)$, i. e.

$$I(f) := \sum_{i=1}^{n} f(P_i) \eta_i \tag{2.20}$$

for a continuous f on $\overline{\Omega}$ and analogous for a continuous f defined on a part of S or for vector- or matrix-valued functions. Then an admissible approximation of (2.18) is given by

$$
\begin{aligned}
(M^m)_{ij} &:= \int_\Omega I(\Theta^m \eta_j \eta_i)\, d\underline{x}, \\
(M_\lambda)_{ij} &:= \int_\Omega I(\varrho_\lambda \eta_j \eta_i)\, d\underline{x},
\end{aligned}
\tag{2.21a}
$$

$$
\begin{aligned}
(K^m)_{ij} :=& \int_\Omega I(\Theta^m \underline{\underline{D}}^m) \nabla \eta_j \cdot \nabla \eta_i - I(\underline{q}^m) \cdot \nabla \eta_i \, \eta_j \, d\underline{x} + \\
&+ \int_{S_3} I(\underline{q}^m \cdot \underline{\hat{n}}) \eta_j \eta_i \, d\underline{\sigma}, \\
b_i^m :=& \, \Delta t \int_{S_2} I(r_1 F^{m+1} + \overline{r_1} F^m) \eta_i \, d\underline{\sigma}.
\end{aligned}
\tag{2.21b}
$$

These integrals can be evaluated exactly. An alternative in (2.21b) is to interpolate the whole integrand. In $\widetilde{M^m}$ and $\widetilde{M_\lambda}$ *mass lumping* has been applied leading to diagonal M^m and M_λ. This enhances the stability properties (possibility of a discrete maximum principle, cf. e.g. [15]), but with respect to M_λ also corresponds to the spatial immobility of s.

Thus the approximation c^{m+1} and s_λ^{m+1} have to fulfill (2.19) and

$$
\begin{aligned}
(M^{m+1} + r_1 \, \Delta t \, K^{m+1}) c^{m+1} + \sum_{\lambda=1}^{\ell} M_\lambda s_\lambda^{m+1} &= \\
= (M^m - \overline{r_1} \, \Delta t \, K^m) c^m + \sum_{\lambda=1}^{\ell} M_\lambda s_\lambda^m + b^{m+1}&,
\end{aligned}
\tag{2.22}
$$

where c^m, $s_\lambda^m \in S_h$ are known from the previous time step. For $m = 0$ they are obtained from (2.10), e.g. by

$$
c^0 := I(c_0), \quad s_\lambda^0 := I(s_0(\lambda, .)), \quad \lambda = 1, \ldots, \ell, \tag{2.23}
$$

if the initial data are smooth enough or by L^2-projection to S_h otherwise.

It remains to approximate the algebraic equations (2.9b). Due to the spatial immobility of s it is reasonable to use mass lumping also here, i. e.

$$\int_\Omega I(\frac{1}{k_\lambda}\frac{s_\lambda^{m+1}-s_\lambda^m}{\Delta t}\eta_i)\,d\underline{x} =$$
$$= \int_\Omega I(f(\lambda,.,t^m + r_3^\lambda\,\Delta t\,, r_2^\lambda c^{m+1} + \overline{r_2^\lambda} c^m, r_3^\lambda s_\lambda^{m+1} + \overline{r_3^\lambda} s_\lambda^m)\eta_i)\,d\underline{x}$$

for $i = 1,\ldots,n$. This is equivalent to (2.9b) for the nodal values, i. e.

$$\begin{aligned}\frac{1}{k_\lambda}\frac{s_{\lambda,i}{}^{m+1}-s_{\lambda,i}{}^m}{\Delta t} =& f(\lambda, P_i, t^m + r_3^\lambda\,\Delta t\,, r_2^\lambda c_i^{m+1} + \overline{r_2^\lambda} c_i^m,\\ & r_3^\lambda s_{\lambda,i}{}^{m+1} + \overline{r_3^\lambda} s_{\lambda,i}{}^m) \quad \text{for} \quad i = 1,\ldots,n.\end{aligned} \tag{2.24}$$

Thus the full discrete problem consists of the equations (2.22), (2.19) and (2.24). The first part are linear equations with the sparse matrix

$$A^{m+1} = M^{m+1} + r_1\,\Delta t\,K^{m+1},$$

the second part are nonlinear but diagonal equations. The Dirichlet conditions can be incorporated into (2.22) such that the sparsity pattern of A^{m+1} is preserved (cf. e.g. [14]); we assume that this has been done.

In the following we will encounter the problem to solve linear equations with the matrix A^{m+1} or a modification of it. Basically we will think in terms of a *direct method* with a storage scheme adapted to the sparsity of A^{m+1} like a storage of the *band* or the *profile* after the application of a *band width reduction algorithm*, e.g. the Cuthill-McKee algorithm. The *Gauss algorithm* also preserves the profile or band, if it can be performed without row exchange. This is the case if A^{m+1} is a M-Matrix (cf. e.g. [2]). We will discuss this aspect and its requirements for our problem in chapter 3. If row exchange is necessary the bandwidth of the upper triangular matrix to be constructed by the Gauss algorithm may be doubled. Let k ($\ll n$) be the bandwidth of A^{m+1}, then we have the following (rough) estimates of the complexity:

$$LU \text{ decomposition: } O(k^2\, n) \text{ operations,} \tag{2.25a}$$
$$\text{forward/backward substitution: } O(kn) \text{ operations.} \tag{2.25b}$$

The work to *assemble* the matrix and right hand side is not taken into account here. If only the right hand side is changed, then only (2.25b) is necessary. Note in this aspect that $A^{m+1} = A^m$, if the water flow region is stationary, i. e. $\Theta^{m+1} = \Theta^m$ etc. and the time step size Δt is not changed. For a fine discretization (n very large) it may be of advantage to use adapted iterative methods with corresponding storage schemes like preconditioned conjugate gradient methods (cf. e.g. [1]). For flows, where the convective part dominates the diffusive/dispersive part considerably, the Galerkin approach above has to be modified. Possible measures are to use upwinding (cf. e.g. [6]) or the modified method of characteristics (cf. e.g. [3]). Also in these cases we are lead to a sparse linear system like (2.22) such that the considerations of the following chapter are also valid here.

An *order of convergence* result estimates the deviation of c^m from $c(.,t^m)$ and s_λ^m from $s_\lambda(.,t^m)$ in some function space norms in terms of powers of Δt and h. On the other hand, the computational complexity of the discrete problem is given by the number of time steps, i. e. roughly $O(1/\Delta t)$, and the number of nodes, i. e. roughly $O(1/h)$ for $N = 1$ and $O(1/h^2)$ for $N = 2$. Hence one aims towards an estimate in terms of possibly high powers of Δt and h in a reasonable norm, say the norm in $L^2(\Omega)$:

$$\|u\|_2 := (\int_\Omega u(\underline{x})^2\, d\underline{x})^{1/2} \tag{2.26}$$

We will not persue this subject rigorously in this paper, but only give some formal indications. If we restrict ourselves to the linear diffusion-convection equation

(1.2), (1.6), we reach a discretization analogous to (2.14) with $\varrho_\lambda \equiv 0$. The best possible order of convergence estimate requires certain smoothness of the solution c of (1.2), (1.6), i. e. of the initial and boundary data and the coefficients. It reads:

$$\begin{aligned} \|c(.,t^m) - c^m\|_2 &= O(h^2 + \Delta t) \quad \text{for} \quad r_1 \neq \frac{1}{2}, \\ \|c(.,t^m) - c^m\|_2 &= O(h^2 + \Delta t^2) \quad \text{for} \quad r_1 = \frac{1}{2} \end{aligned} \tag{2.27}$$

For $r_1 < \frac{1}{2}$ stability requirements lead to a restriction of Δt in terms of h^2 (cf. e.g. [15]). As a rule of thumb, order of convergence estimates corresponds to *order of consistency* estimates, provided the discretization is *stable* in a certain sense. The consistency error is obtained by substituting the exact solution in the discrete set of equations (i. e. for (1.1) with $G \equiv 0$, (1.6): (2.14) with $\varrho_\lambda \equiv 0$).

From (2.27) the choice $r_1 = \frac{1}{2}$, the *Crank-Nicolson scheme*, seems to be optimal. On the other hand, the validity of a *discrete maximum principle* is desirable, leading to non-negative solutions for non-negative data. Usually the corresponding requirements for $r_1 = 1$, the *fully implicit* or *backward Euler scheme*, are lower than for $r_1 = \frac{1}{2}$ (cf. e.g. [15]). Therefore we usually prefer $r_1 = 1$ and thus have $h^2 + \Delta t$ as the best possible order of convergence estimate.

Turning now to nonlinear problems like (2.14), (2.24) for smooth nonlinearities f the order of consistency $h^2 + \Delta t$ remains unchanged. This is independent of the choice of the implicity parameters r_2^λ, r_3^λ. Only if one wants to preserve the order of consistency $h^2 + \Delta t^2$ one is restricted to $r_1 = r_2^\lambda = r_3^\lambda = \frac{1}{2}$. The computational complexity of the discrete equations will depend heavily on the selection of r_2^λ and r_3^λ, as discussed in chapter 3. One has to keep in mind here that this is also the case for the stability requirements leading to more restrictive step sizes Δt for "more explicit" versions.

In some typical adsorption models the nonlinearities are nonsmooth, e.g. if f is of explicit type and the corresponding isotherm φ is of the form

$$\varphi(\lambda, c) = A_\lambda c^{p_\lambda} \quad \text{for} \quad A_\lambda > 0, \quad 0 < p_\lambda < 1, \tag{2.28}$$

the *Freundlich isotherm* (cf. e.g. [8]). Then due to the reduced regularity of the solution, only weaker order of convergence estimates are to be expected (cf. [11] for the case $\Lambda = \{1\}$, $k = \infty$). In this paper we concentrate our discussion on smooth nonlinearities.

3 Approximate solution of the discrete equations: nonlinearities of explicit type

In this chapter and the following we discuss iterative schemes to obtain an approximate solution of the discrete equations (2.22) (modified by (2.19)) and (2.24). To simplify the notation we ignore the modifications for Dirichlet data by setting $S_1 = \emptyset$, i. e. $\overline{n} = n$. Furthermore we assume in the following that in an equilibrium case, i. e. $k_\lambda = \infty$, we have $r_3^\lambda > 0$, as otherwise we cannot expect a uniquely solvable discrete problem.

We start here with the simplest situation, in which *all nonlinearities are of explicit type*, i. e. of the form (2.4). In this case the equations (2.24) can be resolved explicitly for ${s_{\lambda,i}}^{m+1}$, both for $k_\lambda < \infty$ and $k_\lambda = \infty$. Substitution into (2.22) leads to

$$\begin{aligned}(M^{m+1} + r_1\,\Delta t\,K^{m+1})\,c^{m+1} &= (M^m - \overline{r_1}\,\Delta t\,K^m)\,c^m + b^{m+1} + \\ + \sum_{\lambda=1}^{\ell} M_\lambda &\left(\frac{1}{1/(\Delta t\,k_\lambda) + r_3^\lambda}\left(s_\lambda^m - \varphi(\lambda, ., t^m + r_3^\lambda\,\Delta t\,, r_2^\lambda c^{m+1} + \overline{r_2^\lambda} c^m)\right)\right),\end{aligned} \tag{3.1a}$$

$$\begin{aligned}s_\lambda^{m+1} = &\frac{1}{1/(\Delta t\,k_\lambda) + r_3^\lambda}\,\varphi(\lambda, ., t^m + r_3^\lambda\,\Delta t\,, r_2^\lambda c^{m+1} + \overline{r_2^\lambda} c^m) + \\ &+ \frac{1/(\Delta t\,k_\lambda) - \overline{r_3^\lambda}}{1/(\Delta t\,k_\lambda) + r_3^\lambda}\,s_\lambda^m, \quad \lambda = 1, \ldots, \ell.\end{aligned} \tag{3.1b}$$

Hereby, for $c \in \mathbb{R}^n$, $\varphi(\lambda, ., t, c)$ denotes the vector $(\varphi(\lambda, t, P_i, c_i))_{i=1,\ldots,n} \in \mathbb{R}^n$.

We will use the abbreviations

$$
\begin{aligned}
g_\lambda^{m+1}(\underline{x}, c) &:= \frac{1}{1/(\Delta t\, k_\lambda) + r_3^\lambda}\,(s_\lambda^m - \\
&\quad - \varphi(\lambda, \underline{x}, t^m + r_3^\lambda\,\Delta t\,, r_2^\lambda c + \overline{r_2^\lambda} c^m(\underline{x}))) \\
&\quad \text{for} \quad c \in \mathbb{R}, \quad \underline{x} \in \Omega, \\
g^{m+1}(c) &:= \sum_{\lambda=1}^{\ell} M_\lambda (g_\lambda^{m+1}(P_i, c_i))_i \quad \text{for} \quad c \in \mathbb{R}^n, \\
A^{m+1} &:= M^{m+1} + r_1\,\Delta t\, K^{m+1}, \\
\overline{b}^{m+1} &:= (M^m - \overline{r_1}\,\Delta t\, K^m)\, c^m + b^{m+1}.
\end{aligned} \tag{3.3}
$$

I.e., the problem has reduced to the nonlinear equations for c^{m+1}:

$$
A^{m+1} c^{m+1} = \overline{b}^{m+1} + g^{m+1}(c^{m+1}). \tag{3.4}
$$

After its resolution (3.1b) can be used to determine s_λ^{m+1}. The formulas are uniform for equilibrium and non-equilibrium with the interpretation

$$
\frac{1}{\Delta t\, k_\lambda} = 0 \quad \text{for} \quad k_\lambda = \infty. \tag{3.5}
$$

(3.1a) can be interpreted as the discretization of the following semilinear equation for c^{m+1} alone:

$$
\begin{aligned}
\Theta^{m+1} c^{m+1} - \Theta^m c^m - \Delta t\, r_1 L^{m+1} c^{m+1} - \Delta t\, \overline{r_1} L^m c^m &= \\
= \sum_{\lambda=1}^{\ell} \varrho_\lambda g_\lambda^{m+1}(., c^{m+1}) \quad \text{in} \quad \Omega.&
\end{aligned} \tag{3.6}
$$

To simplify the notation we will suppress the superscript $m+1$ in (3.4) in the following, but keep in mind that usually these terms vary from time level to time level. In general it is sufficient to have a *locally convergent* iteration scheme for (3.1a), as a good initial guess is provided by the solution at the old time level, i. e.

$$c^{(0)} := c^m. \tag{3.7}$$

We will compare the *fixed point (or Picard) iteration* with *Newton's method.* In the first method the old *iteration* $c^{(\nu)}$ is substituted in the nonlinearity in (3.1a) to obtain a linear problem for the new iteration $c^{(\nu+1)}$:

$$Ac^{(\nu+1)} = \bar{b} + g(c^{(\nu)}). \tag{3.8}$$

For each iteration a set of n linear equations of the sparse structure discussed in chapter 2 has to be solved, but with the same matrix A.

Setting

$$d^{(\nu)} := Ac^{(\nu)} - \bar{b} - g(c^{(\nu)}) \tag{3.9}$$

for the *defect* of the νth iteration and

$$\Delta c^{(\nu+1)} := c^{(\nu+1)} - c^{(\nu)} \tag{3.10}$$

for the $(\nu+1)$th *correction*, an equivalent form for (3.8) is

$$A\,\Delta c^{(\nu+1)} = -d^{(\nu)}. \tag{3.11}$$

In Newton's method the nonlinearity is linearized at the old iteration. This leads to

$$(A - Dg(c^{(\nu)}))\,\Delta c^{(\nu+1)} = -d^{(\nu)}. \tag{3.12}$$

Here $Dg(\tilde{c})$ denotes the Jacobi matrix of g with respect to c at $\tilde{c}$. $Dg(\tilde{c})$ is a diagonal matrix, as M_λ is a diagonal matrix by mass lumping and thus $(g)_i$ only depends on $\tilde{c_i}$ for $i = 1, \ldots, n$. I. e., for each iteration the matrix A has to be updated by a diagonal matrix.

A method in between (3.11) and (3.12) is the *chord method*, in which a matrix R is fixed, supposed to be an "approximation" of $Dg(., c^{(\nu)})$ and the iteration procedure reads

$$(A - R)\, \Delta c^{(\nu+1)} = -d^{(\nu)}. \tag{3.13}$$

Consider for a while the usual completion of an iteration step

$$c^{(\nu+1)} := c^{(\nu)} + \Delta c^{(\nu+1)} \tag{3.14}$$

and a stopping of the iteration, if for a given $\varepsilon > 0$

$$\|d^{(\nu)}\| \leq \varepsilon, \tag{3.15}$$

where $\|\cdot\|$ is some norm in $\mathbb{R}^n$. Comparing the complexity of the different schemes, (3.11) and (3.13) require $O(kn)$ operations per iteration and once $O(k^2 n)$ operations, whereas (3.12) requires $O(k^2 n)$ operations per iteration. The work to update righthand side (and matrix) is roughly the same.

Nevertheless Newton's method may be advantageous: It is well known that there is quadratic convergence of the $c^{(\nu)}$, if there exists a solution c^{m+1} of (3.4), $A - Dg(c^{m+1})$ is regular and $c^{(0)}$ is close enough to c^{m+1} (cf. e.g. [12]). *Quadratic convergence* means an estimate of the type

$$\|c^{m+1} - c^{(\nu)}\| \leq K\gamma^{2^\nu} \tag{3.16}$$

for constants $K > 0$, $0 < \gamma < 1$. On the other hand an existence and uniqueness proof for c^{m+1} and a convergence proof for (3.8) based on the contraction principle is only possible if Δt is small enough (corresponding to the local convergence above) and $k_\lambda < \infty$ or $r_2^\lambda = 0$ for all $\lambda = 1, \ldots, \ell$. Then the Lipschitz constant of g can be made as small as desired if all $\Delta t\, k_\lambda$ are sufficiently small. The obtained *convergence* usually is only *linear*, i. e. only an estimate of the type

$$\|c^{m+1} - c^{(\nu)}\| \leq K\gamma^\nu \tag{3.17}$$

is possible. There is an extensive literature on these and other iteration schemes, summarized in [12], which we do not want to repeat here. As an example, we indicate how monotonicity properties can be used to clarify these questions. The following results are based on inverse monotonicity.

A matrix $A \in \mathbb{R}^{n,n}$ is called *inverse monotone* if A is regular and $A^{-1} \geq 0$ (defined componentwise). There are several easily accessible criteria for the subclass of M-matrices:

A matrix $A \in \mathbb{R}^{n,n}$ is called a *M-matrix* if A is regular, $A^{-1} \geq 0$ and $a_{ij} \leq 0$ for $i, j = 1, \ldots, n$, $i \neq j$.

We consider an equation of the type (3.4) and assume the existence of $Q, P \in \mathbb{R}^{n,n}$ such that

$$Q \leq Dg(c) \leq P \quad \text{for all} \quad c \in \mathbb{R}^{n,n}. \tag{3.18}$$

The following assertions hold true (cf. [2]):

(A) If $Q = -P$, A and $A - P$ are inverse monotone, then for each $c^{(0)}$ the sequence defined by (3.8) converges to the unique solution of (3.4).

(B) If $R \leq \frac{1}{2}(P+Q)$, $A-P$ and $A-R$ are inverse monotone, then for each $c^{(0)}$ the sequence defined by (3.13) converges to the unique solution of (3.4).

(C) If $A-P$ and $A-R$ for $R=\frac{1}{2}(P+Q)$ are inverse monotone, then a unique solution $\tilde{c}$ of (3.4) exists. If $\|d^{(0)}\|$ is small enough, then there is quadratic convergence of the sequence defined by (3.12).

(D) If $A-P$ and $A-Q$ are inverse monotone and g is *concave*, i. e.

$$g(c_1)-g(c_2) \geq Dg(c_1)(c_1-c_2) \quad \text{for} \quad c_1, c_2 \in \mathbb{R}^n \tag{3.19}$$

then for each $c^{(0)}$ the sequence defined by (3.12) converges to the unique solution of (3.4) and $\tilde{c} \leq c^{(\nu+1)} \leq c^{(\nu)} \leq \ldots \leq c^{(1)}$.

Under the assumptions from e.g. (C) also the stopping criterion (3.15) seems to be reasonable, as

$$|c^{(\nu)}-\tilde{c}| \leq (A-P)^{-1}\,|d^{(\nu)}|, \tag{3.20}$$

where $|c| := (|c_i|)_i \in \mathbb{R}^n$ for $c \in \mathbb{R}^n$.

We will discuss (A) to (D) under the assumption:

- *The matrix A from (3.4) is a M-matrix.* (3.21)

If there is no convective transport (i. e. $\underline{q} \equiv 0$), this holds true for one space dimension ($N=1$) (cf. e.g. [2]), and also for $N=2$ provided all angles of the triangulation $\mathcal{T}_h$ are at most $\pi/2$ (cf. e.g. [15]). These properties remain preserved in the presence of convective flow as long as $h\|\underline{q}\|$ is small enough. As flows in field or in column studies are rather slow, this requirement on the triangulation often is acceptable.

Nearly all isotherms from applications fulfill the next assumption:

- $0 \le \dfrac{\partial}{\partial c}\varphi(\lambda, \underline{x}, t, c) \le K$
  for all $c \in \mathbb{R}, \quad \lambda = 1, \ldots, \ell; \quad (\underline{x}, t) \in Q_T,$ (3.22)
  and some constant $K > 0$.

Here and in the following we assume that φ and f are defined for all $c, s \in \mathbb{R}$ (and not only for $c, s \ge 0$). This can be achieved by extending the definition in the way that the desired properties hold true. If it can be assured that the solution is nonnegative, the specific extension is irrelevant. If an a priori bound of the solution is known, then it is also possible to change φ for large c to fit the requirements of (3.22).

Because of (3.22) we can choose in (3.18) $P = 0$ and $Q = -K\sum_{\lambda=1}^{\ell}(r_2^\lambda/(\frac{1}{\Delta t\, k_\lambda} + r_3^\lambda))\, M_\lambda$. Thus:

◇ *If all nonlinearities f are of explicit type and (3.21), (3.22) are satisfied, then the following holds true:*
For each $c^{(0)}$ the sequence defined by the chord method (3.13) converges to the unique solution of (3.4), provided

$$R \le -\frac{1}{2} K \sum_{\lambda=1}^{\ell} \frac{r_2^\lambda}{1/(\Delta t\, k_\lambda) + r_3^\lambda} M_\lambda .$$

The sequence defined by Newton's method (3.12) converges quadratically to the unique solution of (3.4), provided the initial defect is small enough. The convergence is global (i. e. for every $c^{(0)}$) and monotone from above, if all $\varphi(\lambda, \underline{x}, t, .)$ are convex.

Hereby we use that (cf. e.g. [2]).

If A is a M-matrix and $B \ge A$ such that $b_{ij} \le 0$ for $i \ne j$,
then B is a M-matrix. (3.23)

On the contrary in (A) we have to choose $P = -Q$ and Q as above. We have to show that $A + Q$ is inverse monotone, which we can do only in special cases requiring at least that $\Delta t\, k_\lambda$ is small or $r_2^\lambda = 0$. This does not lead to any improvement of the local convergence results for the Picard iteration mentioned above.

Concerning the choice of r_2^λ and r_3^λ, we see from (2.30) that for

- $r_2^\lambda = 0, \quad r_3^\lambda \in [0,1]$ *for* $\lambda = 1, \ldots, \ell$ *("implicit-explicit" in (2.9b)):* *The discrete problem (3.1) for* c^{m+1} *is linear.*

Furthermore the discretization arising from $r_2^\lambda = 0$, $r_3^\lambda \in (0,1]$ is nearly identical with the fully explicit version $r_2^\lambda = r_3^\lambda = 0$ for the following reduced rate parameters:

$$
\begin{aligned}
k_\lambda < \infty: &\quad \widetilde{k_\lambda} = k_\lambda \frac{1}{1 + k_\lambda r_3^\lambda \,\Delta t}, \\
k_\lambda = \infty: &\quad \widetilde{k_\lambda} = \frac{1}{r_3^\lambda \,\Delta t} \quad (< \infty).
\end{aligned}
\tag{3.24}
$$

The distinction lies in the instance of time, at which φ is evaluated, i. e. the discretizations are identical, if φ is time independent.

On the other hand we can also interpret certain discretizations of non-equilibrium components as discretizations of equilibrium components. The update formula (3.1b) is equivalent with

$$
\widetilde{r_3^\lambda} s_\lambda^{m+1} + (1 - \widetilde{r_3^\lambda}) s_\lambda^m = \varphi(\lambda, ., t^m + r_3^\lambda \,\Delta t\,, r_2^\lambda c^{m+1} + \overline{r_2^\lambda} c^m), \tag{3.25}
$$

where $\widetilde{r_3^\lambda} := r_3^\lambda + 1/(\,\Delta t\, k_\lambda)$.

Thus:

- $r_2^\lambda > r_3^\lambda, \quad k_\lambda := 1/(\,\Delta t\,(r_2^\lambda - r_3^\lambda))$ *leads to the same discretization as*
  $$r_3^\lambda = r_2^\lambda \ (r_2^\lambda \text{ from above}), \quad k_\lambda = \infty, \tag{3.26}$$
  up to the time evaluation of φ.

4 Approximate solution of the discrete equations: general non-linearities

We now return to the general case of arbitrary rate functions, i. e. (2.22), (2.24). A popular approach is to iterate between the transport equations (2.22) and the reaction equations (2.24), i. e. for a given $c^{(\nu)}$ for $\nu \geq 0$ determine $s_\lambda^{(\nu+1)}$ from

$$
\begin{aligned}
&\frac{1}{\Delta t\, k_\lambda}({s_{\lambda,i}}^{(\nu+1)} - {s_{\lambda,i}}^m) = \\
&= f(\lambda, P_i, t^m + r_3^\lambda\, \Delta t\,, r_2^\lambda c_i^{(\nu)} + \overline{r_2^\lambda} c_i^m, r_3^\lambda {s_{\lambda,i}}^{(\nu+1)} + \overline{r_3^\lambda} {s^m_\lambda}_i) \qquad (4.1)\\
&\quad \text{for} \quad \lambda = 1, \ldots, \ell; \quad i = 1, \ldots, n.
\end{aligned}
$$

In general these are $\ell \times n$ nonlinear scalar equations to solve. Only if f is of explicit type (2.4) or $k_\lambda = \infty$ and the isotherm φ is known ${s_{\lambda,i}}^{(\nu+1)}$ is given explicitly. If we substitute s_λ^{m+1} in (2.22) by $s_\lambda^{(\nu+1)}$ we reach at the following set of linear equations for $c^{(\nu+1)}$:

$$
Ac^{(\nu+1)} = \bar{b} - \sum_{\lambda=1}^{\ell} M_\lambda(s_\lambda^{(\nu+1)} - s_\lambda^m). \qquad (4.2)
$$

If all f are of explicit type, then $s_\lambda^{(\nu+1)}$ can be eliminated as above and (4.2) reduces to the fixed point iteration (3.8) for the equation (3.1a), while (4.1) transforms to an update of $s_\lambda^{(\nu+1)}$ analogously to (3.1b), which only has to be performed once at the (approximate) solution c^{m+1}.

The iteration is constructed in such a way that $c^{(\nu)}$ and $s^{(\nu)}$ fulfill the overall mass balance equation (2.22), but there is a lag in the reaction equations (2.24). Another interpretation evolves if the solution of (4.1) is denoted by $s^{(\nu)}$ and correspondingly $s^{(\nu)}$ appears in (4.2). Then the reaction equations are fulfilled by $(c^{(\nu)}, s^{(\nu)})$, but there is a lag in s in the transport equations. We will call this scheme the *fixed point iteration I*. Possible ways to start the iteration are, as above,

$$
c^{(0)} = c^m
$$

or

$$Ac^{(0)} = \bar{b}. \tag{4.3}$$

The second alternative corresponds to a start in s with $s_\lambda^{(0)} := s_\lambda^m$ and then a step according to (4.2), i. e. first the reaction is ignored.

In the applied literature the notations *one-step-methods* and *two-step-methods* are used. In a one-step-method the discrete nonlinear equations are solved approximately, usually with fixed point iteration I. Two-step-methods are defined by one iteration of the fixed point iteration I. There are two variants:

(A) $c^{(0)} := c^m$,
$s_\lambda^{(1)}$ by (4.1): reaction step
$c_\lambda^{(1)}$ by (4.2): transport step
$c^{m+1} := c^{(1)},\ s_\lambda^{m+1} := s_\lambda^{(1)}$

(B) $Ac^{(0)} = \bar{b}$: transport step
$s_\lambda^{(1)}$ by (4.1): reaction step
$c^{m+1} := c^{(0)},\ s_\lambda^{m+1} := s_\lambda^{(1)}$

Variant (B) is used e.g. in [9].

If one wants to avoid solving nonlinear scalar equations in each iteration, there is no direct way to do this, if $k_\lambda = \infty$ and no isotherm $\varphi(\lambda,.)$ for $f(\lambda,.,.)$ is known. For components in non-equilibrium, however, there is the following obvious variant of (4.1):
For given $c^{(\nu)}$ and $s_\lambda^{(\nu)}$, $\nu \geq 0$ determine $s_\lambda^{(\nu+1)}$ for $k_\lambda = \infty$ from (4.1) and for $k_\lambda < \infty$ from

$$\begin{aligned} &\frac{1}{\Delta t\, k_\lambda}({s_{\lambda,i}}^{(\nu+1)} - {s_{\lambda,i}}^m) = \\ &= f(\lambda, P_i, t^m + r_3^\lambda\,\Delta t\,, r_2^\lambda c_i^{(\nu)} + \overline{r_2^\lambda} c_i^m, r_3^\lambda {s_{\lambda,i}}^{(\nu)} + \overline{r_3^\lambda} {s_{\lambda,i}}^m) \\ &\quad \text{for} \quad i = 1, \ldots, n. \end{aligned} \tag{4.4}$$

If *kinetic approximation* is used, i. e. an equilibrium description is approximated by a non-equilibrium description, then (4.4) could be applied to this non-equilibrium description.

Here we need $c^{(0)}$ and $s_\lambda^{(0)}$ (for λ such that $k_\lambda < \infty$) to start, e.g.

$$c^{(0)} := c^m, \quad s_\lambda^{(0)} = s_\lambda^m. \tag{4.5}$$

We will call this scheme the *fixed point iteration II*.

Finally we consider again *Newton's method*. For given $c^{(\nu)}$, $s_\lambda^{(\nu)}$, $\nu \geq 0$, the next iterates are determined by linearizing at $c^{(\nu)}$, $s_\lambda^{(\nu)}$, i. e.

$$Ac^{(\nu+1)} = \overline{b} - \sum_{\lambda=1}^{\ell} M_\lambda(s_\lambda^{(\nu+1)} - s_\lambda^m), \tag{4.6a}$$

$$\begin{aligned} \frac{1}{\Delta t\, k_\lambda}({s_{\lambda,i}}^{(\nu+1)} - {s_{\lambda,i}}^m) &= \widetilde{f} + \widetilde{\frac{\partial f}{\partial c}} r_2^\lambda (c_i^{(\nu+1)} - c_i^{(\nu)}) + \\ &+ \widetilde{\frac{\partial f}{\partial s}} r_3^\lambda ({s_{\lambda,i}}^{(\nu+1)} - {s_{\lambda,i}}^m) \qquad \lambda = 1, \ldots, \ell, \quad i = 1, \ldots, n. \end{aligned} \tag{4.6b}$$

Hereby $\tilde{}$ denotes the evaluation of the corresponding function at the argument

$$(\lambda, P_i, t^m + r_3^\lambda\, \Delta t\, , r_2^\lambda c_i^{(\nu)} + \overline{r_2^\lambda} c_i^m, r_3^\lambda {s_{\lambda,i}}^{(\nu)} + \overline{r_3^\lambda} {s_{\lambda,i}}^m).$$

The appearing λ, i and ν are determined by the context of the formula such that no ambiguities should arise.

Similar to the explicit case the algebraic equations (4.6b) can be solved explicitly for ${s_{\lambda,i}}^{(\nu+1)}$. Substitution into (4.6a) leads to

$$\begin{aligned} Ac^{(\nu+1)} &= \bar{b} - \\ &- \sum_{\lambda=1}^{\ell} M_\lambda \Big(\frac{1}{1/(\Delta t\, k_\lambda) - r_3^\lambda \widetilde{\partial f/\partial s}} (\tilde{f} + \frac{\widetilde{\partial f}}{\partial c} r_2^\lambda (c_i^{(\nu+1)} - c_i^{(\nu)}) + \\ &\quad + \frac{\widetilde{\partial f}}{\partial s} r_3^\lambda ({s_{\lambda,i}}^m - {s_{\lambda,i}}^{(\nu)}))\Big)_i \end{aligned} \tag{4.7}$$

$$\begin{aligned} {s_{\lambda,i}}^{(\nu+1)} &= \frac{1}{1/(\Delta t\, k_\lambda) - r_3^\lambda \widetilde{\partial f/\partial s}} \cdot \\ &\cdot \left(\tilde{f} + \frac{\widetilde{\partial f}}{\partial c} r_2^\lambda (c_i^{(\nu+1)} - c_i^{(\nu)}) - \frac{\widetilde{\partial f}}{\partial s} r_3^\lambda {s_{\lambda,i}}^{(\nu)} + \frac{1}{\Delta t\, k_\lambda} {s_{\lambda,i}}^m \right), \\ &\lambda = 1, \ldots, \ell, \quad i = 1, \ldots, n. \end{aligned} \tag{4.8a}$$

Using (3.10), the linear equations (4.7) can be rewritten as

$$\begin{aligned} (A - D(c^{(\nu)}, s^{(\nu)}))\, \Delta c^{(\nu+1)} &= -\left\{ Ac^{(\nu)} - \bar{b} + \sum_{\lambda=1}^{\ell} M_\lambda (s_\lambda^{(\nu)} - s_\lambda^m) \right\} - \\ - \sum_{\lambda=1}^{\ell} M_\lambda &\left(\frac{1}{1/(\Delta t\, k_\lambda) - r_3^\lambda \widetilde{\partial f/\partial s}} \left(\tilde{f} + \frac{1}{\Delta t\, k_\lambda} ({s_{\lambda,i}}^m - {s_{\lambda,i}}^{(\nu)}) \right) \right)_i, \end{aligned} \tag{4.8b}$$

where

$$D(c^{(\nu)}, s^{(\nu)}) := -\sum_{\lambda=1}^{\ell} r_2^\lambda M_\lambda \operatorname{diag}\left(\left(\frac{\widetilde{\partial f/\partial c}}{1/(\Delta t\, k_\lambda) - r_3^\lambda \widetilde{\partial f/\partial s}} \right)_i \right) \tag{4.9}$$

($\operatorname{diag}((a_i)_i)$ is the diagonal matrix with the entries a_i).

Note that the defect in the transport equations, the term in braces in (4.8b), vanishes for $\nu \geq 1$ according to (4.6a). Again the iteration can be started by (4.5).

We have implicitly assumed that $1/(\Delta t\, k_\lambda) - r_3^\lambda \widetilde{\partial f/\partial s} \neq 0$. A sufficient condition for this is $\widetilde{\partial f/\partial s} \leq 0$ and $\widetilde{\partial f/\partial s} < 0$ if $k_\lambda = \infty$. We will assume this later on.

Compared to the explicit case, the computational complexity is only slightly increased in general: The distinction is given by the appearance of the iterates $s^{(\nu)}$, defined by the explicit formula (4.8a). If for a $\lambda \in \{1, \ldots, \ell\}$ f is of explicit type, (4.8a) reduces to

$$
\begin{aligned}
s_{\lambda,i}{}^{(\nu+1)} &= \frac{1}{1/(\Delta t\, k_\lambda) + r_3^\lambda} \cdot \\
&\cdot \left(\widetilde{\varphi} + \frac{\partial \widetilde{\varphi}}{\partial c} r_2^\lambda (c_i^{(\nu+1)} - c_i^{(\nu)}) + (\frac{1}{\Delta t\, k_\lambda} - \overline{r_3^\lambda}) s_{\lambda,i}{}^m \right).
\end{aligned}
\tag{4.10}
$$

In the case of convergence this is identical with (3.1b). Furthermore the corresponding term in (4.7) becomes independent of $s_\lambda^{(\nu)}$ and (3.1b) need only be evaluated once at the last computed iterate. If all components are of explicit type, the scheme reduces to (3.12).

The use of Newton's method is not so popular in the applied literature. An example is [10].

Now we want to discuss these schemes according to the results (A)–(D). Again we assume that A is a M-matrix (see (3.21)) and

- *If $k_\lambda = \infty$ there is an isotherm φ such that*

$$0 = f(\lambda, \underline{x}, t, c, s) \iff s = \varphi(\lambda, \underline{x}, t, c) \tag{4.11a}$$

  and

$$0 \leq \frac{\partial}{\partial c} \varphi(\lambda, \underline{x}, t, c) \leq K \tag{4.11b}$$

for all $c, s \in \mathbb{R}$, $\lambda = 1, \dots, \ell$, $(\underline{x}, t) \in \overline{Q_T}$, *and some constant* $K > 0$.

If $k_\lambda < \infty$, *then*

$$\begin{aligned} 0 \le \frac{\partial}{\partial c} f(\lambda, \underline{x}, t, c, s) \le K, \\ -K \le \frac{\partial}{\partial s} f(\lambda, \underline{x}, t, c, s) \le 0 \end{aligned} \tag{4.12}$$

for all $c, s \in \mathbb{R}$, $\lambda = 1, \dots, \ell$, $(\underline{x}, t) \in \overline{Q_T}$, *and some constant* $K > 0$.

These conditions are fulfilled in most of the adsorption and ion exchange models (cf. e.g. [8] and [4]).

To simplify notation assume that $k_\lambda = \infty$ corresponds exactly to $\lambda = 1, \dots, \overline{\ell}$ for some $\overline{\ell} \in \{0, \dots, \ell\}$, i. e. for $\overline{\ell} = 0$ there are no equilibrium components.

The following formulation is equivalent to (2.22), (2.24):

$$\begin{aligned} Ac^{m+1} &= \overline{b} + \sum_{\lambda=1}^{\overline{\ell}} M_\lambda \left(\frac{1}{r_3^\lambda} \left(s_\lambda^m - \varphi(\lambda, ., t^m + r_3^\lambda \, \Delta t \, , r_2^\lambda c^{m+1} + \overline{r_2^\lambda} c^m) \right) \right) + \\ &+ \sum_{\lambda=\overline{\ell}+1}^{\ell} M_\lambda (- \Delta t \, k_\lambda f(\lambda, ., t^m + r_3^\lambda \, \Delta t \, , r_2^\lambda c^{m+1} + \overline{r_2^\lambda} c^m , r_3^\lambda s_\lambda^{m+1} + \overline{r_3^\lambda} s_\lambda^m)), \end{aligned} \tag{4.13a}$$

$$\begin{aligned} s_\lambda^{m+1} &= s_\lambda^m + \Delta t \, k_\lambda f(\lambda, ., t^m + r_3^\lambda \, \Delta t \, , r_2^\lambda c^{m+1} + \overline{r_2^\lambda} c^m , r_3^\lambda s_\lambda^{m+1} + \overline{r_3^\lambda} s_\lambda^m), \\ &\lambda = \overline{\ell} + 1, \dots, \ell. \end{aligned} \tag{4.13b}$$

We can regard (4.13) again as a problem of type (3.4), i. e.

$$\widetilde{A} \widetilde{c} = \widetilde{b} + \widetilde{g}(\widetilde{c}), \tag{4.14}$$

now with the unknown vector $\widetilde{c} = (c^{m+1}, s_{\bar{\ell}+1}^{m+1}, \ldots, s_{\ell}^{m+1})^T \in \mathbb{R}^{n\times(\ell-\bar{\ell}+1)}$ and

$$\widetilde{A} = \begin{pmatrix} A & & & \\ & I & & \\ & & \ddots & \\ & & & I \end{pmatrix}, \tag{4.15}$$

where I is the $n \times n$ identity matrix. The definition of the nonlinearity $\widetilde{g}$ and of $\widetilde{b}$ can be seen from (4.13).

The formulation (4.13) has been chosen because the fixed point iteration applied to (4.14) is identical to the fixed point iteration II, defined in (4.1) for $k_\lambda = \infty$, (4.4) and (4.2): To see this, note that because of (4.11a) the definition of $s_\lambda^{(\nu+1)}$ for $k_\lambda = \infty$ is equivalent to the explicit form

$$r_3^\lambda s^{(\nu+1)} + \overline{r_3^\lambda} s^m = \varphi(\lambda, ., t^m + r_3^\lambda \, \Delta t\, , r_2^\lambda c^{(\nu)} + \overline{r_2^\lambda} c^m)$$

and thus, if we use this identity to *define* $s^{(\nu+1)}$, we can substitute the corresponding part in $\widetilde{g}(c^{(\nu)})$ by

$$-\sum_{\lambda=1}^{\bar{\ell}} M_\lambda (s_\lambda^{(\nu+1)} - s_\lambda^m),$$

as in (4.2). The same applies for the non-equilibrium components, as can be seen by a resubstitution of (4.13b) into (4.13a). With similar arguments, we can see that Newton's method applied to (4.14) leads to (4.6), under the assumption that the rate function f for equilibrium components are of explicit form.

To discuss the results (A)–(D) with respect to (4.14), we need estimates of the type (3.18). The Jacobi matrix is of the structure

$$D\tilde{g}(\tilde{c}) = \begin{pmatrix} -(S+C) & -B_{\bar{\ell}+1} & \cdots & -B_{\ell} \\ \widetilde{A_{\bar{\ell}+1}} & \widetilde{B_{\bar{\ell}+1}} & & \\ \vdots & & \ddots & \\ \widetilde{A_{\ell}} & & & \widetilde{B_{\ell}} \end{pmatrix}, \tag{4.16}$$

where

$$\begin{aligned} \widetilde{A_\lambda} &:= \Delta t\, k_\lambda r_2^\lambda \frac{\partial}{\partial c} f(\lambda, .), \quad A_\lambda := M_\lambda \widetilde{A_\lambda}, \quad \lambda = \bar{\ell}+1, \ldots, l, \\ S &:= \sum_{\lambda=\bar{\ell}+1}^{\ell} A_\lambda, \\ \widetilde{B_\lambda} &:= \Delta t\, k_\lambda r_3^\lambda \frac{\partial}{\partial s} f(\lambda, .), \quad B_\lambda := M_\lambda \widetilde{B_\lambda}, \quad \lambda = \bar{\ell}+1, \ldots, l, \\ C &:= \sum_{\lambda=1}^{\ell} \frac{r_2^\lambda}{r_3^\lambda} M_\lambda \frac{\partial}{\partial c} \varphi(\lambda, .). \end{aligned}$$

The Jacobi matrices and therefore all matrices in this partition are diagonal and, because of (4.11), (4.12) we obtain as possible choices

$$\begin{aligned} P &:= \begin{pmatrix} 0 & r_3^{\bar{\ell}+1} E_{\bar{\ell}+1} & \cdots & r_3^{\ell} E_{\ell} \\ r_2^{\bar{\ell}+1} \widetilde{E_{\bar{\ell}+1}} & 0 & & 0 \\ \vdots & & \ddots & 0 \\ r_2^{\ell} \widetilde{E_{\ell}} & 0 & & 0 \end{pmatrix}, \\ Q &:= \begin{pmatrix} -(F+G) & & \\ & -r_3^{\bar{\ell}+1} \widetilde{E_{\bar{\ell}+1}} & \\ & & -r_3^{\ell} \widetilde{E_{\ell}} \end{pmatrix}, \end{aligned} \tag{4.17}$$

where

$$\widetilde{E_\lambda} := \Delta t\, k_\lambda K I, \quad E_\lambda := M_\lambda \widetilde{E_\lambda},$$

$$F := \sum_{\lambda=\bar{\ell}+1}^{\ell} r_2^\lambda E_\lambda,$$

$$G := \sum_{\lambda=1}^{\ell} \frac{r_2^\lambda}{r_3^\lambda} K M_\lambda.$$

As the discretization matrix A (of (3.3)) is a M-matrix, also $\widetilde{A}$ (of (4.15)) is a M-matrix. Also $\widetilde{A}-P$ has the property $(\widetilde{A}-P)_{ij} \leq 0$ for $i \neq j$, but in general it is not inverse monotone. This can only be concluded under more specific assumptions:

A sufficient condition for a matrix $A = (a_{ij})$ to be a M-matrix is given by (cf. e.g. [2]):

$$a_{ij} \leq 0 \quad \text{for} \quad i \neq j, \quad a_{ii} > 0, \quad A \text{ is irreducible,} \tag{4.18a}$$

$$\text{and} \quad A \text{ is diagonally dominant, i. e.} \sum_{j=1}^{n} a_{ij} \geq 0 \quad \text{for} \quad i = 1, \ldots, n. \tag{4.18b}$$

Let us assume that (4.18) is fulfilled by the matrix $K = K^{m+1}$ in (3.3). Then also $A = A^{m+1}$ of (3.3) satisfies these properties and is even strictly diagonally dominant, i. e.

$$\sum_{j=1}^{n} a_{ij} > 0 \quad \text{for} \quad i = 1, \ldots, n.$$

The same properties carry over to $\widetilde{A}$ of (4.15). $\widetilde{A}-P$ then fulfills (4.18a), and also (4.18b), if in addition $\Delta t\, k_\lambda$ is small enough for $\lambda = \bar{\ell}+1, \ldots, \ell$. Therefore, using (3.23), (B) and (C):

◇ a) *If (4.11), (4.12) are fulfilled, K from (2.21) satisfies (4.18) and $\Delta t\, k_\lambda$ is small enough for the non-equilibrium components, then the chord method based on fixed point iteration II, i. e.*

$$(A-R)c^{(\nu+1)} = \bar{b} - Rc^{(\nu)} - \sum_{\lambda=1}^{\ell} M_\lambda (s_\lambda^{(\nu+1)} - s_\lambda^m), \tag{4.19a}$$

(4.1) for $\lambda = 1, \dots, \bar{\ell}$ and

$$\frac{1}{\Delta t\, k_\lambda}(s_\lambda^{(\nu+1)} - s_\lambda^m) - R_\lambda(s_\lambda^{(\nu+1)} - s_\lambda^{(\nu)}) = \tilde{f}, \qquad (4.19b)$$

for $\lambda = \bar{\ell}+1, \dots, \ell$ converges for each $(c^{(0)}, s^{(0)})$ to the unique solution of (2.22), (2.24) provided that

$$R \leq -\frac{1}{2}(F+G), \quad R_\lambda \leq -\frac{1}{2} r_3^\lambda \widetilde{E_\lambda}, \quad \lambda = \bar{\ell}+1, \dots, \ell.$$

b) *Under the assumptions of a) and if all equilibrium components are of explicit form, then Newton's method converges quadratically, provided the initial defect is small enough.*

For the fixed point iteration II itself, (A) does not lead to any advantage compared with the use of Banach's fixed point theorem directly due to the requirement $P = -Q$:

◇ *If $\partial f/\partial c(\lambda, .)$ and $\partial f/\partial s(\lambda, .)$ are bounded and $\Delta t\, k_\lambda$ is small enough for the non-equilibrium components, and if $r_2^\lambda = 0$ for the equilibrium components, then the fixed point iteration II converges for each $(c^{(0)}, s_\lambda^{(0)})$ to the unique solution of (2.22), (2.24).*

None of the results above hold true independent of $\Delta t\, k_\lambda$. This will be different for the fixed point iteration I. We have to find an equivalent form of (2.22), (2.24), whose Picard iteration scheme is identical with (4.1), (4.2). Consider a $\lambda \in \{1, \dots, \ell\}$. Define for $c, s \in \mathbb{R}$

$$l(c,s) := s - \Delta t\, k_\lambda f(\lambda, P_i, t^m + r_3^\lambda \Delta t\,, r_2^\lambda c + \overline{r_2^\lambda} c_i^m, r_3^\lambda s + \overline{r_3^\lambda} s_{\lambda,i}{}^m). \qquad (4.20)$$

For fixed $c \in \mathbb{R}$ $l(c,.)$ is strictly increasing because of (4.12) and $l(c,-\infty) = -\infty$, $l(c,+\infty) = +\infty$. Therefore there exists a unique $s \in \mathbb{R}$ such that

$$l(c,s) = s_{\lambda,i}{}^m.$$

This defines a mapping $h_{\lambda,i} : \mathbb{R} \to \mathbb{R}$ by

$$l(c, h_{\lambda,i}(c)) = s_{\lambda,i}{}^m \tag{4.21}$$

and $h_\lambda : \mathbb{R}^n \mapsto \mathbb{R}^n$ by $h_\lambda((c_i)) = (h_{\lambda,i}(c_i))_i$. Therefore an equivalent formulation for (2.22), (2.24) is given by

$$\begin{aligned} Ac^{m+1} &= \bar{b} + \\ &+ \sum_{\lambda=1}^{\bar{\ell}} M_\lambda \left(\frac{1}{r_3^\lambda} \left(s_\lambda^m - \varphi(\lambda, ., t^m + r_3^\lambda \, \Delta t \, , r_2^\lambda c^{m+1} + \overline{r_2^\lambda} c^m) \right) \right) + \\ &\quad + \sum_{\lambda=\bar{\ell}+1}^{\ell} M_\lambda (s_\lambda^m - h_\lambda(c^{m+1})). \end{aligned} \tag{4.22}$$

The vector c^{m+1} remains to be the only unknown in (4.22). With arguments similar as above we see that the fixed point iteration (3.8) applied to (4.22) is identical with the fixed point iteration I. Furthermore, Newton's method applied to (4.22) gives rise to a new scheme. The linearization of the last term in (4.22) reads:

$$\sum_{\lambda=\bar{\ell}+1}^{\ell} M_\lambda \left(s_\lambda^m - h_\lambda(c^{(\nu)}) - \frac{\partial h_\lambda}{\partial c}(c^{(\nu)}) \, \Delta c^{(\nu+1)} \right).$$

From (4.21) we see

$$\frac{\partial h_{\lambda,i}}{\partial c}(c) = r_2^\lambda \, \Delta t \, k_\lambda \frac{\partial f}{\partial c}(\lambda, c, h_{\lambda,i}(c)) / \left(1 - r_3^\lambda \, \Delta t \, k_\lambda \frac{\partial f}{\partial s}(\lambda, c, h_{\lambda,i}(c)) \right), \tag{4.23}$$

where the argument (λ, c, s) is an abbreviation of $(\lambda, P_i, t^m + r_3^\lambda \Delta t, r_2^\lambda c + \overline{r_2^\lambda} c_i^m, r_3^\lambda s + \overline{r_3^\lambda} s_{\lambda,i}{}^m)$. We define

$$\widetilde{s_\lambda} := h_\lambda(c^{(\nu)}),$$
$$s_{\lambda,i}{}^{(\nu+1)} := \widetilde{s_{\lambda,i}} + \frac{\partial h_{\lambda,i}}{\partial c}(c_i^{(\nu)})\,\Delta c_i^{(\nu+1)}.$$

It follows from (4.23) that

$$\frac{1}{\Delta t\, k_\lambda}(s_{\lambda,i}{}^{(\nu+1)} - \widetilde{s_{\lambda,i}}) =$$
$$= r_2^\lambda \frac{\partial f}{\partial c}(\lambda, c_i^{(\nu)}, \widetilde{s_{\lambda,i}})\,\Delta c_i^{(\nu+1)} + r_3^\lambda \frac{\partial f}{\partial s}(\lambda, c_i^{(\nu)}, \widetilde{s_{\lambda,i}})(s_{\lambda,i}{}^{(\nu+1)} - \widetilde{s_{\lambda,i}})\,.$$

and thus by (4.21)

$$\begin{aligned}\frac{1}{\Delta t\, k_\lambda}&(s_{\lambda,i}{}^{(\nu+1)} - s_i^m) = \\ &= f(\lambda, c_i^{(\nu)}, \widetilde{s_{\lambda,i}}) + r_2^\lambda \frac{\partial f}{\partial c}(\lambda, c_i^{(\nu)}, \widetilde{s_{\lambda,i}})(c_i^{(\nu+1)} - c_i^{(\nu)}) + \\ &\quad + r_3^\lambda \frac{\partial f}{\partial s}(\lambda, c_i^{(\nu)}, \widetilde{s_{\lambda,i}})(s_{\lambda,i}{}^{(\nu+1)} - \widetilde{s_{\lambda,i}}).\end{aligned} \tag{4.24}$$

Assume that for $\lambda = 1, \dots, \overline{\ell}$ the rate functions are of explicit form. Then we can proceed as above with the equilibrium components to conclude that Newton's method applied to (4.22) is identical with the following *Newton–fixed point scheme*:

For given $c^{(\nu)}$ determine $\widetilde{s_\lambda}$ for $\lambda = \overline{\ell} + 1, \dots, \ell$ to fulfill

$$\frac{1}{\Delta t\, k_\lambda}(\widetilde{s_{\lambda,i}} - s_i^m) = f(\lambda, P_i, t^m + r_3^\lambda \Delta t, r_2^\lambda c_i^{(\nu)} + \overline{r_2^\lambda} c_i^m, r_3^\lambda \widetilde{s_{\lambda,i}} + \overline{r_3^\lambda} s_{\lambda,i}{}^m). \tag{4.25a}$$

Then $(c^{(\nu+1)}, s^{(\nu+1)})$ are determined by linearization at $(c^{(\nu)}, \widetilde{s})$, i. e.

$$Ac^{(\nu+1)} = \overline{b} - \sum_{\lambda=1}^{\ell} M_\lambda (s_\lambda^{(\nu+1)} - s_\lambda^m), \tag{4.25b}$$

for $\lambda = 1, \ldots, \overline{\ell}$

$$r_3^\lambda s_i^{(\nu+1)} = \varphi\left(\lambda, P_i, t^m + r_3^\lambda\,\Delta t\,, r_2^\lambda c_i^{(\nu)} + \overline{r_2^\lambda} c_i^m\right) + \\ + r_2^\lambda \frac{\partial \varphi}{\partial c}\left(\lambda, P_i, t^m + r_3^\lambda\,\Delta t\,, r_2^\lambda c_i^{(\nu)} + \overline{r_2^\lambda} c_i^m)(c_i^{(\nu+1)} - c_i^{(\nu)}\right), \tag{4.25c}$$

and for $\lambda = \overline{\ell} + 1, \ldots, \ell$

$$\frac{1}{\Delta t\, k_\lambda}(s_{\lambda,i}{}^{(\nu+1)} - s_i^m) = \\ = f(\lambda, P_i, t^m + r_3^\lambda\,\Delta t\,, r_2^\lambda c_i^{(\nu)} + \overline{r_2^\lambda} c_i^m, r_3^\lambda \widetilde{s_{\lambda,i}} + \overline{r_3^\lambda} s_{\lambda,i}{}^m) + \\ + r_2^\lambda \frac{\partial f}{\partial c}(\lambda, P_i, t^m + r_3^\lambda\,\Delta t\,, r_2^\lambda c_i^{(\nu)} + \overline{r_2^\lambda} c_i^m, r_3^\lambda \widetilde{s_{\lambda,i}} + \overline{r_3^\lambda} s_{\lambda,i}{}^m)(c_i^{(\nu+1)} - c_i^{(\nu)}) + \\ + r_3^\lambda \frac{\partial f}{\partial s}(\lambda, P_i, t^m + r_3^\lambda\,\Delta t\,, r_2^\lambda c_i^{(\nu)} + \overline{r_2^\lambda} c_i^m, r_3^\lambda \widetilde{s_{\lambda,i}} + \overline{r_3^\lambda} s_{\lambda,i}{}^m)(s_{\lambda,i}{}^{(\nu+1)} - \widetilde{s_{\lambda,i}}). \tag{4.25d}$$

In (4.25a) again the nonlinear scalar equations of the fixed point iteration appear. The linear problem actually has only the unknown $c^{(\nu+1)}$, as can be seen from the original formulation or from a reduction as from (4.6) to (4.7) or (4.8b). The only distinction is that in the evaluation abbreviated with $\tilde{\cdot}$ $s_\lambda^{(\nu)}$ has to be substituted by $\widetilde{s_\lambda}$ for $\lambda = \overline{\ell} + 1, \ldots, l$.

The nonlinearity in (4.22) has the same properties as in (3.1a) for the situation where all nonlinearities are of explicit type. Therefore we can conclude from (B) and (C):

◇ a) *If A from (3.3) is a M-matrix and (4.11), (4.12) are fulfilled, then the chord method based on the fixed point iteration I, i. e. (4.1) and*

$$(A - R)c^{(\nu+1)} = \overline{b} - Rc^{(\nu)} - \sum_{\lambda=1}^{\ell} M_\lambda(s_\lambda^{(\nu+1)} - s_\lambda^m) \tag{4.26}$$

converges for each $c^{(\nu)}$ *to the unique solution of (2.22), (2.24), provided*

$$R \leq -\frac{1}{2}K\Big(\sum_{\lambda=1}^{\bar{\ell}} \frac{r_2^\lambda}{r_3^\lambda} M_\lambda + \sum_{\lambda=\bar{\ell}+1}^{\ell} r_2^\lambda \, \Delta t \, k_\lambda M_\lambda\Big). \tag{4.27}$$

b) *Under the assumptions of* a) *and if all equilibrium components are of explicit form, then the Newton-fixed point scheme converges quadratically, provided the initial defect is small enough.*

If in addition to (4.12) a negative upper bound of $\partial f/\partial s$ is known, i. e.

$$\frac{\partial}{\partial s} f(\lambda, \underline{x}, t, c, s) \leq -L$$

for all $c, s \in \mathrm{IR}$, $(\underline{x}, t) \in \overline{Q}_T$ and $\lambda \in \{1, \ldots, \ell\}$ such that $k_\lambda < \infty$, then the bound in (4.27) can be substituted by

$$R \leq -\frac{1}{2}K\Big(\sum_{\lambda=1}^{\bar{\ell}} \frac{r_2^\lambda}{r_3^\lambda} M_\lambda + \sum_{\lambda=\bar{\ell}+1}^{\ell} \frac{r_2^\lambda}{1/(\Delta t \, k_\lambda) + r_3^\lambda L} M_\lambda\Big).$$

This requirement can be fulfilled uniformly for large k_λ.

For the fixed point iteration I itself, again only a convergence result is available for sufficiently small $\Delta t \, k_\lambda$ or $r_2^\lambda = 0$.

We continue with some remarks concerning the choice of the implicity factors r_2^λ, r_3^λ. Consider the *implicit-explicit case*:

- $r_2^\lambda = 0$, $r_3^\lambda \in [0, 1]$ *for* $\lambda = 1, \ldots, \ell$.

If all nonlinearities are of explicit type, then the discrete problem (3.1) for c^{m+1} is linear, as mentioned in chapter 3, and thus fixed point iteration (3.8) and Newton's method (3.12) are identical and converge in one iteration. For general nonlinearities the matrix in (4.8b) is identical with A for each iteration such that the computational complexity of Newton's method in the implicit-explicit case is roughly the same as the complexity of the fixed point iteration (3.8). The fixed point iteration I ((4.1), (4.2)) and the fixed point iteration II ((4.1), (4.2), (4.4)), if some of the equilibrium components are not of explicit type, still contain the problem to solve scalar nonlinear equations in each iteration.

In general there are the following *variants of Newton's method*: In the decoupled form (4.7), (4.8a) linear expansions appear: In (4.7)

$$\tilde{f} + \widetilde{\frac{\partial f}{\partial s}} r_3^\lambda ({s_{\lambda,i}}^m - {s_{\lambda,i}}^{(\nu)})$$

is the expansion of

$$f(\lambda, P_i, t^m + r_3^\lambda\,\Delta t\,, r_2^\lambda c_i^{(\nu)} + \overline{r_2^\lambda} c_i^m, {s_{\lambda,i}}^m) =: \hat{f}. \tag{4.28}$$

Similarly

$$\tilde{f} + \widetilde{\frac{\partial f}{\partial c}} r_2^\lambda (c_i^{(\nu+1)} - c_i^{(\nu)})$$

is the expansion of

$$f(\lambda, P_i, t^m + r_3^\lambda\,\Delta t\,, r_2^\lambda c_i^{(\nu+1)} + \overline{r_2^\lambda} c_i^m, r_3^\lambda {s_{\lambda,i}}^{(\nu)} + \overline{r_3^\lambda} {s_{\lambda,i}}^m) =: \overline{f}. \tag{4.29}$$

Therefore the following iteration, called *modified Newton's method*, seems to be reasonable: Given $c^{(\nu)}$, $s_\lambda^{(\nu)}$, $\lambda = 1, \ldots, \ell$, determine $c^{(\nu+1)}$, $s_\lambda^{(\nu+1)}$ as the solutions of the linear equation

$$\begin{aligned} Ac^{(\nu+1)} = \overline{b} - \\ - \sum_{\lambda=1}^{\ell} M_\lambda \left(\frac{1}{1/(\Delta t\, k_\lambda) - r_3^\lambda \widetilde{\partial f/\partial s}} \left(\hat{f} + \widetilde{\frac{\partial f}{\partial c}}\, r_2^\lambda (c_i^{(\nu+1)} - c_i^{(\nu)}) \right) \right)_i \end{aligned} \tag{4.30}$$

and of the update formulas

$$s_{\lambda,i}{}^{(\nu+1)} = \frac{1}{1/(\Delta t\, k_\lambda) - r_3^\lambda \widetilde{\partial f/\partial s}} \left(\widetilde{f} - \frac{\widetilde{\partial f}}{\partial s} r_3^\lambda s_{\lambda,i}{}^{(\nu)} + \frac{1}{\Delta t\, k_\lambda} s_{\lambda,i}{}^m \right), \qquad (4.31)$$

$$\lambda = 1, \dots, \ell, i = 1, \dots, n.$$

A similar modification is possible for the Newton-fixed point scheme, written in the decoupled form analogously to (4.7), (4.8a) (supplemented by (4.25a)).

We return to the implicit-explicit discretization ($r_2^\lambda = 0$ for $\lambda = 1, \dots, \ell$). Analogously to (3.24) the first iterates $c^{(1)}$, $s_\lambda^{(1)}$, $\lambda = 1, \dots, \ell$, for the modified Newton's method, being started by (4.5), are nearly identical with the solution of the fully explicit version $r_2^\lambda = r_3^\lambda = 0$ for the following reduced "rate parameters":

$$\begin{aligned} k_\lambda < \infty: \quad & \widetilde{k_\lambda} = k_\lambda \frac{1}{1 - k_\lambda r_3^\lambda\, \Delta t\, \widetilde{\partial f/\partial s}}, \\ k_\lambda = \infty: \quad & \widetilde{k_\lambda} = \frac{1}{-r_3^\lambda\, \Delta t\, \widetilde{\partial f/\partial s}} \quad (< \infty), \end{aligned} \qquad (4.32)$$

where

$$\frac{\widetilde{\partial f}}{\partial s} = \frac{\partial f}{\partial s}(\lambda, P_i, t^m + r_3^\lambda\, \Delta t\,, c_i^m, s_{\lambda,i}{}^m).$$

Again the only distinction lies in the instance of time, at which f is evaluated.

Finally we discuss the relation between the fixed point iteration I and Newton's method. The scalar nonlinear equations (4.1) also have to be solved approximately by an iteration scheme. We discuss the application of Newton's method for these equations. Let $\nu \geq 0$ be fixed. We denote the iterates by $s_\lambda^{[j]} := s_\lambda^{[\nu,j]}$ and start the iteration with

$$s_\lambda^{[0]} := s_\lambda^{(\nu)}, \quad \lambda = 1, \dots, \ell.$$

Then

$$\frac{1}{\Delta t\, k_\lambda}(s_{\lambda,i}^{[j+1]} - s_{\lambda,i}{}^m) = \widetilde{f} + \widetilde{\frac{\partial f}{\partial s}} r_3^\lambda (s_{\lambda,i}^{[j+1]} - s_{\lambda,i}^{[j]}), \qquad \lambda = 1, \ldots, \ell, \quad i = 1, \ldots, n, \tag{4.33}$$

where here $\widetilde{\cdot}$ is the abbreviation for evaluation at

$$(\lambda, P_i, t^m + r_3^\lambda\, \Delta t\,, r_2^\lambda c_i^{(\nu)} + \overline{r_2^\lambda} c_i^m, r_3^\lambda s_{\lambda,i}^{[j]} + \overline{r_3^\lambda} s_{\lambda,i}{}^m). \tag{4.34}$$

If we do only *one* Newton step in (4.33) and set

$$s_\lambda^{(\nu+1)} := s_\lambda^{[1]}$$

we get the following iteration scheme for $c^{(\nu)}$, $s_\lambda^{(\nu)}$:

$$\frac{1}{\Delta t\, k_\lambda}(s_{\lambda,i}{}^{(\nu+1)} - s_{\lambda,i}{}^m) = \widetilde{f} + \widetilde{\frac{\partial f}{\partial s}} r_3^\lambda (s_{\lambda,i}{}^{(\nu+1)} - s_{\lambda,i}{}^{(\nu)}), \quad i = 1, \ldots, n, \tag{4.35a}$$

$$Ac^{(\nu+1)} = \overline{b} - \sum_{\lambda=1}^{\ell} M_\lambda (s_\lambda^{(\nu+1)} - s_\lambda^m), \quad \lambda = 1, \ldots, \ell, \tag{4.35b}$$

where $\widetilde{\ }$ here again abbreviates the evaluation at

$$(\lambda, P_i, t^m + r_3^\lambda\, \Delta t\,, r_2^\lambda c_i^{(\nu)} + \overline{r_2^\lambda} c_i^m, r_3^\lambda s_{\lambda,i}{}^{(\nu)} + \overline{r_3^\lambda} s_{\lambda,i}{}^m). \tag{4.36}$$

This iteration scheme is identical with Newton's method for the implicit-explicit case ($r_2^\lambda = 0$ for $\lambda = 1, \ldots, \ell$) up to the different evaluation of f and $\partial f / \partial s$. Therefore there is also a decoupled form analogous to (4.8a) and (4.8b) or (4.7) with $r_2^\lambda = 0$, but $\widetilde{\ }$ interpreted by (4.36). The remarks above about the implicit-explicit case also apply here. Again there is a *modified version* given by (4.30) and (4.8a) with $r_2^\lambda = 0$, but $\widetilde{\ }$ defined by (4.36). For this one the same relation to the fully explicit version $r_2^\lambda = r_3^\lambda = 0$ exists as indicated in (4.32).

5 Adsorption and physical non-equilibrium

In the model (2.1a, b) considered so far, non-equilibrium effects are due to slow (chemical) reactions. Often such effects are observed, although the reactions are believed to be in equilibrium. A popular model attributes this fact to the existence of adsorption sites, which are situated in aggregates and thus only accessible by diffusion. If the concentrations in the aggregate are assumed to be microscopically homogeneous and the diffusion rate into the aggregates is proportional to the difference in concentrations dissolved, we reach the following model (cf. e.g. [5]):

$$\frac{\partial}{\partial t}(\Theta_1 c_1 + \Theta_2 c_2)(\underline{x},t) + \varrho_1(\underline{x})\frac{\partial}{\partial t}s_1(\underline{x},t) + \varrho_2(\underline{x})\frac{\partial}{\partial t}s_2(\underline{x},t) - (Lc_1)(\underline{x},t) = 0, \tag{5.1a}$$

$$\frac{\partial}{\partial t}(\Theta_2 c_2 + \varrho_2 s_2)(\underline{x},t) = \alpha(c_1(\underline{x},t) - c_2(\underline{x},t)), \tag{5.1b}$$

$$\frac{\partial}{\partial t}s_\lambda(\underline{x},t) = k_\lambda f(\lambda,\underline{x},t,c_\lambda(\underline{x},t),s_\lambda(\underline{x},t)), \quad \lambda = 1,2, \quad (\underline{x},t)\in Q_T. \tag{5.1c}$$

Here the index 1 refers to the *inter-aggregate*, i. e. the *mobile phase*, and index 2 to the *intra-aggregate*, i. e. *immobile phase*. $\alpha > 0$ is the rate parameter describing the exchange between the phases, $k_\lambda > 0$ are the rate parameters for the chemical reactions, where we allow for $k_\lambda = \infty$ with the same interpretation as in chapter 2. The following considerations can be easily extended to the case of multiple inter- and intra-aggregate adsorption sites, where $s_i(\underline{x},t)$ has to be substituted by $⨍_{\Lambda_i} s(\lambda,\underline{x},t)\,d\lambda$ in (5.1a) and (5.1b) and λ in (5.1c) now ranges in $\Lambda := \Lambda_1 \cup \Lambda_2$.

If $k_2 = \infty$, then (5.1a–c) can be reduced to a problem of the form (2.1a, b). Set

$$\beta(\underline{x},t,c) := \Theta_2(\underline{x},t)c + \varrho_2(\underline{x})\varphi_2(\underline{x},t,c) \quad \text{for} \quad (\underline{x},t)\in Q_T,\ c \geq 0, \tag{5.2}$$

where φ_2 is the adsorption isotherm for f_2 (compare (2.4)). We assume that for each fixed $(\underline{x}, t) \in Q_T$ $\varphi_2(\underline{x}, t, .)$ is increasing and $\varphi_2(\underline{x}, t, 0) = 0$. Then $\beta(\underline{x}, t, .)$ is an bijective mapping from $c \geq 0$ to $w \geq 0$. Let $\gamma(\underline{x}, t, w)$ denote the inverse function. Then in the new variables c_1, s_1 and

$$w(\underline{x}, t) := \beta(\underline{x}, t, c_2(\underline{x}, t)) \tag{5.3}$$

The equations (5.1a–c) transform to

$$\begin{aligned}
&\frac{\partial}{\partial t}(\Theta_1 c_1)(\underline{x}, t) + \varrho_1(\underline{x})\frac{\partial}{\partial t} s_1(\underline{x}, t) + \frac{\partial}{\partial t} w(\underline{x}, t) - (Lc_1)(\underline{x}, t) = 0, \\
&\frac{\partial}{\partial t} s_1(\underline{x}, t) = k_1 f(1, t, c_1(\underline{x}, t), s_1(\underline{x}, t)), \\
&\frac{\partial}{\partial t} w(\underline{x}, t) = \alpha(c_1(\underline{x}, t) - \gamma(\underline{x}, t, w(\underline{x}, t)).
\end{aligned} \tag{5.4}$$

This is a model of the form (2.1a, b) with $\Lambda = \{1, 2\}$. All considerations of the preceeding chapters can be applied to (5.4).

Alternatively we will discuss the discretization of (5.1a–c) directly, allowing also for finite k_2. For the sake of simplicity we will only consider the discretization in time. From chapter 2 it is clear how to perform the discretization in space. A scheme analogous to (2.9a, b) reads:

$$\begin{aligned}
&\frac{\Theta_1^{m+1} c_1^{m+1} - \Theta_1^m c_1^m}{\Delta t} + \frac{\Theta_2^{m+1} c_2^{m+1} - \Theta_2^m c_2^m}{\Delta t} + \\
&+ \varrho_1 \frac{s_1^{m+1} - s_1^m}{\Delta t} + \varrho_2 \frac{s_2^{m+1} - s_2^m}{\Delta t} - r_1 L^{m+1} c_1^{m+1} - \overline{r_1} L^m c_1^m = 0,
\end{aligned} \tag{5.5a}$$

$$\begin{aligned}
&\frac{\Theta_2^{m+1} c_2^{m+1} - \Theta_2^m c_2^m}{\Delta t} + \varrho_2 \frac{s_2^{m+1} - s_2^m}{\Delta t} = \\
&= \alpha(r_4 c_1^{m+1} + \overline{r_4} c_1^m - r_5 c_2^{m+1} - \overline{r_5} c_2^m),
\end{aligned} \tag{5.5b}$$

$$\frac{1}{k_\lambda}\frac{s_\lambda^{m+1}-s_\lambda^m}{\Delta t}=$$
$$= f(\lambda,.,t^m + r_3^\lambda\,\Delta t\,, r_2^\lambda c_\lambda^{m+1} + \overline{r_2^\lambda} c_\lambda^m\,, r_3^\lambda s_\lambda^{m+1} - \overline{r_3^\lambda} s_\lambda^m),$$
$$\lambda = 1,2, \tag{5.5c}$$

where r_1, r_2^λ, r_3^λ, r_4, $r_5 \in [0,1]$ are implicity factors and $\overline{r_1}$ etc. have to be understood according to (2.11). The initial and boundary conditions are analogous to (2.9c), (2.10) and will not be considered explicitly.

The equation (5.5b) can be resolved for c_2^{m+1}:

$$c_2^{m+1} = \delta[(\Theta_2^m - \alpha\,\Delta t\,\overline{r_5})c_2^m - \varrho_2(s_2^{m+1} - s_2^m) + \alpha\,\Delta t\,(r_4 c_1^{m+1} + \overline{r_4} c_1^m)]. \tag{5.6}$$

Substitution into (5.5a) leads to

$$\frac{\Theta_1^{m+1}c_1^{m+1} - \Theta_1^m c_1^m}{\Delta t} - r_1 L^{m+1}c_1^{m+1} - \overline{r_1}L^m c_1^m = -\varrho_1\frac{s_1^{m+1}-s_1^m}{\Delta t} -$$
$$-\widetilde{\varrho_2}\frac{s_2^{m+1}-s_2^m}{\Delta t} + \alpha(\Theta_2^m r_5 + \Theta_2^{m+1}\overline{r_5})\delta c_2^m - \widetilde{\alpha}(r_4 c_1^{m+1} + \overline{r_4}c_1^m), \tag{5.7}$$

where

$$\delta := \frac{1}{\Theta_2^{m+1} + \alpha\,\Delta t\,r_5},$$
$$\widetilde{\varrho_2} := \varrho_2(1 - \Theta_2^{m+1}\delta), \quad \widetilde{\alpha} := \alpha\Theta_2^{m+1}\delta.$$

We have $\widetilde{\varrho_2} < \varrho_2$ and $\widetilde{\varrho_2} > 0$ for $r_5 > 0$, $\widetilde{\varrho_2} = 0$ for $r_5 = 0$. Analogously $\widetilde{\alpha} > 0$, and $\widetilde{\alpha} < \alpha$ for $r_5 > 0$, $\widetilde{\alpha} = \alpha$ for $r_5 = 0$. The factor in front of c_2^m coincides with $\widetilde{\alpha}$ for Θ_2 constant in time. Therefore on the time-discrete level the diffusion into aggregates can be expressed equivalently by a reduction of the part of the bulk density corresponding to the intra-aggregate sites, a linear source in terms of c_2 from the old time level and a linear sink in terms of c_1. These sources and sinks are now compared with (2.9). Their spatial discretization leads to the following additional term in the definition of A^{m+1} in (3.3)

$$r_4 \alpha \, \Delta t \int_\Omega I(\Theta_2^{m+1} \delta \varphi_j \varphi_i) \, d\underline{x}$$

and corresponding terms in $\overline{b}^{m+1}$. I.e., we apply mass lumping, as the terms under consideration stem from the spatially immobile exchange reaction. These modifications of A^{m+1} and $\overline{b}^{m+1}$ do not conflict with any of the preceeding schemes and the results developed.

In (5.7) we have already eliminated c_2^{m+1}. We have to do the same with (5.5c) to show that (5.5a–c) transforms to a problem of the form (2.9) for c_1^{m+1}, s_1^{m+1}, s_2^{m+1} up to an additional linear sink in (2.9a). Therefore we substitute c_2^{m+1} from (5.6) into (5.5c) for $\lambda = 2$. This leads to

$$\begin{aligned} \frac{1}{k_2} \frac{s_2^{m+1} - s_2^m}{\Delta t} &= f(2, ., t^m + r_3^2 \, \Delta t, \\ &\quad (1 - \overline{\varepsilon} r_2^2) c_2^m + \alpha \, \Delta t \, r_2^2 \delta (r_4 c_1^{m+1} + \overline{r_4} c_1^m) - \varrho_2 r_2^2 \delta (s_2^{m+1} - s_2^m), \\ &\quad r_3^2 s_2^{m+1} + \overline{r_3^2} s_2^m), \end{aligned} \tag{5.8}$$

where

$$\varepsilon := \frac{\Theta_2^m - \alpha \, \Delta t \, \overline{r_5}}{\Theta_2^{m+1} + \alpha \, \Delta t \, r_5}, \qquad \overline{\varepsilon} := 1 - \varepsilon.$$

We have $\varepsilon < 1$ and also $\varepsilon > 0$ for small Δt.

The time-discrete problem for c_1^{m+1}, s_1^m, s_2^{m+1} consists of (5.7), (5.5c) for $\lambda = 1$, and (5.8). After its solution c_2^{m+1} is defined by means of (5.6). Surprisingly the coupling of (5.8) on c_1^{m+1} in (5.8) is weak, as the factor in front of c_1^{m+1} vanishes for $\Delta t \to 0$.

Note that, if $f(2,.)$ is of explicit type in c_2 and s_2, in general it is no longer of explicit type in c_1 and s_2 in (5.8). If $f(2,.)$ is linear, i. e.

$$f(2, \underline{x}, t, c, s) = ac - bc$$

with constants a, b, (5.8) simplifies to

$$\frac{1}{k_2} \frac{s_2^{m+1} - s_2^m}{\Delta t} = a[(1 - \overline{\varepsilon} r_2^2) c_2^m + \alpha \, \Delta t \, r_2^2 \delta (r_4 c_1^{m+1} + \overline{r_4} c_1^m)] - b[\widetilde{r_3^2} s_2^{m+1} + \overline{\widetilde{r_3^2}} s_2^m],$$

where

$$\widetilde{r_3^2} = r_3^2 + \frac{a}{b} \varrho_2 r_2^2 \delta.$$

6 Implementation and numerical case studies

In this final chapter we add some computational details and discuss numerical experiments. We are only interested in non-negative solutions. Usually this property requires smallness of the time step size Δt in relation to the spatial discretization parameter h (compare chapter 2). Therefore it is desirable to control the non-negativity of the solution by an *appropriate step size control* in connection with *damping* in the iteration. I.e., (3.14) is substituted by

$$c^{(\nu+1)} := c^{(\nu)} + \omega\,\Delta c^{(\nu+1)} \tag{6.1}$$

The same applies for the iteration schemes, defined in chapters 4 and 5. The iterates $c^{(\nu+1)}$, $s_\lambda^{(\nu+1)}$ are considered as intermediate values, denoted now by $\widetilde{c^{(\nu+1)}}$, $\widetilde{s_\lambda^{(\nu+1)}}$. Then

$$\begin{aligned}
\Delta c^{(\nu+1)} &:= \widetilde{c^{(\nu+1)}} - c^{(\nu)}, \\
\Delta s_\lambda^{(\nu+1)} &:= \widetilde{s_\lambda^{(\nu+1)}} - s_\lambda^{(\nu)}, \\
c^{(\nu+1)} &:= c^{(\nu)} + \omega\,\Delta c^{(\nu+1)}, \\
s_\lambda^{(\nu+1)} &:= s_\lambda^{(\nu)} + \omega\,\Delta s_\lambda^{(\nu+1)}.
\end{aligned}$$

The parameter ω is chosen as the maximal positive number not exceeding 1, such that all components of $c^{(\nu+1)}$ (and $s_\lambda^{(\nu+1)}$) are non-negative. If this is not possible, the iteration is considered to be "divergent" and it is stopped. Another case of "divergence" is a defect not small enough after a maximal number of iterations. In the case of "divergence", the time step size Δt is reduced, e.g. to $\Delta t/2$, and the time step is repeated. Another reason to repeat a (convergent) time step is a too large variation in the solution, i. e. for schemes in chapter 3 if

$$\|c^{(\nu+1)} - c^m\| \geq 2\,dc_{\text{opt}},$$

dc_{opt} being a control parameter. If a time step is accepted, a new time step size is adapted to $\|c^{(\nu+1)} - c^m\|$ and dc_{opt}. Time steps have to range in an interval $[\Delta t_{\min}, \Delta t_{\max}]$. If no step size control is possible the time stepping is stopped.

If f is not continuously differentiable in c and s, Newton-type schemes cannot be applied directly. An example is an f of explicit type, defined by the Freundlich isotherm (2.28). Usually there are only a few isolated points c or s, at which f is not smooth. Then it is easy to substitute f by a smooth approximation f_ε. If f is of explicit type and $\varphi(\lambda, \underline{x}, t, .)$ is smooth for $c > 0$, then a possible choice is given by:

$$
\begin{aligned}
f_\varepsilon(\lambda, \underline{x}, t, c, s) &:= k_\lambda(\varphi_\varepsilon(\lambda, \underline{x}, t, c) - s), \\
\varphi_\varepsilon(\lambda, \underline{x}, t, c) &:= \begin{cases} \varphi(\lambda, \underline{x}, t, c) & \text{for } c \geq \varepsilon, \\ (\varphi(\lambda, \underline{x}, t, \varepsilon)/\varepsilon\, c & \text{for } 0 \leq c < \varepsilon. \end{cases}
\end{aligned}
\tag{6.2}
$$

The schemes of chapter 3 are realized in the programs

SO(lute)TRA(nsport)—SL(ow)AD(sorption)—1(space dimension)
and —2(space dimension).

Numerical experiments in one dimension are already discussed in [7]. We consider here the following two-dimensional example.

We want to concentrate on the numerical study of the (non-equilibrium) adsorption phenomenon, therefore we take geometry and flow control as simple as possible:

$$
\begin{aligned}
\Omega &= (0, L_1) \times (0, L_2) \quad \text{(cf. Figure 1)}, \\
\underline{q} &= (q_1, q_2) = \text{ constant with } q_2 = 0,
\end{aligned}
\tag{6.3}
$$

i. e. there is water inflow at $x_1 = 0$, constant and uniform in x_2, outflow at $x_1 = L_1$ and no flow for $x_2 = 0$ or $x_2 = L_2$. Thus we take (cf. (1.5), (1.6)) $S_1 := 0$ (no Dirichlet conditions)

$$\begin{aligned} S_2 &:= \{(x_1, x_2) \mid x_1 = 0 \quad \text{and} \quad x_2 \in [0, L_2]\}, \\ S_3 &:= \{(x_1, x_2) \mid x_1 = L_1 \quad \text{and} \quad x_2 \in [0, L_2] \quad \text{or} \\ &\quad x_1 \in (0, L_1) \quad \text{and} \quad x_2 \in \{0, L_2\}\}. \end{aligned} \tag{6.4}$$

We confine to constant Θ and ϱ and take the diffusion/dispersion matrix to be of the (usual) form

$$\Theta D = \Theta \tau D_0 \begin{pmatrix} 1 & 0 \\ 0 & 1 \end{pmatrix} + q_1 \begin{pmatrix} \lambda_L & 0 \\ 0 & \lambda_T \end{pmatrix}. \tag{6.5}$$

with constant τD_0, the diffusion coefficient reduced by toruosity, and constant longitudinal and transversal dispersion lengths λ_L and λ_T. The mass inflow in normal direction at S_2 is given by

$$F(x_2) := q_1 \begin{cases} c_f & \text{for } x_2 \in [H_1, H_2], \\ 0 & \text{otherwise.} \end{cases} \tag{6.6a}$$

where $0 < H_1 < H_2 < L_1$ and c_f is a positive constant, i. e. at $x_1 = 0$, $x_2 \in [H_1, H_2]$ a "plume" is situated. In the last computations we do not consider this *continuous feed*, but rather a *pulse*, i. e.

$$F(x_2, t) = \begin{cases} \text{as defined by (6.6a)} & \text{for } t \in [0, T_1], \\ 0 & \text{for } t > T_1. \end{cases} \tag{6.6b}$$

We always start with a "clean" medium:

$$c_0(\underline{x}) = s_0(\underline{x}) = 0 \quad \text{for} \quad \underline{x} \in \Omega. \tag{6.6c}$$

The adsorption we take to be homogeneous with respect to the adsorption sites, i. e. $\Lambda = \{1\}$, and we omit the corresponding index. We assume a process in non-equilibrium with rate parameter k, described by a rate function of explicit type (cf. (2.4)), with an adsorption isotherm of Freundlich form, where we allow for spatial dependence of the coefficient, i. e.

$$\varphi(\underline{x}, c) := A(\underline{x})c^p \quad \text{with} \quad A(\underline{x}) > 0,\ 0 < p < 1. \tag{6.7}$$

The numerical parameters for the following computations are

$$\begin{aligned}
L_1 &= 5\ [\mathrm{cm}], \quad L_2 = 10\ [\mathrm{cm}] \\
q_1 &= 2.5\ [\mathrm{cm/day}] \\
\Theta &= 0.5 \quad [-], \quad \varrho = 1.5\ [\mathrm{g/cm^3}] \\
\tau D_0 &= 0.025\ [\mathrm{cm^2/day}] \\
\lambda_L &= 0.595\ [\mathrm{cm}], \quad \lambda_T = 0.0595\ [\mathrm{cm}] \\
c_f &= 1\ [100\mu\mathrm{g/cm^3}] \\
H_1 &= 1.875\ [\mathrm{cm}], \quad H_2 = 3.125\ [\mathrm{cm}] \\
k &= 0.5, \\
p &= 0.5,
\end{aligned} \tag{6.8}$$

and A will be specified later on. Thus we have taken the dimensions

$$\begin{aligned}
c &: \quad [100\mu\mathrm{g/cm^3}] \\
s &: \quad [100\mu\mathrm{g/g}].
\end{aligned}$$

The values in (6.8) correspond fairly well to experimental data for contaminants of high water solubility.

The parameters in the numerical procedure are

$$\begin{aligned}
r_1 = r_2 = r_3 &= 1 \quad (\text{in } (2.9)), \\
\varepsilon &= 10^{-12} \quad (\text{in } (6.2)),
\end{aligned} \tag{6.9}$$

and we apply Newton's method ((3.12)). The stopping criterion is (cf. (3.15))

$$\|d^{(\nu)}\|_\infty \leq 2 \cdot 10^{-5}.$$

The parameter dc_{opt} in the step size control is chosen so large that it is not active. Time stepping starts with a rather small $\Delta t_{\min}$: The maximal time step $\Delta t_{\max}$, which eventually corresponds to the time step chosen, is such that the CFL criterion is satisfied in x_1-direction. For the first computations with homogeneous adsorption, i. e. constant A, we use the triangulation displayed in Figure 1 a. It has 625 nodes and a maximal edge length $\Delta x = 0.325$ such that the Peclet criterion is satisfied:

$$\frac{q_1 \, \Delta x}{\Theta_T D_0 + \lambda_L q_1} \leq 2. \tag{6.10}$$

For comparison we consider also the case of no adsorption ($A = 0$), the linear diffusion-convection equation. In Figure 2 and 3 the mass flow field $-\Theta D \nabla c + qc$ and lines of constant concentration c are shown. The vector fields are scaled from Figure to Figure, such that the largest "arrow" has the same length. The values of concentration correspond to a logarithmic scale: from level line to level line the value is doubled, the smallest value is $c = 10^{-3}$. In Figure 2 a we see the infinite speed of propagation, leading to a positive concentration everywhere in the domain also for small times. The direction of the flow field and the ellipsoidal shape of the level lines show the dominance of convection. The concentration profile is monotone in the sense that the level lines in the edges of the domain correspond to $c = 10^{-3}$ and the values increase exponentially up to $c = 0.512$ for the innermost level line. The continuous problem has a unique asymptotic steady state for $t \to \infty$. The same applies to the discrete problem. This steady state is reached very quickly. From $t = 3$ on (cf. Figure 2 b) there is hardly any change.

In the Figures 3 and 4 we consider homogeneous adsorption, where

$$A(\underline{x}) = 1 \quad \text{for} \quad \underline{x} \in \Omega.$$

Now the speed of propagation is finite (cf. [8], [4]). For small times this leads to a concentration front beyond which there is no mass. The adsorbed concentration has the same property and is displayed also in a logarithmic scale, where the smallest value is $s = 3.3 \cdot 10^{-4}$. The concentration profiles are again monotone. The largest concentration displayed as a level line is $c = 0.512$, $s = 0.341$ in Figure 3 and $s = 0.682$ in Figure 4. Obviously due to adsorption the transport process is delayed, nevertheless the continuous problem has the same unique steady state as the problem without adsorption. This can be seen formally by setting the time derivatives to zero in (2.1) and can also be made rigorously mathematically (cf. [7]). This means that for "$t = \infty$" the presence of adsorption cannot be seen any more in the dissolved phase (but of course in the adsorbed phase). The distinction lies in the speed, with which this steady state is approached, i. e. in the amount of time, which has to elapse till the concentration profiles are close to the steady state. Comparison of Figure 4 and 2 b shows that this is not yet the case for $t = 20$ in this example.

Finally we consider spatially inhomogeneous adsorption by setting

$$A(\underline{x}) = \begin{cases} 10 & \text{for } x_1 \in [4,6],\ x_2 \in [0, L_2], \\ 1 & \text{otherwise,} \end{cases}$$

which may be visualized as a layer of high adsorption in the middle of the domain. Here the computations are based on the triangulation in Figure 1 b, containing 1105 nodes, for which the maximal edge length is the same as above. Figures 5 to 9 show the time development for this case. The layer of strong adsorption acts as an obstacle for the mass flow. Correspondingly the maximum of adsorbed concentration is attained there and steep gradients appear at the boundaries $x_1 = 4$ (and $x_2 = 6$) between the different layers. The maximal values of s displayed as level lines are $s = 0.341, 2.731, 5.461, 5.461, 5.461$ for this sequence of times. As the asymptotic steady state for $t \to \infty$ is the same as without adsorption, there must be eventually a "breakthrough" as in Figure 7. The approach to the steady state, however, is still more delayed than above: For $t = 40$ (Figure 9) the steady state is not yet reached. In Figures 10 to 12, the last scenario is considered again, but now for a pulse ((6.6b)) with $T_1 = 8$ instead of continuous feed. Thus the time evolution up to $t = 8$ is identical to Figures 5 and 6 and Figures 10 to 12 show the redistribution after the cut-off of the mass inflow. The layer of large adsorbed concentration now acts as a source for the dissolved concentration in form of a desorption process.

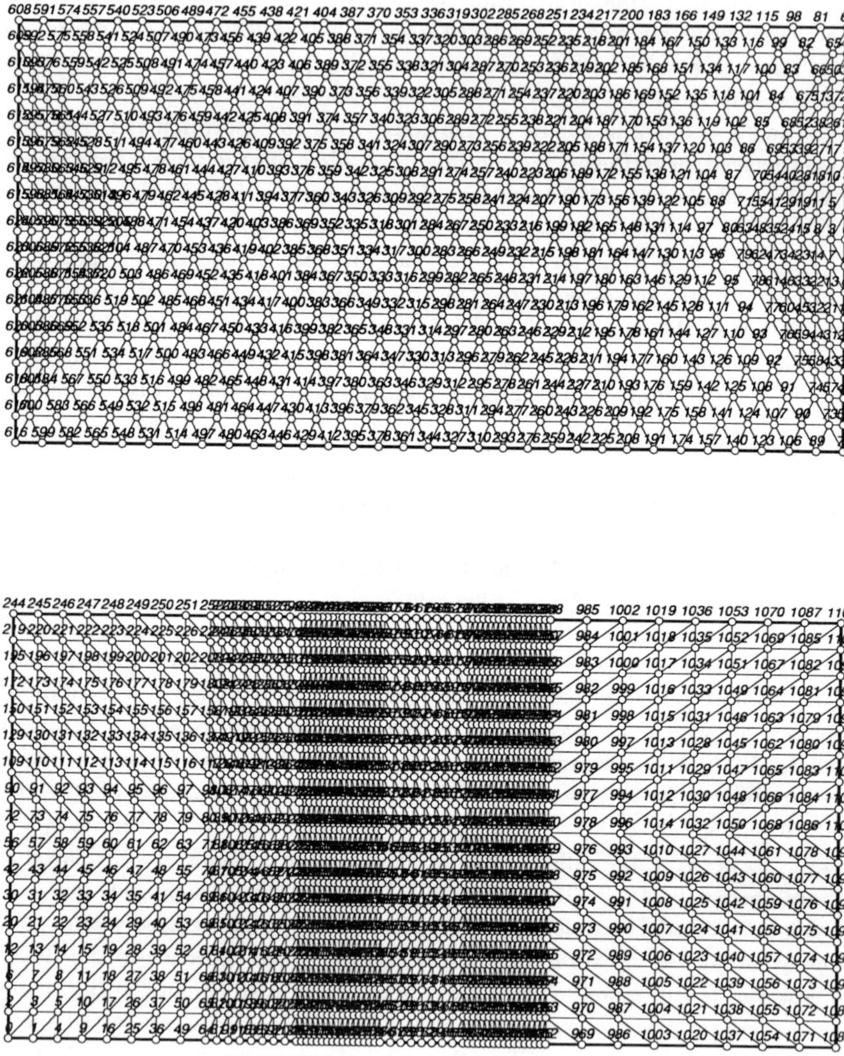

Figure 1.

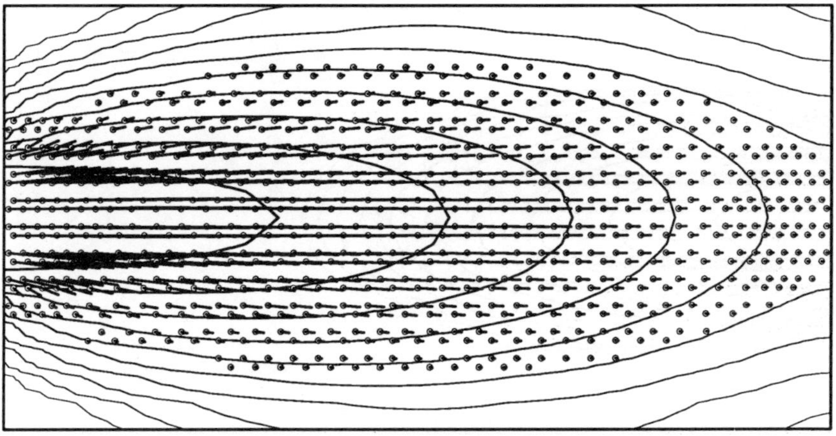

No adsorption: dissolved phase t = 1.0

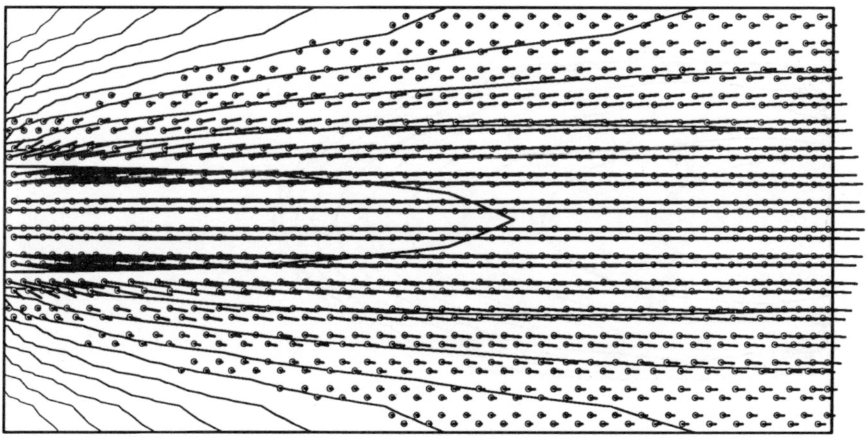

No adsorption: dissolved phase t = 3.0

Figure 2.

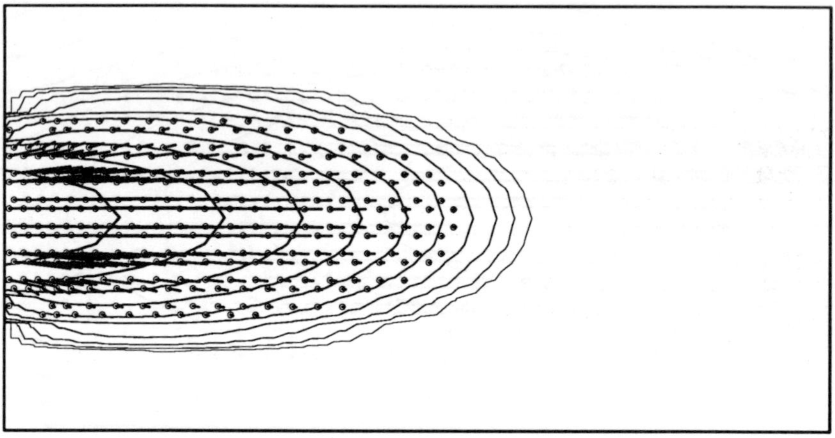

Homogeneous adsorption: dissolved phase t = 1.0

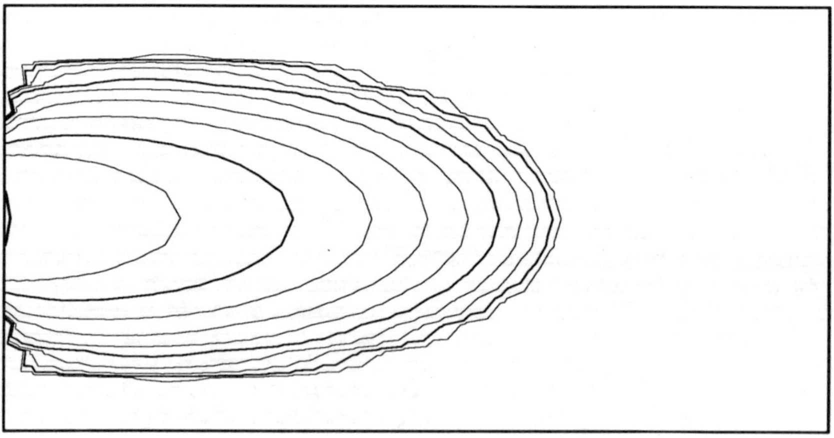

Homogeneous adsorption: adsorbed phase t = 1.0

Figure 3.

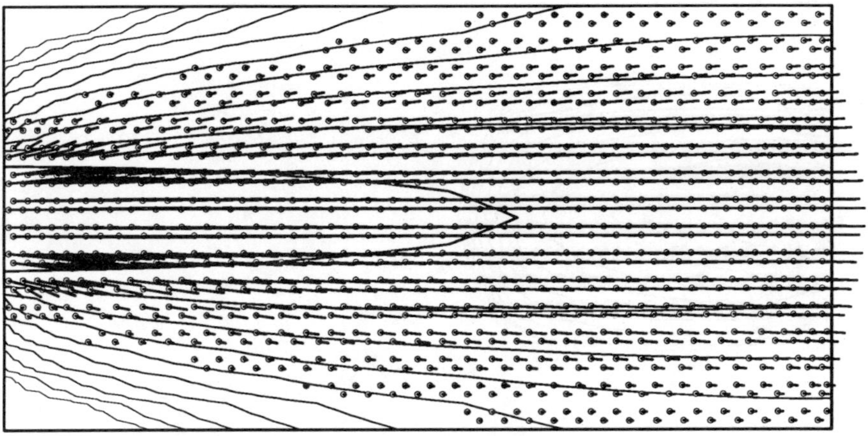

Homogeneous adsorption: dissolved phase t = 20.0

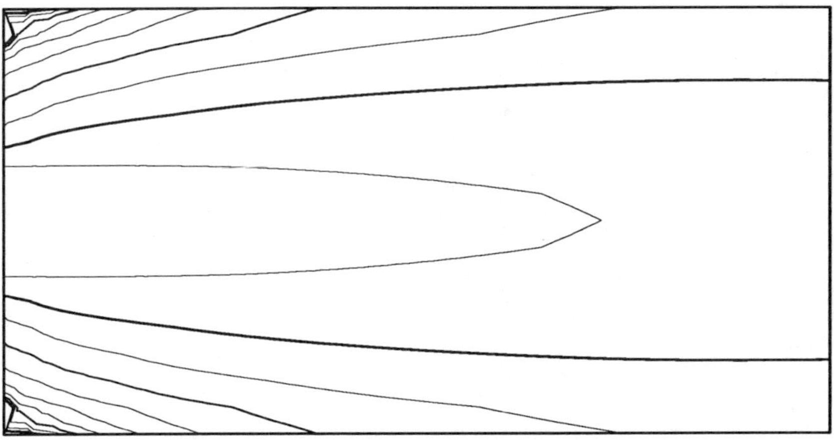

Homogeneous adsorption: adsorbed phase t = 20.0

Figure 4.

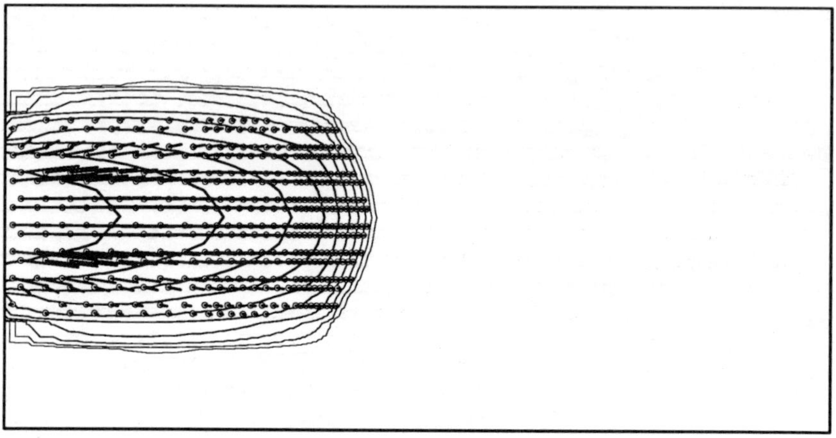

Inhomogeneous adsorption 1 to 10: dissolved phase t = 1.0

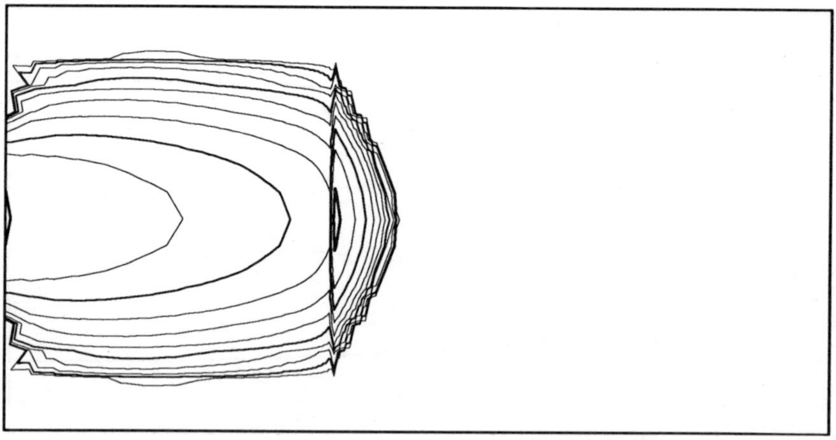

Inhomogeneous adsorption 1 to 10: adsorbed phase t = 1.0

Figure 5.

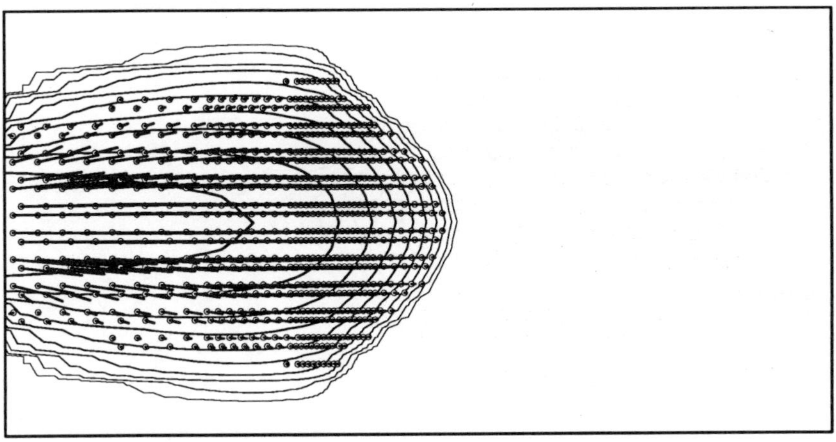

Inhomogeneous adsorption 1 to 10: dissolved phase t = 5.0

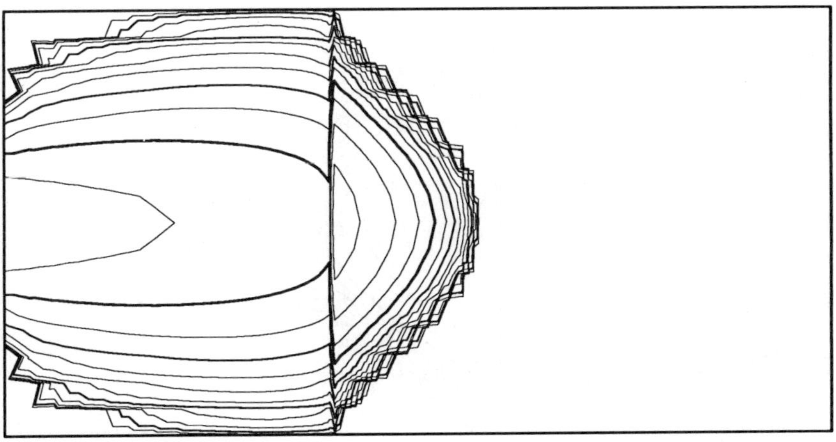

Inhomogeneous adsorption 1 to 10: adsorbed phase t = 5.0

Figure 6.

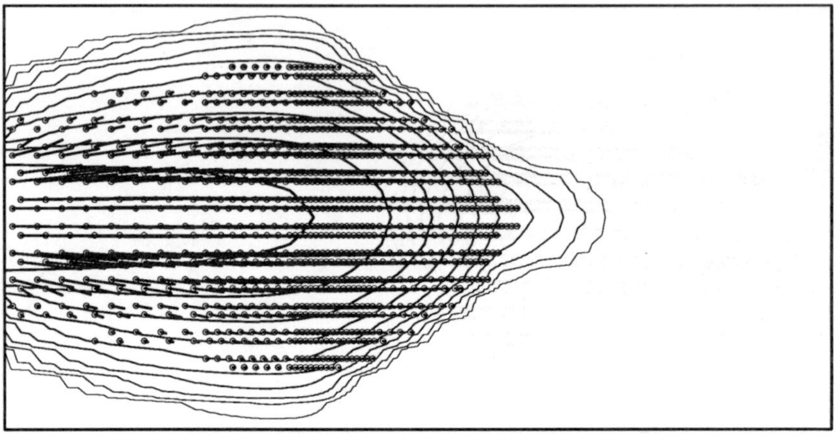

Inhomogeneous adsorption 1 to 10: dissolved phase t = 10.0

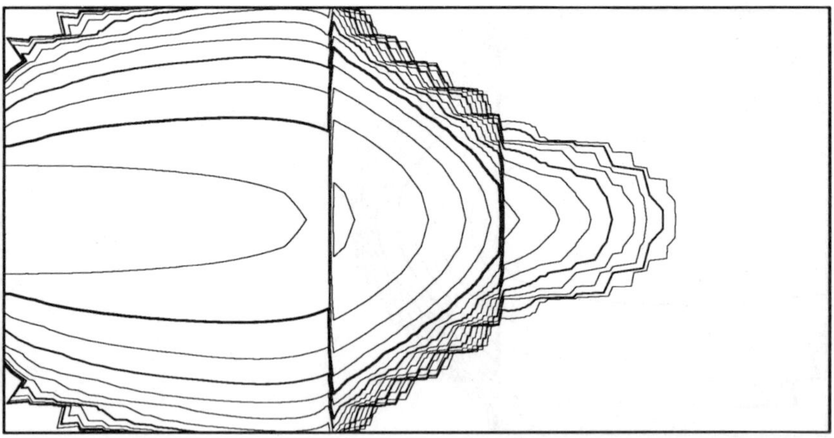

Inhomogeneous adsorption 1 to 10: adsorbed phase t = 10.0

Figure 7.

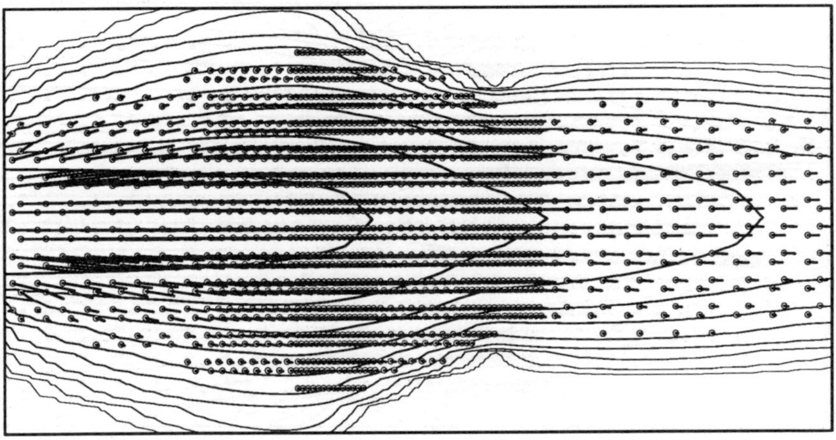

Inhomogeneous adsorption 1 to 10: dissolved phase t = 20.0

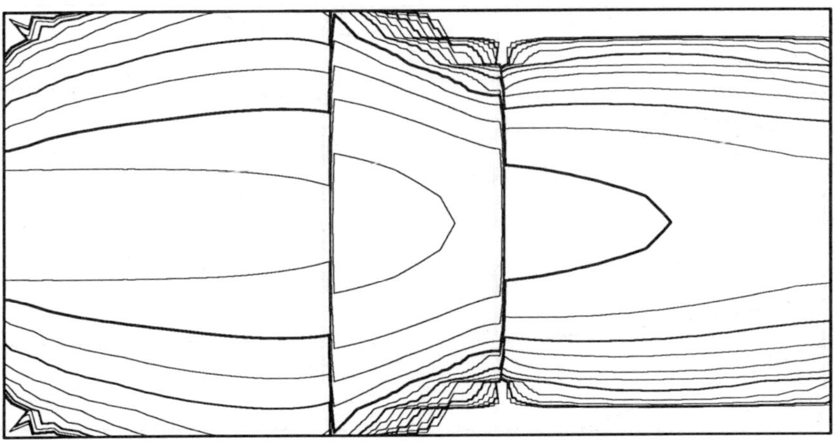

Inhomogeneous adsorption 1 to 10: adsorbed phase t = 20.0

Figure 8.

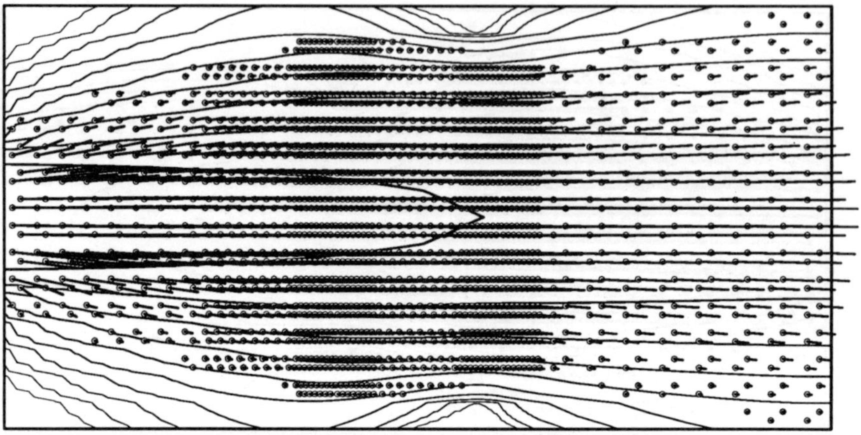

Inhomogeneous adsorption 1 to 10: dissolved phase t = 40.0

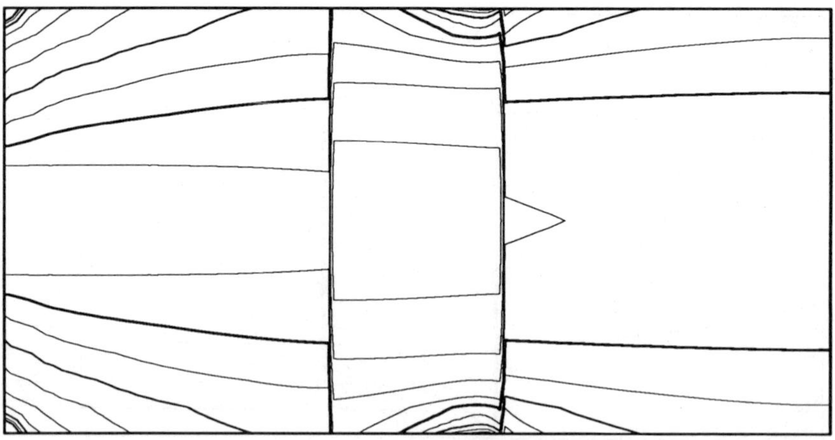

Inhomogeneous adsorption 1 to 10: adsorbed phase t = 40.0

Figure 9.

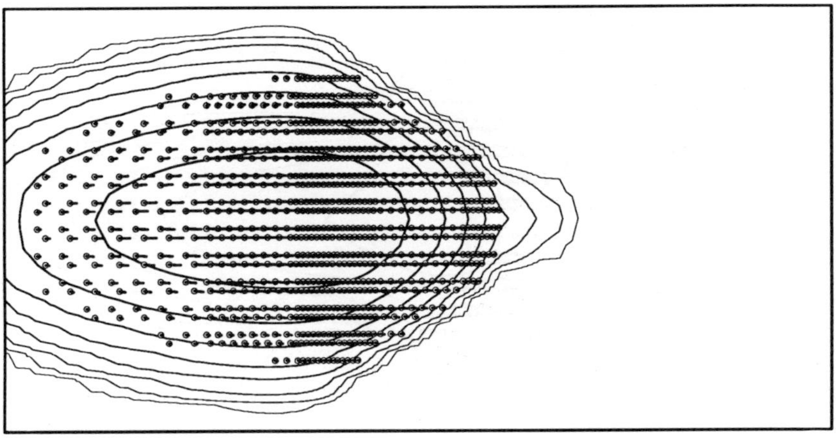

Inhomogeneous adsorption 1 to 10, pulse: dissolved phase t = 10.0

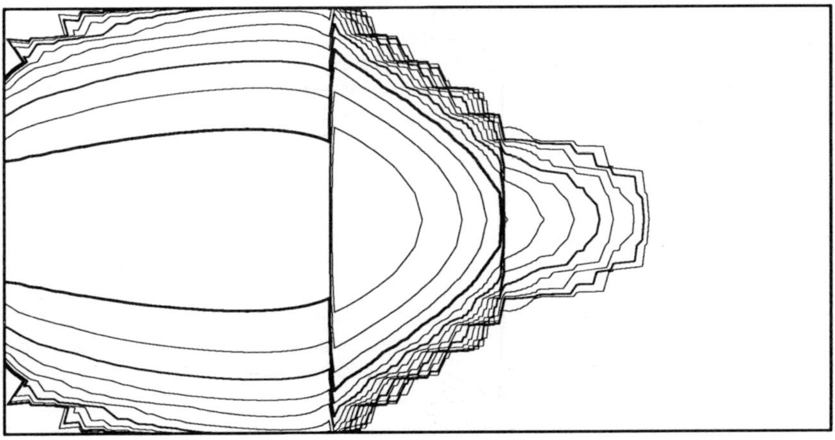

Inhomogeneous adsorption 1 to 10, pulse: adsorbed phase t = 10.0

Figure 10.

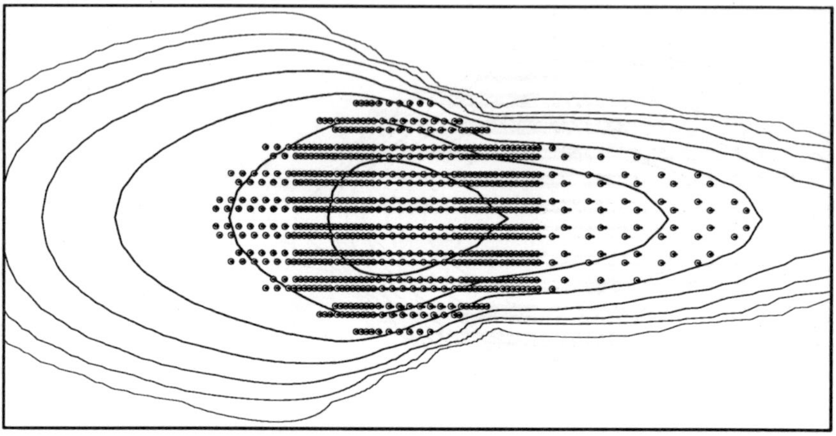

Inhomogeneous adsorption 1 to 10, pulse: dissolved phase t = 20.0

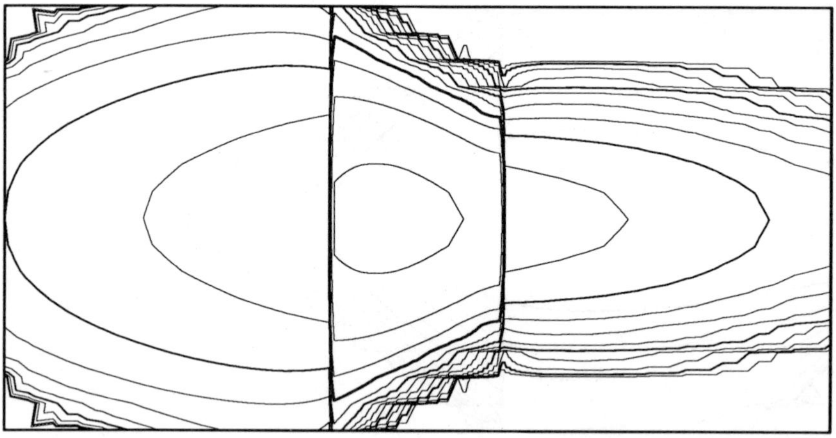

Inhomogeneous adsorption 1 to 10, pulse: adsorbed phase t = 20.0

Figure 11.

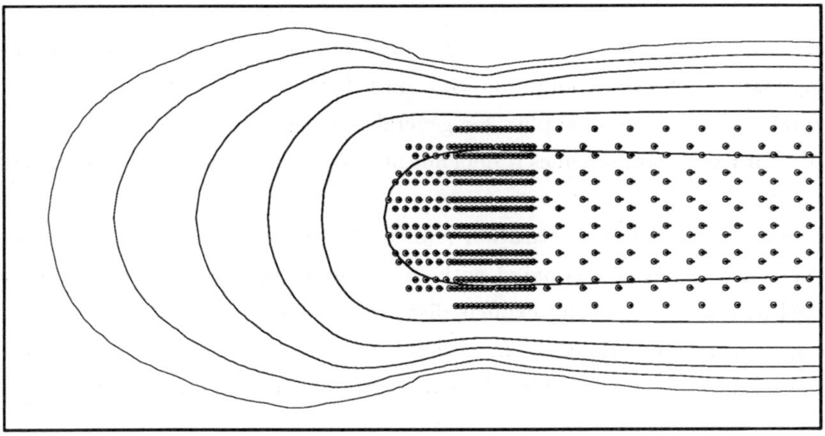

Inhomogeneous adsorption 1 to 10, pulse: dissolved phase t = 40.0

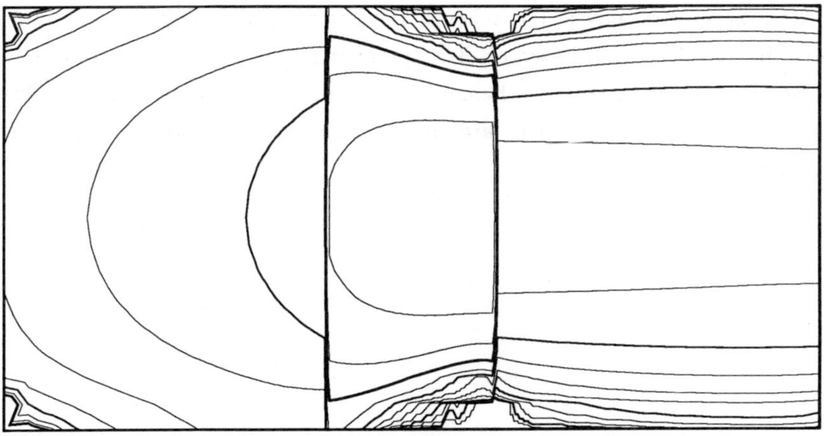

Inhomogeneous adsorption 1 to 10, pulse: adsorbed phase t = 40.0

Figure 12.

7 Conclusions

If all rate functions are of explicit type, there is nearly no distinction in numerical complexity between a model allowing only for equilibrium adsorption and the general case. The same applies for general nonlinearities, if Newton's method is used. For fixed point schemes it depends on the way the scalar nonlinear equations are solved.

The implicit-explicit discretization is of particular interest, as it has a reduced numerical complexity compared with the general implicit case. It can be interpreted as related to a fully explicit discretization with reduced rate parameters. For general rate functions the fixed point iteration with one Newton step for the scalar nonlinear equations coincides with Newton's method for the implicit-explicit discretization.

The usual monotonicity structure of the rate functions allows to solve the discrete equations in a stable manner uniformly in the rate parameter, if an appropriate method is applied: If all rate functions are of explicit type, this applies to Newton's method and the chord method, but not necessarily to the fixed point iteration. In the general case only a new scheme, the Newton-fixed point scheme, and partially the chord method based on fixed point iteration I are appropriate, but not necessarily any of the other schemes. It remains open whether this conclusions drawn by "maximum principle“ arguments can be extended by "energy estimate“ arguments.

8 References

[1] Axelsson, O., Barker, V. A.: *Finite Element Solution of Boundary Value Problems.* Academic Press, Orlando FL, 1984.

[2] Bohl, E.: *Finite Modelle gewöhnlicher Randwertaufgaben.* Teubner, Stuttgart, 1988.

[3] Douglas, J., Russell, T. F.: *Numerical Methods for Convection-Dominated Diffusion Problems Based on Combining the Method of Characteristics with Finite Element or Finite Difference Procedures.* SIAM J. Numer. Anal., 19 (1982), 871–885.

[4] Duijn, C. J. van, Knabner, P.: *Travelling Waves in the Transport of Reactive Solutes through Porous Media: Adsorption and Binary Ion Exchange–Part 1.* Submitted, Preprint "SPP Optimierung“ No. 198, 1990.

[5] Genuchten, M. T. van, Wierenga, P. J.: *Mass Transfer Studies in Sorbing Porous Media I.* Soil Sci. Soc. Amer. J. 40 (1976), 473–480.

[6] Heinrich, J. C., Huyakorn, P. S., Zienkiewicz, O. C., Mitchell, A. R.: *An 'Upwind' Finite Element Scheme for Two-Dimensional Convective Transport Equation.* Int. J. Numer. Math. Engrg. 11 (1977), 131–143.

[7] Knabner, P.: *Mathematische Modelle für Transport und Sorption in porösen Medien.* (In German) Verlag Peter Lang, Frankfurt/Main, 1991.

[8] Knabner, P.: *Models for the Transport of Reacting Solutes in Porous Media—A Mathematical Survey.* In preparation.

[9] Lake, L. W., Helfferich, F.: *Cation exchange and Chemical flooding 2.* Soc. Pet. Eng. J. 18 (1978), 435–444.

[10] Miller, C. W., Benson, L. V.: *Simulation of Solute in a Chemically Reactive Heterogeneous System: Model development and applications.* Water Resour. Res. 19 (1983), 381–391.

[11] Nochetto, R. H., Verdi, C.: *Approximation of Degenerate Parabolic Problems Using Numerical Integration.* SIAM J. Numer. Anal. 25 (1988), 784–814.

[12] Ortega, J. M., Rheinboldt, W. C.: *Iterative Solution of Nonlinear Equations in Several Variables.* Academic Press, New York, 1970.

[13] Raviart, P. A.: *The Use of Numerical Integration in Finite Element Methods for Solving Parabolic Equations.* In: J. J. H. Miller (ed.): Topics in Numerical Analysis, 233–264, Academic Press, London, 1973.

[14] Schwarz, H. R.: *The Finite Element Method.* Academic Press, New York, 1988.

[15] Thomée, V.: *Galerkin Finite Element Methods for Parabolic Problems.* Lecture Notes in Mathematics 1054, Springer Verlag, Berlin, 1984.

FREE BOUNDARY PROBLEMS IN FRESH-SALT GROUNDWATER FLOW

C.J. van Duijn
Delft University of Technology
Department of Technical Mathematics and Informatics
P.O. Box 5031, 2600 GA Delft
The Netherlands
also
Leiden University
Mathematical Institute
P.O. Box 9512, 2300 RA Leiden
The Netherlands

with an introduction on the physical background by

G. de Josselin de Jong
Retired from
Delft University of Technology
Department of Civil Engineering
Private address
Stadhouderslaan 66A
2517 JA Den Haag
The Netherlands.

Introduction

In this contribution density induced, i.e. gravity driven, groundwater flow problems are studied for fluids with a low relative density difference. This allows us to assume that the flow is incompressible, e.g. see van Duijn et al. (1992). This situation occurs in the case of fresh and salt groundwater flow where $(\rho_s - \rho_f)/\rho_f \approx 1/40$. It will also be assumed that the scale of the problems is such that a description in terms of an abrupt transition between the fluids, an *interface* or *free boundary* between the fresh and salt water, is possible.
This work is organized as follows.

In Section 1, written by G. de Josselin de Jong, the necessary physical properties of groundwater flow are discussed. Darcy's law is introduced and differential equations for the pressure and stream function are derived to describe the flow of a fluid with a variable density through a porous material.

In Section 2, some mathematical preliminaries are given. In particular the concepts of test functions and distributions are introduced. Also expressions for the divergence and the curl of a vector field across a discontinuity are derived.

Next, in Section 3, the interface models are set up. Two cases are treated in detail: a two-dimensional fresh-salt problem in a horizontal aquifer and an axially symmetric flow problem in which fresh water, on top of the heavier salt water, flows towards a well (the coning problem). For both cases a numerical method is given.

Finally in Section 4 a simplification will be considered which arises in many practical situations and which is called the Dupuit-approximation. In this approximation, following de Josselin de Jong (1981), it is assumed that the horizontal component of the specific discharge is constant (in vertical direction) in each fluid and has a jump at the interface. It leads to a description in terms of a nonlinear diffusion equation. The mathematical theory for this equation, which has been developed over the past few years, is summarized in this section. Also some extensions, in particular adding convection terms, are discussed.

1 Physical properties of groundwater flow: an introduction

A scientific treatise on the flow of fluids through permeable soils necessarily starts with Darcy, who as an inspector of civil engineering was responsible for the drinking water supply to the fountains of the town of Dyon in France. In order to establish how much water could be expected to percolate through the sand layers in his area, he performed tests and formulated in 1856 a law based on the results, which since then became classic. The test setup consisted of a vertical pipe filled over different lengths with saturated sand, through which water could percolate under various pressure differences on either side of the sand body. He observed a linear relationship between the specific discharge of water and the pressure difference divided by the percolation length. He introduced a coefficient of proportionality, K $[m/s]$, which since then is called the Darcy coefficient of permeability or hydraulic conductivity.

Darcy's law

The law proposed by Darcy expresses the specific discharge, q $[m/s]$, which is defined as the volume of fluid $[m^3]$ passing through a unit cross section $[m^2]$ per unit time $[s]$, in terms of a pressure difference. He wrote the law as

$$q = K \frac{(h_1 - h_2 + l)}{l}, \tag{1.1}$$

where l $[m]$ is the length of the vertical sand column and h_1, h_2 $[m]$ are water heights, representing the pressures p_1, p_2 $[N/m^2]$ measured by mercury manometers on either side of the column. The pressures are converted into water heights by division through γ $[N/m^3]$, the specific weight of water. This presentation is possibly due to the c.g.s system of units, used at the time, in which γ for water equals 1 $[grf/cm^3]$, permitting to disregard conversion factors.

After Darcy performed the first tests, his law has often been verified. It was found to be valid for all kinds of soils and fluids, when the fluid velocity is not extremely high. In the engineering practice of hydrology the velocities are generally small enough for the Darcy coefficient, K, to be constant, when a particular soil is percolated by a particular fluid.

Darcy's law was not developed from basic physical principles, but deduced from experiments. Its direct simplicity was very useful for civil engineers who solved many practical problems with it in the realm of hydrology. This started with Dupuit (1865) who was responsible for the fountains of Paris and used the theory for the design of filtration basins. Attempts to base Darcy's law on basic principles were also initiated by Dupuit, who compared the flow through the pore space of a soil with the flow of fluids through open channels and parallel thin tubes. Although he confirmed the law by applying the open channel concept, the thin tube analogy turned out later to be more appropriate for understanding the physical properties of the underlying mechanism.

Thin tube analogy

At the time Poiseuille (1840, 1841) had developed the theory for the flow of a viscous fluid through small, horizontal tubes of constant, circular cross section with radius r $[m]$. (As a physician his objective was to study the flow of blood through human veins). Assuming laminar flow, with velocities small enough to disregard inertial effects, he determined the velocity distribution to be parabolic by equating the forces on concentric cylindrical subsections of the fluid in the tube. These forces are: a driving force resulting from the pressure difference $(p_1 - p_2)$ $[N/m^2]$ over a length l $[m]$ of the tube and a viscous force originating from the fluid resistance against internal deformation. This resistance is proportional to the dynamic viscosity, μ $[Ns/m^2]$ of the fluid. Integration of the parabolic velocity distribution over the cross section gives the total discharge Q $[m^3/s]$ of fluid to be

$$Q = \frac{\pi\, r^4}{8\,\mu}\,\frac{(p_1 - p_2)}{l}. \tag{1.2}$$

Schematically the soil can be compared with a rigid solid perforated by a system of parallel, circular tubes with different radii, r_i $[m]$. Here i is a number counting the N tubes, that pass through a perpendicular cross section with area A $[m^2]$ of the solid.

According to (1.2) the total discharge of fluid can be written as

$$Q = (\sum_{i=1}^{N} \frac{\pi\, r_i^4}{8})\,(p_1 - p_2)\,/(\mu l). \tag{1.3}$$

The specific discharge q $[m/s]$ is the total discharge Q divided by the cross sectional area A. So from (1.3) there results

$$q = (\sum_{i=1}^{N} \frac{\pi\, r_i^4}{8})\,(p_1 - p_2)\,/(\mu l A). \tag{1.4}$$

The average velocity $\overline{v}$ $[m/s]$ of the fluid over all the tubes is Q divided by their combined cross sectional area's $(\sum \pi\, r_i^2)$ $[m^2]$ such that

$$\overline{v} = (\sum_{i=1}^{N} \frac{\pi r_i^4}{8}(p_1 - p_2)/(\mu l \sum_{i=1}^{N} \pi r_i^2). \tag{1.5}$$

Comparison with Darcy's law

The resemblance with Darcy's law (1.1) is recognized by recalling, that the waterheights h_1, h_2 are obtained from the pressures p_1, p_2 by division through the specific weight $\gamma = \rho g$ $[N/m^3]$ of the fluid, giving

$$q \;=\; K \frac{(p_1 - p_2)}{\rho g \;\; l} \tag{1.6}$$

Here γ is replaced by the density of the fluid ρ $[kg/m^3]$, multipied by g $[m/s^2]$ the acceleration of gravity. Moreover, the vertical column length l in the numerator can be ignored, when the tubes are considered to be horizontal. Comparing (1.6) with (1.4) shows, that the parallel tube model has an analogous Darcy coefficient K, given by

$$K \;=\; (\sum_{i=1}^{N} \frac{\pi\, r_i^4}{8})\, \frac{\rho g}{A\, \mu}. \tag{1.7}$$

The expression (1.7) shows, that the coefficient K consists of two parts: one part being $(\rho g/\mu)$ $[1/(m\;s)]$ which contains the properties of the percolating fluid and a second part $(\sum \pi r_i^4)/8A$ $[m^2]$, representing the influence of the pore structure on the flow.

Intrinsic permeability

The above substitution of the pore space by parallel tubes must be considered as a metaphor. The real pore space in a soil forms an interconnected system of fluid transmitting vessels with intricate tortuosity, consisting of cavities that alternate with constrictions, both bizarrely shaped and randomly oriented. In fact the pores occupy the space, that remains between the soil particles after they have been packed together. Attempts to describe the pore space geometry, starting from the distribution of particle sizes and shapes, have often failed. A fortiori it has proven impossible to describe the flow through such a space, although certain averaging techniques result in macroscopic flow rules of the type of Darcy's law. For this we refer to the contribution of Hornung, elsewhere in this book.

In hydrology it is customary to disregard the details of the flow inside the pores and to consider only averages of flow and pressure. In fact the system of particles and pores is conceptually replaced by a continuum with continuum properties.

The relation (1.7) is written in the form

$$K = \kappa \, \rho g \, / \, \mu, \tag{1.8}$$

where κ is a coefficient, which similar to $(\sum \pi r_i^4)/8A$ has the dimension $[m^2]$ and is called the intrinsic permeability. All the unknown properties of the real flow geometry are dumped in this κ, which therefore represents the percolation transmissivity of the pore space. The value of κ can only be determined experimentally.

Sofar the soil is considered as a rigid perforated solid. In reality a soil skeleton is flexible, reacting with deformations on pressures either applied from the outside or from the inside by fluid pressures in the pores. This interaction forms a soil mechanics problem, which remains outside the scope of this contribution, where the soil skeleton will be assumed to be perfectly rigid and fixed to coordinates x, y, z.

Average velocity

Up to here the specific discharge q is used to describe the flow of fluid through the soil. The underlying reason is that it is the amount of water that can be collected outside the soil as referred to a specific soil cross section. For establishing, however, the velocity of for instance a contaminant dissolved in the fluid, it is necessary to know $\overline{v}$ the average real velocity of the fluid in the pores.

The relation between $\overline{v}$ and q contains the porosity ε of the soil, which is defined as the ratio of pore volume to total volume (pores + particles). For the parallel tube model the pore space is $(\sum \pi r_i^2)l$ and the total volume is $A\,l$. So the porosity is $\varepsilon = (\sum \pi r_i^2)/A$. Comparison of (1.4) with (1.5) shows, that for this model $q = \varepsilon \overline{v}$. According to the analogy one uses a similar relation for a real soil. So in general $\overline{v}$ is related to q by

$$\overline{v} = q \, / \, \varepsilon. \tag{1.9}$$

Hydraulic head

Darcy expressed the relevant waterpressures p_1 and p_2, in terms of water heights $h_1 = p_1/\rho g$ and $h_2 = p_2/\rho g$ above the levels in the sand column, where p_1 and p_2 were measured. These heights may be called pressure heads. Afterwards it was observed, that the law is simplified by introducing another kind of head, h $[m]$, called the hydraulic head. This h is defined as the height of the waterlevel in a standpipe, measured with respect to an arbitrary datum level. Let z $[m]$ be the elevation above the same datum level of a filter that provides the connection between the water in the aquifer and the water in the standpipe. Then the relation between h and the pressure p $[N/m^2]$ in the pore water at elevation z is

$$h = z + (p/\rho g). \tag{1.10}$$

When a fluid does not move, the water pressure in it decreases with elevation. Letting z be defined as the coordinate pointing upwards, $(p + \rho g\, z)$ is constant for water that is stationary. This means, that in a soil there is no flow, i.e. no specific discharge, between two points with the same head. Flow occurs only, when h differs and according to the law proposed by Darcy the discharge is proportional to the hydraulic head difference, divided by the distance of the points where the heads are measured. Written in terms of the hydraulic gradient, grad h, Darcy's law is

$$\boldsymbol{q} = -K \operatorname{grad} h. \tag{1.11}$$

The formulation (1.11) of Darcy's law is only valid, when ρ the density of the fluid is a constant, since for ρ varying in space the gradient of h given by (1.10) is no more representative for the specific discharge $\boldsymbol{q}$. Traditionally the form (1.11) was considered to be the necessary extension of Darcy's original percolation law for treating space problems. Its limitation to ρ constant, is a logical consequence of his law, being deduced from tests that were performed on one kind of fluid (fresh water), only.

Potential

When K is a constant, the relation (1.11) can be written as

$$\boldsymbol{q} = -\operatorname{grad} \Phi, \qquad \text{with } \Phi = Kh. \tag{1.12}$$

In the case that a velocity vector $\boldsymbol{q}$ can be derived by differentiation from a function Φ in the manner of (1.12), this function Φ is called a potential and the corresponding flow is a potential flow. Since potential flow occurs in many other physical processes, its mathematical properties have received much attention.

A property, that is often overlooked when studying groundwater flow, is its rotation $\boldsymbol{\omega}$. This rotation is defined by $2\boldsymbol{\omega} = \text{curl}\,\boldsymbol{q}$, where in hydrodynamics $2\boldsymbol{\omega}$ is called the "vorticity", e.g. see Lamb (1945). When $\boldsymbol{q}$ satisfies (1.12), its vorticity is given by $2\boldsymbol{\omega} = \text{curl}\,\boldsymbol{q} = -\text{curl grad}\,\Phi$. Since the curl grad of a smooth, differentiable function Φ is zero it follows, that

$$2\boldsymbol{\omega} \;=\; \text{curl}\,\boldsymbol{q} \;=\; -\,\text{curl grad}\,\Phi \;=\; 0 \tag{1.13}$$

indicating that, if a potential Φ exists, the flow has no vorticity and hence no rotation. It is then called "irrotational".

The concept, that Darcy's law could be formulated in terms of a potential flow, had many convenient consequences from which the applicability of complex variables in two dimensions is perhaps the most prominent. Also the simple relation $\Phi = Kh$ permitted to identify observations of heads, as measured in stand pipes in the field, to potentials Φ. The advantage of a formulation in terms of potentials was so convincing, that traditionally it was considered obvious to search for such a formulation, also when the aquifer contains several fluids, such as fresh and salt water or their mixtures. Several attempts were made to construct appropriate heads to provide such potentials.

However, it will be shown below, that in the general case of varying fluid or aquifer properties, potentials cannot exist, because the flow then possesses rotation. In order to demonstrate this it is necessary to develop a more general percolation law, which shows the influence of ρg, μ and κ explicitly. This is accomplished by considering the equilibrium of forces acting in a pore fluid.

Generalization of Darcy's law

The essential character of Darcy's law is the action of two forces. The driving force $\boldsymbol{f}^p$ due to pressure differences in the fluid and the resistance force $\boldsymbol{f}^r$ due to the viscous resistance of the fluid, wriggling through the pores.

Both forces have an averaged character in the same manner as $\boldsymbol{v}$ and p are averages. Let an infinitesimal volume $dV = dxdydz$ be considered, large enough to contain so much particles and pores that a reasonable permeability, κ, is produced, but small enough with respect to the entire flow to be acceptable as mathematical small. dV is a representative elementary volume REV as introduced by Bear (1972).

The pressure p in the pore fluid within dV is averaged in such a manner, that p is a continuous function of x, y, z, the location of dV. The x-component f_x^p of the driving force $\boldsymbol{f}^p$ is then defined as the pressure decrease $-(\partial p/\partial x)\,dx$ multiplied by $dy\,dz$ the cross section of dV perpendicular to x. By this definition the remaining forces in the soil skeleton in x-direction are exclusively those that are responsible for its deformation, which is further disregarded in this treatise.

Similar components in y- and z-directions produce, together with the x-component, a driving force $\boldsymbol{f}^p = -\text{grad } p \; dV$. In addition gravity exerts on the specific weight ρg of the fluid a force $(-\rho g\, \boldsymbol{e}_z)\, dV$, where $\boldsymbol{e}_z$ is the unit vector in z-direction pointing vertically upwards. Together this gives for the driving force

$$\boldsymbol{f}^p \;=\; (-\,\text{grad}\, p \;-\; \rho g\, \boldsymbol{e}_z\,)\; dV. \tag{1.14}$$

The driving force is balanced by the resistance force $\boldsymbol{f}^r$ experienced by the fluid, when percolating through the pores. This resistance force is, simlarly as encountered in Poiseuille flow, proportional to the specifc discharge $\boldsymbol{q}$ and to μ the viscosity of the fluid, but inverse proportional to κ the intrinsic permeability of the soil skeleton. Further the force $\boldsymbol{f}^r$ is taken to be proportional to dV, allthough the fluid only occupies the pore space $\varepsilon\, dV$. The porosity ε is generally absorbed in the value of κ, giving

$$\boldsymbol{f}^r \;=\; (\,\frac{\mu}{\kappa}\,\boldsymbol{q}\,)\; dV. \tag{1.15}$$

The balance of $\boldsymbol{f}^r$ and $\boldsymbol{f}^p$ gives after division by dV

$$\frac{\mu}{\kappa}\,\boldsymbol{q} \;=\; -\;\text{grad}\, p \;-\; \rho g\, \boldsymbol{e}_z \tag{1.16}$$

This is the generalized form of Darcy's law.

A relation similar to (1.16) was first published in 1912 by H.A.Lorentz (the creator of the special theory of relativity). As a Dutch physicist, he got involved in problems of the drinking water supply for the large towns in the western part of the country. Since this water was obtained from the rain falling on the dune area near the sea, its pureness was endangered by salt water percolating through underlying sand layers. In order to put the discussion on a sound physical basis, he used the concept of REV and developed the above pertinent differential equation (1.16) by equating the driving and resistance forces.

The relation (1.16) is called the "generalized form of Darcy's law", since it shows explicitly how the properties of fluid and soil are to be accounted for, seperately. It is the logical extension of Darcy's original law, especially in those cases, where the fluid and/or soil properties vary in space. Darcy's pioneering experiments did not incoporate such variations since they were carried out with fresh water only and his factor of proportionality K combines all physical properties involved, see (1.8).

Divergence of the specific discharge; continuity

Besides Darcy's law the second basic property governing the flow through soils is the conservation of fluid volume. This is a simplification, since different factors can affect the volume of a certain amount of fluid. In the first place the volume changes, when the salt concentration increases due to diffusion, dispersion or solution. This volume increase is, however, perceptible only for very high concentrations, which are not considered here. In the second place fluid is almost incompressible. The pressure variations occurring in the practice of hydrology are small enough for their influence to be disregarded. Only when air or gas bubbles are entrapped in the soil or the fluid, important volume changes may occur.

For simplicity this study is limited to completely saturated fluids (i.e. without air or gas bubbles) and salt concentrations that donot affect the volume of the fluid. Together with the postulated rigidity of the soil skeleton the consequence is, that the volume of fluid entering an elementary volume $dV = dxdydz$ is equal to the volume leaving it. This is called "continuity".

Expressed in terms of the discharge vector, continuity requires that the increase of its component q_x in x-direction, $(\partial q_x/\partial x)\, dx$, multiplied by the area, $dydz$, perpendicular to x, together with similar increases in y and z directions, vanishes. Division by the volume $dxdydz$ gives

$$(\partial q_x/\partial x) + (\partial q_y/\partial y) + (\partial q_z/\partial z) = \operatorname{div} \boldsymbol{q} = 0 \qquad (1.17)$$

This equation is the so-called "continuity equation".

Relation for p, when density varies in space

The continuity equation (1.17) combined with the general form of Darcy's law, leads to the elimination of the specific discharge $\boldsymbol{q}$ as follows:

$$\operatorname{div} \boldsymbol{q} = 0 = -\operatorname{grad}(\kappa/\mu)\,(\operatorname{grad} p + \rho g\ \boldsymbol{e}_z) - (\kappa/\mu)\,(\operatorname{div}\operatorname{grad} p + g(\partial\rho/\partial z)) \qquad (1.18)$$

When the terms with variations of (κ/μ) can be disregarded with respect to the other terms this reduces to

$$\operatorname{div}\operatorname{grad} p = \nabla^2 p = -\,g(\partial\rho/\partial z). \qquad (1.19)$$

This equation is due to Knudsen (1962) and permits to determine the distribution of p in a soil body subject to given boundary conditions in terms of the pressure p.

Vorticity in three dimensions

The property of the discharge vector $\boldsymbol{q}$ complementary to div $\boldsymbol{q}$ is its vorticity curl $\boldsymbol{q}$, which is related to the rotation $\boldsymbol{\omega}$ by $2\boldsymbol{\omega} = \text{curl } \boldsymbol{q}$. Using the general form of Darcy's law (1.16) and assuming that the terms with variations of (κ/μ) can be disregarded, there results

$$2\boldsymbol{\omega} \;=\; \text{curl } \boldsymbol{q} \;=\; -\,(\frac{\kappa}{\mu})\,\text{curl}\,(\,\text{grad}\, p \;+\; \rho g\, \boldsymbol{e}_z\,) \qquad (1.20)$$

Since the curl grad p vanishes for miscible fluids and curl $\boldsymbol{e}_z$ is zero, this reduces to

$$2\boldsymbol{\omega} \;=\; -\,(\frac{\kappa}{\mu})\, g\,(\,\text{grad}\,\rho \,\times\, \boldsymbol{e}_z\,) \qquad (1.21)$$

The cross product of the vectors grad ρ and $\boldsymbol{e}_z$ is a vector that is perpendicular to both of them. This means that the vorticity has an axis tangent to the line of ρ constant in the horizontal plane. Its magnitude is $-(\kappa g/\mu)$ multiplied by the horizontal component of grad ρ, which is equal to $\sqrt{(\partial\rho/\partial x)^2 + (\partial\rho/\partial y)^2}$. The direction of $\boldsymbol{\omega}$ is such that the three vectors $-$gradρ, $\boldsymbol{e}_z$ and $\boldsymbol{\omega}$ form a right handed system. As a consequence the sense of rotation is such, that the heavier fluid slides under the lighter one.

From (1.21) it follows directly that pore fluid rotates in a region, where ρ varies. Since potentials or heads produce irrotational flow, they cannot exist in such regions.

Stream function, two dimensions

A function complementary to the pressure, and suitable for describing the vorticity of the specific discharge is the stream function Ψ. In three dimensions the stream-function has a somewhat intricate character. The description is therefore limited here to the two dimensional case, where q_y is zero and all derivatives with respect to y vanish. The continuity equation (1.17) then reduces to

$$(\partial q_x/\partial x) \;+\; (\partial q_z/\partial z) \;=\; 0. \qquad (1.22)$$

Let Ψ be a smooth function and let a vector field $\boldsymbol{q}$ be related to it by the definition

$$(\partial\Psi/\partial x) \;=\; +\,q_z, \qquad (\partial\Psi/\partial z) \;=\; -\,q_x. \qquad (1.23)$$

This vector field satisfies automatically the continuity condition (1.22). In fluid dynamics, where $\boldsymbol{q}$ is interpreted as a velocity, the function Ψ is called the stream function, see for instance Lamb (1945). The reason for this name is the following.

Let a local, orthogonal coordinate system s, n be oriented such, that s has the direction of $\boldsymbol{q}$. Then q_n is zero. According to (1.23) $\partial\Psi/\partial s$ vanishes and this means that Ψ is constant along s. So lines for constant Ψ have in every point the direction of $\boldsymbol{q}$, making them streamlines. Another consequence of this is that impervious boundaries are lines of constant Ψ.

Vorticity in two dimensions

In two dimensions the vorticity $2\boldsymbol{\omega}$ has only one component ω_y. This component is the rotation around the y-axis of the right handed x, y, z coordinate system. It is given by

$$2\omega_y \;=\; (\text{curl}\, \boldsymbol{q})_y \;=\; \partial q_x/\partial z \;-\; \partial q_z/\partial x. \tag{1.24}$$

By use of the stream function. Ψ, related to q_x, q_z by (1.23), this becomes

$$2\omega_y \;= -\partial^2\Psi/\partial z^2 - \partial^2\Psi/\partial x^2 \;=\; -\nabla^2\Psi. \tag{1.25}$$

Application of the general form of Darcy's law (1.16) gives for (1.24) in the case that variations of κ/μ can be disregarded

$$2\omega_y \;= -\,(\frac{\kappa}{\mu})\,(\,\partial^2 p/\partial x\partial z \;-\; \partial^2 p/\partial z\partial x \;-\; \partial(\rho g)/\partial x\,). \tag{1.26}$$

Since p is a continuous, smooth function (1.26) reduces to

$$2\omega_y \;=\; +\,(\frac{\kappa}{\mu})g\,(\,\partial\rho/\partial x\,). \tag{1.27}$$

This relation shows that the vorticity $2\omega_y$ being a right handed screw in y-direction, rotates the fluid in such a manner that the heavier fluid slides under the lighter one.

Taking account of (1.25) it follows that

$$\nabla^2\Psi = \;-\,(\frac{\kappa}{\mu})g\,(\,\partial\rho/\partial x\,). \tag{1.28}$$

This last expression, obtained by de Josselin de Jong (1960), has similar to the Knudsen equation (1.19) the character of a Poisson equation. The relation (1.28) is suited for the determination of the flow distribution in a region where the boundary conditions are given in terms of the specific discharge $\boldsymbol{q}$ or of Ψ for that matter.

Sharp interface

In the case of a sharp interface the value of (ρg) varies abruptly. The lighter fluid with specific weight $(\rho_1 g)$ is separated by a plane from the heavier fluid with $(\rho_2 g)$. The behaviour of the fluids at either side of the plane was described by Edelman (1940) for the two dimensional x, z case as follows.

Let a local coordinate system s, n be considered, with s along the interface and n normal to it, both pointing upwards, i.e. against the direction of gravity. The angle between s and x is called α, such that $\partial z/\partial s = \sin\alpha$. The specific discharge $\boldsymbol{q}$ at the interface is discontinuous being $\boldsymbol{q}_1$ above and $\boldsymbol{q}_2$ below it.

Continuity along the interface requires that no gaps or overlaps are created and this is achieved by prescribing that the normal components of the discharge are equal or

$$q_{n_1} = q_{n_2}. \tag{1.29}$$

When the fluids are miscible the pressures p_1, p_2 on either side of the interface are equal. This implies that $\partial p_1/\partial s = \partial p_2/\partial s$ along the interface. Using the general form of Darcy's law, the components q_{s_1}, q_{s_2} of the discharges, parallel to the interface on either side of it, obey

$$\begin{aligned}\frac{\mu}{\kappa}(q_{s_1} - q_{s_2}) &= -(\partial p_1/\partial s) - \rho_1 g(\partial z/\partial s) + (\partial p_2/\partial s) + \rho_2 g(\partial z/\partial s) \\ &= (\rho_2 - \rho_1) g \sin\alpha. \end{aligned} \tag{1.30}$$

The difference $(q_{s_1} - q_{s_2})$ is called the shear flow. Its direction is always such that the heavier fluid slides under the lighter one. This kind of flow possesses a vorticity that is concentrated in the interface plane, which in that case is called a vortex sheet, see de Josselin de Jong (1960).

The discussion presented in this section only treats those aspects of flows through porous media which are necessary to serve as a physical background for the problems discussed in later sections. A thorough and more extended introduction in the field of physics of porous media flow is given by Bear (1972).

References

Bear, J. (1972). *Dynamics of fluids in porous media*, New York: American Elsevier

Darcy, H. (1856). *Les fontaines publique de la ville de Dijon*, Parijs: Dalmont.

Dupuit, J. (1865). *Traité théorique et pratique de la conduite et de la distribution des eaux*, Paris: Ed. Dunod.

Edelman, J.H. (1940). *Strooming van zoet en zout water*, rapport inzake de watervoorziening van Amsterdam. Bijlage 2: 8-14.

de Josselin de Jong, G. (1960), *Singularity distributions for the analysis of multiple fluid flow through porous media*, J, Geophys. Res. **65**, 3739-3758.

Knudsen, W.C. (1962). *Equations of fluid flow through porous media - incompressible fluids of varying density*, J. Geophys. Res. **67**, 733-737.

Lamb, H. (1932). *Hydrodynamics, 6th ed.*, New York: Dover Publications.

Lorentz, H.A. (1912). *Grondwaterbeweging in de nabijheid van bronnen*, De Ingenieur, **49**, 24-26.

Poiseuille, J.M.L. (1840, 1841), *Recherches expérimentales sur le mouvement des liquides dans les tubes de très petits diamètres*, Comptes Rendues XI, XII.

2 Preliminaries

In this section we discuss a number of mathematical concepts which will appear at various places in later sections. They are introduced here by considering some simple flow situations related to the movement of one or two incompressible fluids through a saturated, rigid and isotropic porous material. Both the fluids and the porous medium are assumed to be homogeneous (i.e. their properties are constant in space).

2.1 Divergence in sense of distributions: one space dimension

Consider an "experiment" as shown in Figure 2.1.

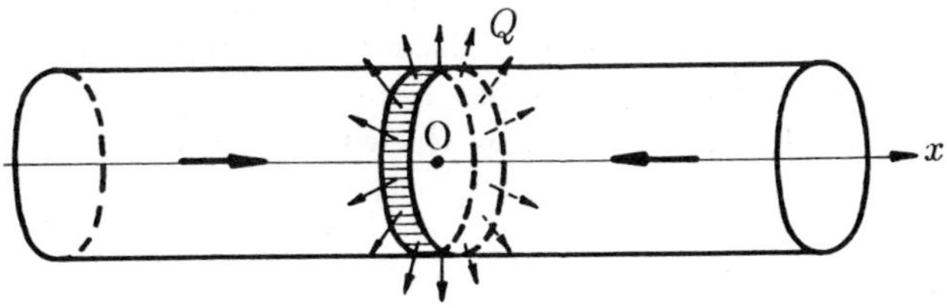

Fig. 2.1: One-dimensional flow towards the origin.

A pipe, having an impermeable wall for $x < 0$ and for $x > 0$, with a constant cross section $A[m^2]$, is filled with a porous material which is saturated by a fluid. At the origin fluid is being withdrawn at the rate $Q[m^3/s]$. Let us suppose that the specific discharge $\boldsymbol{q}[m/s]$ of the fluid is uniformly distributed over each cross section of the pipe and that it is symmetric with respect to the distance to the origin. Then it is given by

$$\boldsymbol{q} = q(x)\boldsymbol{e}_x, \tag{2.1}$$

where

$$q(x) = \begin{cases} -\frac{Q}{2A} & x > 0 \\ +\frac{Q}{2A} & x < 0 \end{cases} \tag{2.2}$$

and where $\boldsymbol{e}_x$ denotes the unit vector in positive x-direction, pointing to the right in Figure 2.1. Using the Heaviside function

$$H(x) = \begin{cases} 1 & x > 0 \\ 0 & x < 0 \end{cases} \tag{2.3}$$

one writes (2.2) as

$$q(x) = \frac{Q}{2A} - \frac{Q}{A}H(x) \quad \text{for } -\infty < x < \infty. \tag{2.4}$$

We now pose the following question:

$$\text{What can be said about div } \boldsymbol{q}? \tag{Q_1}$$

Cleary

$$\text{div } \boldsymbol{q} = \frac{dq}{dx} = 0 \;\; \text{for every } x \neq 0. \tag{2.5}$$

But at $x = 0$, the function q is discontinuous and consequently not differentiable. Thus in the classical sense one cannot evaluate dq/dx for every $-\infty < x < \infty$. However it is possible to generalize the notion of derivative, which allows us to differentiate also discontinuous functions.
For this purpose we introduce test functions on the flow domain. In our trivial example the flow domain is given by the whole real line, the x-axis, which is denoted by $\mathbb{R}$.

Definition 2.1 A function $\xi : \mathbb{R} \to \mathbb{R}$ is called a test function on $\mathbb{R}$ if

(i) ξ is infinitely many times differentiable on $\mathbb{R}$;

(ii) ξ vanishes identically at large distances from the origin.

The class of all such test functions is denoted by $C_0^\infty(\mathbb{R})$.

Example 2.2 Consider the function

$$J(x) = \begin{cases} \exp\left(\frac{-x^2}{1-x^2}\right) & \text{for } |x| < 1, \\ 0 & \text{for } |x| \geq 1. \end{cases}$$

One easily verifies that this function is infinitely many times differentiable with respect to x. Then for any $x_0 \in \mathbb{R}$ and for any $c, \varepsilon > 0$, the shifted and scaled functions

$$\xi(x) = cJ\left(\frac{x - x_0}{\varepsilon}\right), \quad -\infty < x < \infty,$$

are test functions: they are also infinitely many times differentiable on $\mathbb{R}$ and they are zero outside the interval $(x_0 - \varepsilon,\, x_0 + \varepsilon)$.

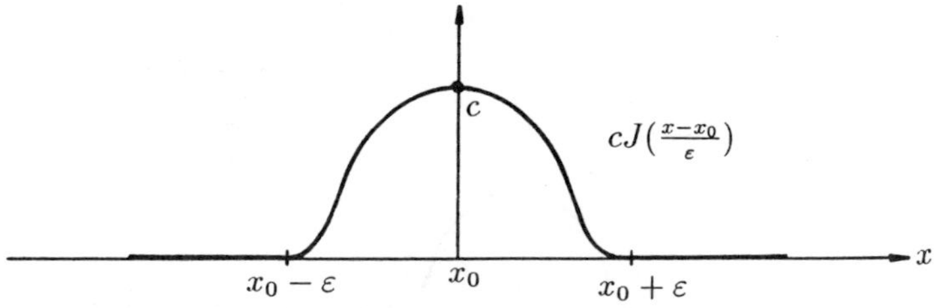

Fig. 2.2: Test function from $C_0^\infty(\mathbb{R})$.

Let us first assume that q is smooth, i.e. continuous differentiable on $\mathbb{R}$, and satisfies

$$\frac{dq}{dx} = f \text{ on } \mathbb{R}. \tag{2.6}$$

Multiplying the equation by an arbitrarily chosen test function ξ from $C_0^\infty(\mathbb{R})$ and integrating the result gives

$$\int_{\mathbb{R}} \frac{dq}{dx}\xi dx = \int_{\mathbb{R}} f\xi dx.$$

Integrating by parts and using $\xi(\pm\infty) = 0$, yields

$$-\int_{\mathbb{R}} q\frac{d\xi}{dx}dx = \int_{\mathbb{R}} f\xi dx \quad \text{for all} \quad \xi \in C_0^\infty(\mathbb{R}). \tag{2.7}$$

With respect to this expression we observe:
(i) When q is smooth, (2.7) implies - in the reverse direction - (2.6). To see this we use again integration by parts to put (2.7) into the form

$$\int_{\mathbb{R}} \left(\frac{dq}{dx} - f\right)\xi dx = 0 \quad \text{for all} \quad \xi \in C_0^\infty(\mathbb{R}). \tag{2.8}$$

We argue by contradiction. Suppose there exists $x_0 \in \mathbb{R}$ at which

$$\left(\frac{dq}{dx} - f\right)(x_0) \neq 0, \text{ say } \left(\frac{dq}{dx} - f\right)(x_0) > 0.$$

Then also $\frac{dq}{dx} - f > 0$ in a small neighbourhood of x_0 (by continuity). Choosing a test function as in Figure 2.3, would lead to the inequality

$$\int_{\mathbb{R}} \left(\frac{dq}{dx} - f\right)\xi dx > 0,$$

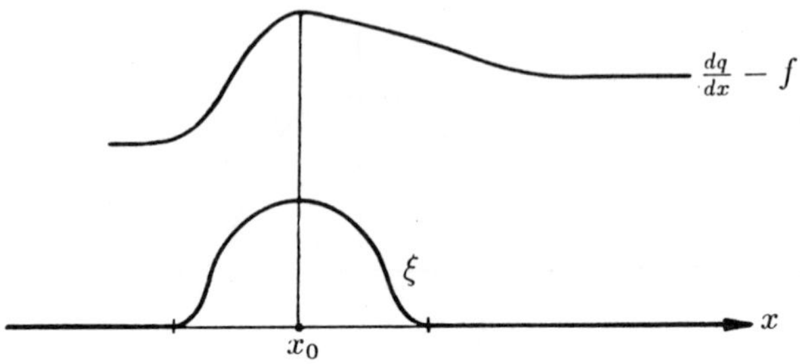

Fig. 2.3: Choice of test function.

which contradicts (2.8). Hence $\left(\frac{dq}{dx} - f\right)(x_0) = 0$.

(ii) When q is non-smooth, say with a discontinuity as in (2.4), expression (2.7) still holds.

One often writes for (2.7), instead of using test functions,

$$\frac{dq}{dx} = f \text{ in sense of distributions,} \tag{2.9}$$

to indicate that for smooth q, expression (2.7) implies (2.6).

To answer the question posed earlier (Q_1), we compute div $\boldsymbol{q}$, with $\boldsymbol{q}$ given by (2.1) and (2.2), in sense of distributions. In view of (2.7) and (2.9) this means, that we have to determine for any $\xi \in C_0^\infty(\mathbb{R})$:

$$\begin{aligned} \int_{\mathbb{R}} \boldsymbol{q} \cdot \text{ grad } \xi \, dx &= \int_{\mathbb{R}} q(x) \tfrac{d\xi}{dx} dx \\ &= \int_{-\infty}^{0} \tfrac{Q}{2A} \tfrac{d\xi}{dx} dx - \int_0^\infty \tfrac{Q}{2A} \tfrac{d\xi}{dx} dx \\ &= \tfrac{Q}{2A}\{\xi(0) - \xi(-\infty) - \xi(\infty) + \xi(0)\}. \end{aligned}$$

Using

$$\xi(\pm\infty) = 0, \tag{2.10}$$

we obtain

$$\int_{\mathbb{R}} \boldsymbol{q} \cdot \text{ grad } \xi \, dx = \frac{Q}{A}\xi(0) \quad \text{for all } \xi \in C_0^\infty(\mathbb{R}). \tag{2.11}$$

Note that the right hand side of this expression only depends on the value of the test function at $x = 0$. This expresses the fact that dq/dx only acts in the origin (see also (2.5)). Instead of (2.11) one writes

$$\text{div } \boldsymbol{q} = \frac{dq}{dx} = -\frac{Q}{A}\delta_0 \quad \text{in sense of distributions,} \tag{2.12}$$

where δ_0 denotes the one-dimensional Dirac-distribution (or delta distribution) at the origin. In many engineering texts one identifies the Dirac distribution with a "function", the socalled delta function $\delta(x-0) = \delta(x)$, which has the following properties: (i) $\delta(x) \geq 0$, (ii) $\delta(x) = 0$ for all $x \neq 0$ and (iii) $\int_{\mathbb{R}} \delta(x)dx = 1$. With this interpretation in mind, equation (2.12) can be considered as the differential form of the continuity equation for the fluid. In our one dimensional setting, the point $x = 0$ acts as a sink of strength Q/A.

Remark 2.3 When a one-dimensional discharge field $\boldsymbol{q} = q(x)\boldsymbol{e}_x$, which is differentiable outside the origin, satisfies (2.11), then it must be of the form

$$q(x) = \begin{cases} q_1 & \text{for } x < 0 \\ q_2 & \text{for } x > 0, \end{cases} \tag{2.13}$$

where q_1 and q_2 are constants. This can be seen by taking appropriate test functions in (2.11) and by integrating by parts. Substitution of (2.13) into expression (2.11) leads to

$$q_1 - q_2 = Q/A. \tag{2.14a}$$

Imposing a symmetry condition on $\boldsymbol{q}$, requires

$$q_1 + q_2 = 0, \tag{2.14b}$$

which shows that q satisfies (2.2). Thus expression (2.11), or equivalently (2.12), and a symmetry condition completely determines the flow field from Figure 2.1.

Writing $\boldsymbol{q}$ as the gradient of a potential Φ, i.e.

$$\boldsymbol{q} = -\text{ grad } \Phi \quad \text{or} \quad q = -\frac{d\Phi}{dx},$$

implies

$$\Phi = \Phi(x) = \frac{Q}{2A}|x| \quad \text{for } -\infty < x < \infty. \tag{2.15}$$

Here the constant of integration has been set equal to zero. From (2.11) it follows that this potential satisfies.

$$\int_{\mathbb{R}} \text{grad } \Phi \cdot \text{ grad } \xi \; dx = -\frac{Q}{A}\xi(0) \quad \text{for all} \quad \xi \in C_0^\infty(\mathbb{R}), \tag{2.16}$$

or equivalently

$$\Delta\Phi = \frac{d^2\Phi}{dx^2} = \frac{Q}{A}\delta_0 \quad \text{in sense of distributions.} \tag{2.17}$$

2.2 Divergence in sense of distributions: two space dimensions

Consider the flow of a fluid towards a well in a vertically confined aquifer of constant thickness $H[m]$. The aquifer is bounded above and below by an impermeable boundary and extends in all horizontal directions. The well is located in the origin of the horizontal plane and penetrates over the full height of the aquifer, see Figure 2.4. Fluid is being withdrawn at a rate $Q[m^3/s]$.

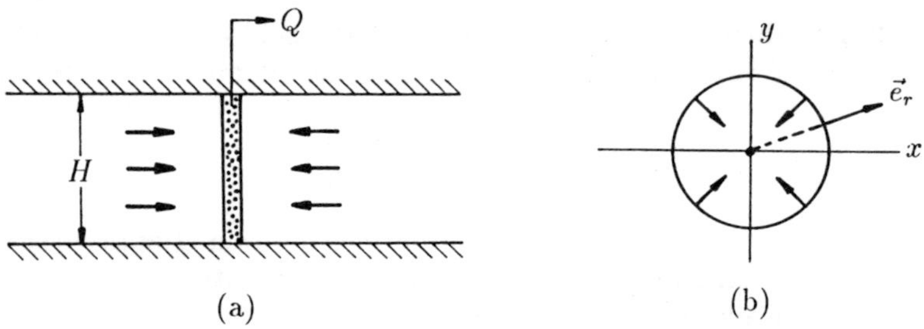

Fig. 2.4: Two dimensional flow; (a) cross section, (b) top view.

Assuming that the flow is uniform over the height of the aquifer, we obtain a two dimensional problem with the x,y plane (also denoted by $\mathbb{R}^2$) as flow domain.
Let $D_r(O), r > 0$, be the disk centered at the origon O with radius r and let $\partial D_r(O)$ denote its boundary. Using the incompressibility of the fluid gives for the continuity equation

$$Q + H \int_{\partial D_r(O)} \boldsymbol{q} \cdot \boldsymbol{e}_r \, ds = 0, \tag{2.18}$$

where again $\boldsymbol{q}$ denotes the fluid specific discharge and where $\boldsymbol{e}_r$ is the radial unit vector in $\mathbb{R}^2$. Assuming that the flow is axially symmetric, i.e.

$$\boldsymbol{q} = q(r)\boldsymbol{e}_r, \tag{2.19}$$

in which r is the distance towards the well, one finds from (2.18)

$$q(r) = -\frac{Q}{2\pi H r}, \quad r > 0. \tag{2.20}$$

Again we ask ourselves

$$\text{What can be said about div } \boldsymbol{q}? \tag{Q_2}$$

Clearly

$$\text{div } \boldsymbol{q} = \frac{1}{r}\frac{d}{dr}(rq(r)) = 0 \quad \text{for all} \quad r > 0. \tag{2.21}$$

However at $r = 0$, $\boldsymbol{q}$ has a singularity. Therefore again div $\boldsymbol{q}$ cannot be evaluated in the classical sense at the origin of the flow domain. As in the previous section we shall introduce the concept of divergence in sense of distributions, which allows us to consider the divergence of vector fields which are only integrable. For this we need again test functions, now defined on $\mathbb{R}^2$. Here the class of all such test functions is denoted by $C_0^\infty(\mathbb{R}^2)$.

Example 2.4 Consider the function J from Example 2.2. Let $X_0 = (x_0, y_0) \in \mathbb{R}^2$ and let $c, \varepsilon > 0$. Then the functions

$$\xi(X) = cJ(\rho/\varepsilon), \text{ with } \rho = \sqrt{(x - x_0)^2 + (y - y_0)^2} \text{ and } X = (x, y) \in \mathbb{R}^2,$$

are test functions: they are infinitely many times differentiable in $\mathbb{R}^2$ and they vanish outside the disc $D_\varepsilon(X_0)$.

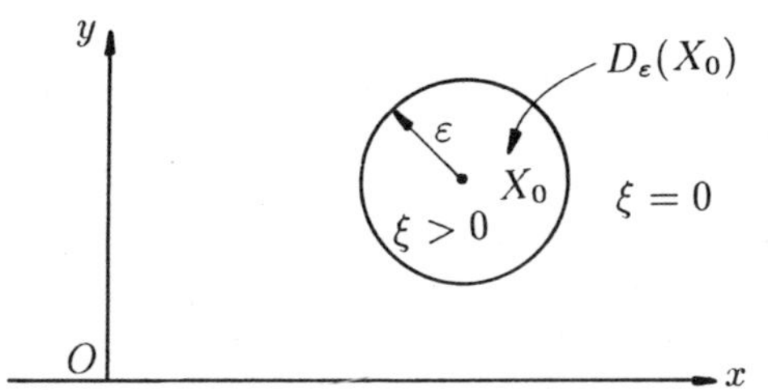

Fig. 2.5: Test function from $C_0^\infty(\mathbb{R}^2)$.

Remark 2.5 If the flow domain is a subset Ω of $\mathbb{R}^2$ (for instance a strip), then we consider test functions defined on Ω such that

(i) ξ is infinitely many times differentiable in Ω;

(ii) ξ vanishes identically in a neighbourhood of the boundary $\partial\Omega$ of Ω and at infinity (if Ω is an unbounded set).

These test functions are denoted by $C_0^\infty(\Omega)$.

To compute the divergence of $\boldsymbol{q}$ in sense of distributions, we proceed as in Section 2.1. Thus for any $\xi \in C_0^\infty(\mathbb{R}^2)$ we consider

$$\int_{\mathbb{R}^2} \boldsymbol{q} \cdot \text{ grad } \xi \, dA = -\frac{Q}{2\pi H} \int_{\mathbb{R}^2} \frac{1}{r} \boldsymbol{e}_r \cdot \text{ grad } \xi \, dA$$

$$= -\frac{Q}{2\pi H} \int_{\mathbb{R}^2} \frac{1}{r} \frac{\partial \xi}{\partial r} dA.$$

Using polar coordinates (r,ϕ) we obtain

$$\int_{\mathbb{R}^2} \boldsymbol{q} \cdot \text{ grad } \xi \, dA = -\frac{Q}{2\pi H} \int_0^{2\pi} \left\{ \int_0^\infty \frac{1}{r} \frac{\partial \xi}{\partial r} r dr \right\} d\phi,$$

which implies

$$\int_{\mathbb{R}^2} \boldsymbol{q} \cdot \text{ grad } \xi \, dA = +\frac{Q}{H} \xi(O) \quad \text{for all } \xi \in C_0^\infty(\mathbb{R}^2). \tag{2.22}$$

This equation is the two dimensional version of (2.11). Note that also here the right hand side only depends on the value of the test function at the origin. Again this expresses the fact that div $\boldsymbol{q}$ only acts in the origin (see also (2.21)). Instead of (2.22) one writes, without test functions,

$$\text{div } \boldsymbol{q} = -\frac{Q}{H} \delta_0 \text{ in sense of distributions,} \tag{2.23}$$

where now δ_0 denotes the two-dimensional Dirac-distribution at the origin. If we interpret this distribution as a "function" $\delta(X_0) = \delta(X)$ (with $X = (x,y)$ denoting a point in $\mathbb{R}^2$) satisfying (i) $\delta(X) \geq 0$, (ii) $\delta(X) = 0$ for all $X \neq 0$ and (iii) $\int_{\mathbb{R}^2} \delta(X) dX = 1$, then (2.23) can be considered as a differential equation expressing the continuity of the fluid. The origin O acts here as a sink of strength Q/H.

As in Remark 2.3, where we considered the one dimensional case, we show here that if a flow field $\boldsymbol{q}$ satisfies (2.22), and in addition is axially symmetric with $\boldsymbol{q} = q(r)\boldsymbol{e}_r$, then q must satisfy (2.20). Thus also in this case the flow field from Figure 2.4 is completely determined by the axially symmetric solution of the integrated equation (2.22). The argument is as follows. Suppose $\boldsymbol{q}$ is smooth outside the origin. Then choosing $\xi \in C_0^\infty(\mathbb{R}^2)$ in (2.22), such that $\xi = 0$ at small and at large distances from the origin, and using the divergence theorem, it follows that

$$0 = \int_{\mathbb{R}^2} \boldsymbol{q} \cdot \text{ grad } \xi \, dA = -\int_{\mathbb{R}^2} \xi \text{ div } \boldsymbol{q} \, dA. \tag{2.24}$$

Next we apply the symmetry of $\boldsymbol{q}$ and we restrict ourselves to axially symmetric test functions $\xi = \xi(r)$. This leads to

$$\int_0^\infty \xi(r) \frac{d}{dr}(rq) dr = 0 \quad \text{for all} \quad \xi \in C_0^\infty((0,\infty)).$$

As in Section 2.1 we use a continuity argument and obtain

$$\frac{d}{dr}(rq) = 0 \quad \text{for all} \quad r > 0,$$

which implies

$$q(r) = \frac{C}{r} \quad \text{for} \quad r > 0, \tag{2.25}$$

where C is the constant of integration. Substituting (2.25) into the left-hand side of (2.22) gives

$$\int_0^{2\pi} \int_0^\infty \frac{C}{r} \frac{\partial \xi}{\partial r} r dr d\varphi = -C 2\pi \xi(O), \tag{2.26}$$

where again we used that ξ vanishes at large distances from the origin. Hence

$$C = -\frac{Q}{2\pi H},$$

resulting in the desired flow (2.20).

Also in this two-dimensional case one can express $\boldsymbol{q}$ in terms of a potential Φ, which is given by

$$\Phi = \Phi(r) = +\frac{Q}{2\pi H} \ln r. \tag{2.27}$$

This potential satisfies the equation

$$\int_{\mathbb{R}^2} \operatorname{grad} \Phi \cdot \operatorname{grad} \xi \ dA = -\frac{Q}{H} \xi(O) \quad \text{for all} \quad \xi \in C_0^\infty(\mathbb{R}^2). \tag{2.28}$$

Equivalently one writes

$$\Delta \Phi = \frac{\partial^2 \Phi}{\partial x^2} + \frac{\partial^2 \Phi}{\partial y^2} = \frac{Q}{H} \delta_0 \ \text{ in sense of distributions.} \tag{2.29}$$

Remark 2.6 A three dimensional example arises, when considering the flow towards a point sink in an unbounded medium. Then the discharge is given by

$$\boldsymbol{q} = q(r)\boldsymbol{e}_r \text{ with } q(r) = -\frac{Q}{4\pi r^2}, \tag{2.30}$$

where $r = \sqrt{x^2 + y^2 + z^2}$ is the distance between a point $X = (x, y, z)$ and the origin -the position of the sink- and where $\boldsymbol{e}_r$ denotes the three dimensional radial unit vector. The equation for $\boldsymbol{q}$ is

$$\int_{\mathbb{R}^3} \boldsymbol{q} \cdot \text{ grad } \xi \; dV = Q\xi(O) \quad \text{for all} \quad \xi \in C_0^\infty(\mathbb{R}^3), \tag{2.31}$$

or equivalently

$$\text{div } \boldsymbol{q} = -Q\delta_0 \text{ in sense of distributions,} \tag{2.32}$$

where δ_0 is the three dimensional Dirac distribution at the origin. Again interpreting this distribution as a "function" $\delta(X - 0) = \delta(X)$ (with $(X = (x,y,z)$ denoting a point in $\mathbb{R}^3$) satisfying (i) $\delta(X) \geq 0$, (ii) $\delta(X) = 0$ for all $X \neq 0$ and (iii) $\int_{\mathbb{R}^3} \delta(X)dX = 1$, equation (2.32) can be considered as the continuity equation for the fluid. The origin O now acts as a sink of strength Q. Again we can express $\boldsymbol{q}$ in terms of a potential Φ, which is now given by

$$\Phi(r) = \frac{Q}{4\pi r}. \tag{2.33}$$

This potential satisfies the equation

$$\int_{\mathbb{R}^3} \text{grad } \Phi \cdot \text{ grad } \xi \; dV = -Q\xi(O) \text{ for all } \xi \in C_0^\infty(\mathbb{R}^3), \tag{2.34}$$

or equivalently

$$\Delta\Phi = \frac{\partial^2\Phi}{\partial x^2} + \frac{\partial^2\Phi}{\partial y^2} + \frac{\partial^2\Phi}{\partial z^2} = Q\delta_0 \text{ in sense of distributions.} \tag{2.35}$$

2.3 Application to two fluid case

Let us suppose that a three dimensional region Ω is occupied by a porous medium in which two fluids are present. A smooth surface S separates Ω into two subregions Ω_1 and Ω_2, such that domain Ω_1, is saturated by one fluid and domain Ω_2 by the other. Let again $\boldsymbol{q}$ denote the specific discharge of the fluids. We write

$$\boldsymbol{q} = \begin{cases} \boldsymbol{q}_1 & \text{on } \Omega_1, \\ \boldsymbol{q}_2 & \text{on } \Omega_2, \end{cases} \tag{2.36}$$

where the vector fields $\boldsymbol{q}_i$ are continuously differentiable on the sub domains Ω_i, up to the surface S. We shall also assume that the fluids are incompressible. This leads to the differential equations

$$\text{div } \boldsymbol{q}_i = 0 \quad \text{on } \Omega_i \ (i = 1, 2). \tag{2.37}$$

Without further conditions on the flow field $\boldsymbol{q}$, we shall determine the divergence across the surface S. To do this we need again test functions, taken from the class $C_0^\infty(\Omega)$, which enable us to compute the divergence in sense of distributions. As in the previous sections, this means that we have to consider the integrals

$$\int_\Omega \boldsymbol{q} \cdot \text{ grad } \xi \; dV \quad \text{with} \quad \xi \in C_0^\infty(\Omega). \tag{2.38}$$

First we write

$$\int_\Omega \boldsymbol{q} \cdot \text{ grad } \xi \; dV = \int_{\Omega_1} \boldsymbol{q}_1 \cdot \text{ grad } \xi \; dV + \int_{\Omega_2} \boldsymbol{q}_2 \cdot \text{ grad } \xi \; dV.$$

Because $\boldsymbol{q}_i$ is differentiable on Ω_i and satisfies (2.37), we can apply the divergence theorem to the integrals on the right hand side. This results in

$$\begin{aligned} \textstyle\int_{\Omega_i} \boldsymbol{q}_i \cdot \text{ grad } \xi \; dV &= \textstyle\int_{\Omega_i} \text{ div } (\xi \boldsymbol{q}_i) \; dV - \int_{\Omega_i} \xi \text{ div } \boldsymbol{q}_i \; dV \\ &= \textstyle\int_{\Omega_i} \text{ div } (\xi \boldsymbol{q}_i) \; dV = \int_{\partial\Omega_i} \xi \boldsymbol{q}_i \cdot \boldsymbol{n} \; dV, \end{aligned}$$

where $\boldsymbol{n}$ denotes the outward unit normal to the boundary $\partial\Omega_i$. Since ξ vanishes near the boundary of $\partial\Omega$, we are left with two ($i = 1, 2$) integrals across the surface S. Hence for every test functions $\xi \in C_0^\infty(\Omega)$ there remains

$$\int_\Omega \boldsymbol{q} \cdot \text{ grad } \xi \; dV = \int_S \boldsymbol{n} \cdot (\boldsymbol{q}_1 - \boldsymbol{q}_2) \; \xi \; dA. \tag{2.39}$$

Here $\boldsymbol{n}$ is the unit normal to the surface S, pointing into domain Ω_2. Thus if the flow $\boldsymbol{q}$ satisfies in addition to the differential equations (2.37) the interface continuity condition

$$\boldsymbol{n} \cdot (\boldsymbol{q}_1 - \boldsymbol{q}_2) = 0 \quad \text{along } S, \tag{2.40}$$

then

$$\int_{\Omega} \boldsymbol{q} \cdot \text{ grad } \xi \, dV = 0 \quad \text{for all } \xi \in C_0^\infty(\Omega), \tag{2.41a}$$

which means that

$$\text{div } \boldsymbol{q} = 0 \text{ in sense of distributions.} \tag{2.41b}$$

Conversely, if a vector field $\boldsymbol{q}$, which is continuously differentiable in Ω_1 and in Ω_2 and continuous up to both sides of S, satisfies (2.41), then $\boldsymbol{q}$ also satisfies (2.37) and the interface condition (2.40). This follows by choosing different test functions ξ in (2.41). First let $\xi \in C_0^\infty(\Omega_1)$, integrate (2.41a) by parts and apply a continuity argument as in Section 2.1. This leads to div $\boldsymbol{q}_1 = 0$ in Ω_1. Next choose $\xi \in C_0^\infty(\Omega_2)$ and repeat the procedure. This yields div $\boldsymbol{q}_2 = 0$ in Ω_2. Then we can apply the argument below (2.38) which gives here

$$0 = \int_S \boldsymbol{n} \cdot (\boldsymbol{q}_1 - \boldsymbol{q}_2) \, \xi \, dA \quad \text{for all } \xi \in C_0^\infty(\Omega). \tag{2.42}$$

Now suppose there is a point $X_0 \in S$ where

$$\boldsymbol{n} \cdot (\boldsymbol{q}_1 - \boldsymbol{q}_2)(X_0) \neq 0, \quad \text{say } > 0. \tag{2.43}$$

By a continuity argument, there exists $\varepsilon > 0$ such that also $\boldsymbol{n} \cdot (\boldsymbol{q}_1 - \boldsymbol{q}_2) > 0$ on the intersection of S with the ball $B_\varepsilon(X_0)$ (i.e. a ball, centered at X_0 with radius ε), see the shaded part of S in Figure 2.7.

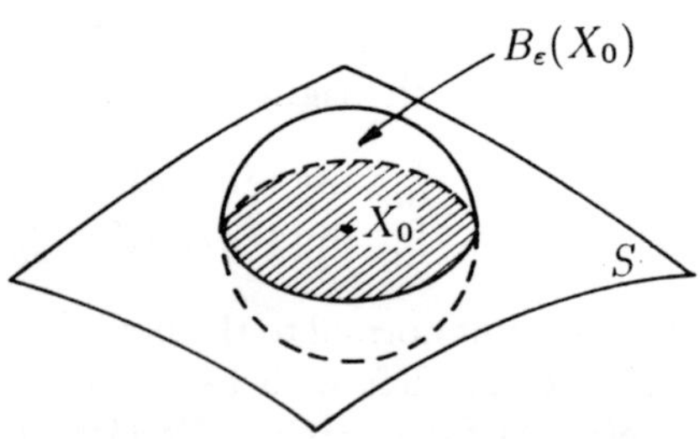

Fig. 2.7: Intersection of ball and surface.

Then for $\xi \in C_0^\infty(\Omega)$ chosen such that $\xi(X_0) > 0$, $\xi \geq 0$ in $B_\varepsilon(X_0)$ and $\xi = 0$ outside $B_\varepsilon(X_0)$, we obtain

$$\int_S \boldsymbol{n} \cdot (\boldsymbol{q}_1 - \boldsymbol{q}_2) \, \xi \, dA > 0,$$

which contradicts (2.42). Thus points $X_0 \in S$ where (2.43) holds do not exist. Therefore we conclude that both descriptions are equivalent. In other words

$$(2.37) \text{ and } (2.40) \Longleftrightarrow (2.41).$$

Expression (2.39) can also be written as a differential equation in sense of distributions. In case of a bounded surface S this equation is obtained as follows.
Divide S into N small surface elements by choosing a partition $\{A_k\}_{k=1}^N$ satisfying $\cup_{k=1}^N A_k = S$ and $A_k \cap A_j = \emptyset$ for every $k \neq j$. Within each element A_k we choose a point X_k such that

$$\int_{A_k} \boldsymbol{n} \cdot (\boldsymbol{q}_1 - \boldsymbol{q}_2)\ \xi\ dA = \boldsymbol{n} \cdot (\boldsymbol{q}_1 - \boldsymbol{q}_2)(X_k)\xi(X_k)\Delta A_k,$$

where ΔA_k denotes the measure of element A_k. The existence of such points follows from the mean value theorem for integrals. In principle they depend on the choice of test function ξ. However when N is large and ΔA_k small (below we let $N \to \infty$ and $\Delta A_k \to 0$) we disregard this dependence. Thus instead of (2.39) we have for all $\xi \in C_0^\infty(\Omega)$

$$\int_\Omega \boldsymbol{q} \cdot \text{ grad } \xi\ dV = \sum_{k=1}^{N} \boldsymbol{n} \cdot (\boldsymbol{q}_1 - \boldsymbol{q}_2)(X_k)\xi(X_k)\Delta A_k.$$

By comparing this expression with (2.31) we observe that each point X_k acts as a sink of strength $\boldsymbol{n} \cdot (\boldsymbol{q}_1 - \boldsymbol{q}_2)(X_k)\Delta A_k$ in the three dimensional space Ω. As in (2.32) we may also write

$$\text{div } \boldsymbol{q} = -\sum_{k=1}^{N} \boldsymbol{n} \cdot (\boldsymbol{q}_1 - \boldsymbol{q}_2)(X_k)\Delta A_k \delta_{X_k} \quad \text{in sense of distributions,}$$

where δ_{X_k} denotes the three dimensional Dirac-distribution at the point $X_k \in S$. Letting $N \to \infty$ in the right-hand side, we obtain the differential form of expression (2.39):

$$\text{div } \boldsymbol{q} = -\int_{X_s \in S} \boldsymbol{n} \cdot (\boldsymbol{q}_1 - \boldsymbol{q}_2)(X_s)\delta_{X_s} dA \quad \text{in sense of distributions,} \tag{2.44}$$

or as a "function",

$$\text{div } \boldsymbol{q}(X) = -\int_{X_s \in S} \boldsymbol{n} \cdot (\boldsymbol{q}_1 - \boldsymbol{q}_2)(X_s)\delta(X - X_s)\ dA, \tag{2.45}$$

where $X \in \Omega \backslash S$.

2.4 Vorticity in sense of distributions

In this section we consider again the situation where two fluids are present in the porous medium. However, instead of the incompressibility conditions (2.37), we assume here that no vorticity is present in the fluids, i.e.

$$\text{curl}\ \boldsymbol{q} = \boldsymbol{\nabla} \times \boldsymbol{q} = \mathbf{0} \quad \text{on} \quad \Omega_i (i = 1, 2). \tag{2.46}$$

Without further conditions on the flow field, we shall now determine the vorticity contribution from the surface S. This we do by computing the vorticity in sense of distributions. On the analogy of the divergence, we compute here

$$\int_\Omega \boldsymbol{q} \cdot \text{curl}\ \boldsymbol{\xi}\ dV, \tag{2.47}$$

where $\boldsymbol{\xi} = (\xi_x, \xi_y, \xi_z)$ is an arbitrary smooth vector function whose components are taken from the class $C_0^\infty(\Omega)$. Again we write

$$\int_\Omega \boldsymbol{q} \cdot \text{curl}\ \boldsymbol{\xi}\ dV = \int_{\Omega_1} \boldsymbol{q}_1 \cdot \text{curl}\ \boldsymbol{\xi}\ dV + \int_{\Omega_2} \boldsymbol{q}_2 \cdot \text{curl}\ \boldsymbol{\xi}\ dV. \tag{2.48}$$

In the right-hand side we use the identity

$$\begin{aligned} \text{div}\ (\boldsymbol{\xi} \times \boldsymbol{q}) &= \boldsymbol{q} \cdot \text{curl}\ \boldsymbol{\xi} - \boldsymbol{\xi} \cdot \text{curl}\ \boldsymbol{q} \\ &= \boldsymbol{q} \cdot \text{curl}\ \boldsymbol{\xi} \qquad\qquad \text{in } \Omega_1 \cup \Omega_2 \end{aligned}$$

as a consequence of (2.46). Thus expression (2.48) becomes

$$\int_\Omega \boldsymbol{q} \cdot \text{curl}\ \boldsymbol{\xi}\ dV = \int_{\Omega_1} \text{div}\ (\boldsymbol{\xi} \times \boldsymbol{q})\ dV + \int_{\Omega_2} \text{div}\ (\boldsymbol{\xi} \times \boldsymbol{q})\ dV.$$

The application of the divergence theorem and the fact that $\boldsymbol{\xi}$ vanishes near the boundary of Ω yields

$$\int_\Omega \boldsymbol{q} \cdot \text{curl}\ \boldsymbol{\xi}\ dV = \int_S \{\boldsymbol{\xi} \times (\boldsymbol{q}_1 - \boldsymbol{q}_2)\} \cdot \boldsymbol{n}\ dA, \tag{2.49}$$

where $\boldsymbol{n}$ is the unit normal to the surface S pointing into domain Ω_2 and where $\boldsymbol{q}_1$ and $\boldsymbol{q}_2$ denote the specific discharge just above and below the interface, respectively.

Using the vector identity $\boldsymbol{A} \cdot (\boldsymbol{B} \times \boldsymbol{C}) = (\boldsymbol{A} \times \boldsymbol{B}) \cdot \boldsymbol{C}$, gives

$$\{\boldsymbol{\xi} \times (\boldsymbol{q}_1 - \boldsymbol{q}_2)\} \cdot \boldsymbol{n} = -\{\boldsymbol{n} \times (\boldsymbol{q}_1 - \boldsymbol{q}_2)\} \cdot \boldsymbol{\xi},$$

and equation (2.49) becomes

$$\int_\Omega \boldsymbol{q} \cdot \operatorname{curl} \boldsymbol{\xi}\, dV = -\int_S \{\boldsymbol{n} \times (\boldsymbol{q}_1 - \boldsymbol{q}_2)\} \cdot \boldsymbol{\xi}\, dA, \tag{2.50}$$

for all test vectors $\boldsymbol{\xi}$ with components in $C_0^\infty(\Omega)$.

Instead of (2.50) one also writes, compare (2.39) and (2.44)

$$\operatorname{curl} \boldsymbol{q} = \int_{X_s \in S} \{\boldsymbol{n} \times (\boldsymbol{q}_1 - \boldsymbol{q}_2)\}(X_s)\delta_{X_s} dA \text{ in sense of distributions} \tag{2.51}$$

or as a "function"

$$\operatorname{curl} \boldsymbol{q}(X) = \int_{X_s \in S} \{\boldsymbol{n} \times (\boldsymbol{q}_1 - \boldsymbol{q}_2)\}(X_s)\delta(X - X_s) dA, \tag{2.52}$$

for every $X \in \Omega \backslash S$.

3 Interface models

Having introduced some basic physical and mathematical concepts in Sections 1 and 2, we now turn to the description of the movement of fresh and salt groundwater. We shall treat here two flow problems. One arises in a two-dimensional model for fresh-salt water flow in a horizontally extended coastal aquifer and the other in an axially symmetric model describing the influence of a fresh water well, say for drinking water purposes, on the salt distribution (the coning problem).

In both cases we suppose that the scale of the problem is sufficient large, so that an abrupt change in specific weight, from fresh water with γ_f to salt water with γ_s, can be assumed. Thus the models discussed below are interface models, leading to free boundary problems in their mathematical formulation.

In many engineering texts similar flow problems, and quite often much more complicated situations, are being studied: see for instance the books by Bear (1972, 1979), Bear & Verruijt (1987) and Pinder & Gray (1977).
Here we intend to keep the problem as "simple and clean" as possible, thus without many technical (in practice very often important) complications, in order to demonstrate the underlying mathematical structure.

The two-dimensional problem is discussed in detail in a paper by Chan Hong et al. (1989) and the coning problem in a paper by Bruining et al. (1991).

3.1 Formulation of the flow problem

Let the strip

$$\Omega = \{(x, z) : -\infty < x < \infty, 0 < z < H\}$$

denote a vertical cross-section of a horizontally extended aquifer of constant thickness $H > 0$, which is bounded above and below by impervious layers. It consists of a homogeneous and isotropic porous medium, characterized by a porosity $\varepsilon \in (0, 1)$ and a permeability $\kappa \in (0, \infty)$, and it is saturated either by fresh water or by salt water. These fluids are assumed to have the same viscosity $\mu \in (0, \infty)$, see Verruijt (1980), but a different specific weight γ such that

$$\gamma = \begin{cases} \gamma_f & \text{for fresh water,} \\ \gamma_s & \text{for salt water,} \end{cases}$$

where $0 < \gamma_f < \gamma_s < \infty$. Under normal conditions: $\gamma_f = 1000[\text{kg}/m^3]$ and $\gamma_s = 1025[\text{kg}/m^3]$.

We also suppose that the interface separating the fluids can be parametrized as a function of the horizontal coordinate x and that the heavier fluid (salt water) is always below the lighter fluid (fresh water). In mathematical terms this means that there exists a function $u : \mathbb{R} \to [0,H]$ such that

$$S = \{(x,z) \in \Omega : -\infty < x < \infty, \quad z = u(x)\}$$

denotes the interface and

$$\gamma(x,z) = \begin{cases} \gamma_f & \text{on} \quad \Omega_f := \{(x,z) \in \Omega : -\infty < x < \infty, \ z > u(x)\} \\ \gamma_s & \text{on} \quad \Omega_s := \{(x,z) \in \Omega : -\infty < x < \infty, \ z < u(x)\} \end{cases} \tag{3.1}$$

or, using the Heaviside function (2.3),

$$\gamma(x,z) = (\gamma_s - \gamma_f)H(u(x) - z) + \gamma_f \quad \text{on } \Omega. \tag{3.2}$$

Finally we require that the fluids are at rest at infinity, i.e. as $x \to \pm\infty$.

Then for a given fresh-salt distribution, thus for a given interface function u, the flow problem is described by the equations

(i) Darcy's law

$$\frac{\mu}{\kappa}\boldsymbol{q} + \text{ grad } p + \gamma \boldsymbol{e}_z = 0 \quad \text{on } \Omega; \tag{3.3}$$

(ii) Continuity

$$\text{div } \boldsymbol{q} = 0 \quad \text{in } \Omega; \tag{3.4}$$

(iii) Boundary condtions

$$\begin{aligned} &\boldsymbol{q} \cdot \boldsymbol{n} = 0 \quad \text{on } \partial\Omega, \\ &|\boldsymbol{q}|(x,z) \to 0 \ \text{ as } x \to \pm\infty \text{ for every } z \in (0,H). \end{aligned} \tag{3.5}$$

Here $\boldsymbol{n}$ denotes the outward unit normal to $\partial\Omega$ and $|\cdot|$ the Eucledian norm. Because γ is discontinuous at the interface between the fluids, equations (3.3) and (3.4) need to be interpreted in a week or distributional sense, similar to what we did in Sections 2.3 and 2.4. Because of the two dimensional description given here, we have to adjust the definition of curl. We consider the curl of a vector function $\mathbf{a} : \mathbb{R}^2 \to \mathbb{R}^2$ as the scalar

$$\text{curl } \mathbf{a} = \frac{\partial a_x}{\partial z} - \frac{\partial a_z}{\partial x}, \tag{3.6}$$

and the curl of a function $\xi : \mathbb{R}^2 \to \mathbb{R}$ as the vector

$$\mathbf{curl}\ \xi = \left(-\frac{\partial \xi}{\partial z}, \frac{\partial \xi}{\partial x}\right). \tag{3.7}$$

Using these definitions, we consider instead of (3.3) and (3.4) the expressions

$$\left.\begin{aligned} &\text{curl}\ (\tfrac{\mu}{\kappa}\boldsymbol{q} + \gamma \boldsymbol{e}_z) = 0 \qquad (3.8)\\ &\text{div}\ \boldsymbol{q} = 0 \qquad (3.9)\end{aligned}\right\} \text{ in sense of distributions,}$$

or equivalently

$$\left.\begin{aligned} &\textstyle\int_\Omega (\tfrac{\mu}{\kappa}\boldsymbol{q} + \gamma \boldsymbol{e}_z)\cdot \mathbf{curl}\ \xi\ dA = 0 \qquad (3.10)\\ &\textstyle\int_\Omega \boldsymbol{q}\cdot\ \text{grad}\ \xi\ dA = 0 \qquad (3.11)\end{aligned}\right\} \text{ for all } \xi \in C_0^\infty(\Omega).$$

Writing

$$\boldsymbol{q} = \begin{cases} \boldsymbol{q}_f & \text{on } \Omega_f \\ \boldsymbol{q}_s & \text{on } \Omega_s \end{cases}$$

and assuming that the restrictions $\boldsymbol{q}_f$ and $\boldsymbol{q}_s$ are differentiable on the subdomains and continuous up to the interface, one deduces from (3.10) and (3.11), by the methods from Sections 2.3 and 2.4, that

$$\text{div}\ \boldsymbol{q}_i = 0 \ \text{ and } \ \text{curl}\ \boldsymbol{q}_i = 0 \ \text{ in } \Omega_i \ \ (i = f, s) \tag{3.12}$$

and

$$(\boldsymbol{q}_f - \boldsymbol{q}_s)\cdot \boldsymbol{n} = 0 \text{ and } (\boldsymbol{q}_f - \boldsymbol{q}_s)\cdot \boldsymbol{s} + \frac{\kappa}{\mu}(\gamma_f - \gamma_s)\boldsymbol{e}_z \cdot \boldsymbol{s} = 0 \ \text{ along } S. \tag{3.13}$$

Here the orthogonal unit vectors $\boldsymbol{n}$ and $\boldsymbol{s}$ are defined in Figure 3.1.

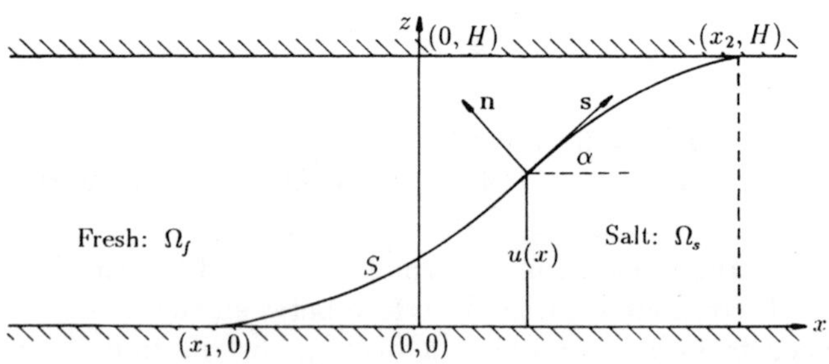

Fig. 3.1: Fresh-salt distribution in horizontal aquifer.

Thus we conclude that the distributional formulation (3.8), (3.9) of the flow problem leads exactly to the same equations and interface conditions as obtained on physical grounds in Section 1.

Remark 3.1. The shear flow condition (the second statement in (3.13), also follows from (2.50) after replacing $\boldsymbol{q}$ by $\frac{\mu}{\kappa}\boldsymbol{q}+\gamma\boldsymbol{e}_z$ and after taking for $\boldsymbol{\xi}$ the vector $\xi\boldsymbol{e}_y$, where $\boldsymbol{e}_y$ is the unit vector perpendicular to Figure 3.1 and $\xi \in C_0^\infty(\Omega)$.

Expressions (3.10) and (3.11) can be combined into one equation by introducing the stream function (see also (1.23))

$$\boldsymbol{q} = \mathbf{curl}\ \Psi \quad \text{in } \Omega. \tag{3.14}$$

Whenever such a function exists we conclude that:

(i) Expression (3.11) is satisfied. This results immediately after integrating by parts.

(ii) From the boundary conditions

$$\Psi(x,0) = \Psi(x,H) = 0 \text{ for every } -\infty < x < \infty. \tag{3.15}$$

(iii) Expression (3.10) leads to the Ψ-equation

$$\int_\Omega (\frac{\mu}{\kappa}\mathbf{curl}\ \Psi + \gamma\boldsymbol{e}_z)\cdot \mathbf{curl}\ \xi\ dA = 0,$$

or

$$\int_\Omega \{\frac{\mu}{\kappa} \text{ grad } \Psi \cdot \text{ grad } \xi + \gamma\frac{\partial\xi}{\partial x}\} = 0 \text{ for all } \xi \in C_0^\infty(\Omega), \tag{3.16}$$

which is the same as saying

$$-\Delta\Psi = \frac{\kappa}{\mu}\frac{\partial\gamma}{\partial x} \text{ in sense of distributions.} \tag{3.17}$$

Substituting (3.2) into (3.16) gives for every $\xi \in C_0^\infty(\Omega)$

$$\begin{aligned} \textstyle\int_\Omega \text{ grad } \Psi \cdot \text{ grad } \xi\ dA &= -\Gamma \textstyle\int_{\mathbb{R}}\{\int_0^{u(x)} \frac{\partial\xi}{\partial x}dz\}dx \\ &= +\Gamma \textstyle\int_{\mathbb{R}} \xi(x,u(x))\frac{du}{dx}(x)dx, \end{aligned} \tag{3.18}$$

where $\Gamma = \frac{\kappa}{\mu}(\gamma_s - \gamma_f)$. If we now compare this expression with (2.39) and (2.49), where volume and surface integrals are considered, we find that the stream function satisfies the differential equation

$$-\Delta\Psi(x,z) = \Gamma\int_{x'\in\mathbb{R}} \frac{du}{dx}(x')\delta(x-x',z-u(x'))dx' \text{ for all } (x,z)\in\Omega. \tag{3.19}$$

Since $-\Delta\Psi = \text{curl}\ \boldsymbol{q}$, this equation clearly shows that the vorticity is concentrated in the interface plane, which indeed can be considered as a vortex sheet, see also de Josselin de Jong (1960).

In the numerical approach to this problem we use the weak formulation (3.18) to obtain a finite element formulation. But first we consider the time evolution of the interface.

3.2 The time evolution

Let $u = u(x,t)$ denote the height of the fresh-salt interface at a certain time $t > 0$. Then the corresponding stream function can in principle be found by solving equation (3.16) with the boundary conditions (3.15). To stress the time dependence we write $\Psi = \Psi(x,z,t)$. From the stream function the displacement of the interface is calculated as follows.
Introduce

$$Q(x,t) = \int_0^{u(x,t)} \boldsymbol{q} \cdot \boldsymbol{e}_x \, dz, \tag{3.20}$$

as the total discharge of salt water in the horizontal, positive x-direction. With the definition (3.7) and the boundary condition $\Psi(x,0,t) = 0$, we find that

$$Q(x,t) = -\Psi(x, u(x,t), t). \tag{3.21}$$

Continuity of the salt water requires, see de Josselin de Jong (1981),

$$\varepsilon \frac{\partial u}{\partial t} + \frac{\partial Q}{\partial x} = 0. \tag{3.22}$$

We combine equations (3.21) and (3.22) to find the *interface motion equation*

$$\varepsilon \frac{\partial u}{\partial t}(x,t) = \frac{\partial}{\partial x}\{\Psi(x, u(x,t), t)\}, \tag{3.23}$$

in the region $(x,t) \in \mathbb{R} \times \mathbb{R}^+$. Note that the derivation of (3.23) is only formal because we have not enough information about the regularity of u and Ψ.
Thus the time dependent flow problem is described by the differential equations (3.18) and (3.23), subject to the boundary conditions (3.15) and the intial condition

$$u(x,0) = u_0(x) \text{ for } -\infty < x < \infty. \tag{3.24}$$

Now suppose the initial distribution is of the form as in Figure 3.1, i.e. there exist numbers $-\infty < a_f < a_s < \infty$ such that

$$u_0(x) = \begin{cases} 0 & -\infty < x \leq a_f \\ \in (0,H) & a_f < x < a_s \\ H & a_s \leq x < \infty. \end{cases} \tag{3.25}$$

Based on the nature of the problem we conjecture that for every $t > 0$, the interface height $u(\cdot, t)$ also is of the form as in Figure 3.1. More precisely, we conjecture that for $i = f, s$ there exist functions $S_i : [0,\infty) \to \mathbb{R}$, with $S_i(0) = a_i$ and $S_f < S_s$, so that the set

$$M := \{(x,t) \in \mathbb{R} \times \mathbb{R}^+ : u(x,t) \in (0,H)\},$$

can be written as

$$M = \{(x,t) \in \mathbb{R} \times \mathbb{R}^+ : S_f(t) < x < S_s(t)\}.$$

Equivalently one can define, for any $t \geq 0$,

$$S_f(t) = \sup\{x \in \mathbb{R} : u(s,t) = 0 \text{ for all } s \leq x\}. \tag{3.26a}$$

and

$$S_s(t) = \inf\{x \in \mathbb{R} : u(x,t) = H \text{ for all } s \geq x\}. \tag{3.26b}$$

Next we concentrate on the differential equations for S_f and S_s. Below we give a formal derivation concerning S_f only. The equation for S_s is found similarly. First we introduce for a given position and time the salt water weight

$$m = \gamma_s \varepsilon \int_{-\infty}^{x} u(s,t)\, ds \geq 0. \tag{3.27}$$

Conversely, expression (3.27) defines for a given $m > 0$ a curve $(L(t;m),t)$ in the x,t-plane for which the total salt water weight on the left is constant and equal to m. To obtain an equation for this level curve one has to differentiate equation (3.27) with respect to t. This gives

$$0 = \int_{-\infty}^{L(t;m)} \frac{\partial u}{\partial t}(s,t)ds + u(L(t;m),t)\dot{L}(t;m),$$

where the dot denotes differentiation with respect to time. In the integral we substitute equation (3.23). Then using the boundary condition $\Psi(x,0,t) = 0$ and the fact that $u(x,t) = 0$ for $x \leq S_f(t)$ we obtain the differential equation

$$\varepsilon \dot{L} = -\frac{\Psi(L, u(L,t), t)}{u(L,t)} \quad \text{for } t > 0. \tag{3.28}$$

At $t = 0$ we have the initial condition

$$L(0;m) = L_0,$$

where L_0 is defined by

$$m = \gamma_s \varepsilon \int_{-\infty}^{L_0} u_0(s)ds.$$

By the positivity of u for $L > S_f(t)$, we have for all $t \geq 0$ and for any $0 < m_1 < m_2 < \infty$ that $L(t; m_1) < L(t; m_2)$ and that

$$L(t; m) \downarrow S_f(t) \text{ if and only if } m \downarrow 0. \tag{3.29}$$

Next we assume that also

$$\dot{S}_f(t) = \lim_{m \downarrow 0} \dot{L}(t; m).$$

Then using (3.28), we have for any $t > 0$

$$\varepsilon \dot{S}_f(t) = -\lim_{m \downarrow 0} \frac{\Psi(L(t; m), u(L(t; m), t), t)}{u(L(t; m), t)}$$

and by (3.29)

$$\varepsilon \dot{S}_f(t) = -\lim_{x \downarrow S_f(t)} \frac{\Psi(x, u(x, t), t)}{u(x, t)}. \tag{3.30}$$

For S_s we find for any $t > 0$.

$$\varepsilon \dot{S}_s(t) = \lim_{x \uparrow S_s(t)} \frac{\Psi(x, u(x, t), t)}{H - u(x, t)}. \tag{3.31}$$

Equations (3.30) and (3.31) are not a part of the original problem. However, they will be used in the numerical algorithm.

Remark 3.2. With equation (3.21), one can write (3.30) as

$$\varepsilon \dot{S}_f(t) = \lim_{x \downarrow S_f(t)} \frac{Q(x, t)}{u(x, t)} \quad \text{for } t > 0,$$

where $Q(x, t)/u(x, t)$, for $x > S_f(t)$, denotes the average velocity in the horizontal direction of the salt water. Thus equation (3.30) tells us that the speed of the point S_f, the toe of the interface, is given by the average velocity of the salt water at that point.

Remark 3.3. If the height of the interface u is non-decreasing on $\mathbb{R}$, then equation (3.18) implies that $\Delta\Psi \leq 0$. In view of the boundary conditions this means that $\Psi > 0$ in Ω. In particular, $\Psi(x, u, t) > 0$. Based on (3.30) and (3.31) we therefore expect that S_f is decreasing and that S_s is increasing. Thus the set M expands in time.

3.3 The numerical approach and an example

The problem to be solved is

$$(CP)\begin{cases} \varepsilon\dfrac{\partial u}{\partial t} = \dfrac{\partial}{\partial x}\{\Psi(x,u(x,t),t)\} & \text{for } -\infty < x < \infty,\ t > 0 \\ u(x,0) = u_0(x) & \text{with } -\infty < x < \infty, \end{cases} \qquad (3.32)$$

where the stream function Ψ is apriori unknown and has to be determined, in terms of $u(\cdot,t)$, from equation (3.18) and the boundary conditions (3.15). We propose a numerical approach which is based on an explicit time integration scheme for Problem (CP). Below we give the main steps of the procedure: remaining details can be found in Chan Hong et al. (1989).
We first have to solve the problem for the stream function (for given u) and then to proceed with the integration in time.
For convenience we introduce the dimensionless variables

$$x := x/H\ ,\ z := z/H\ ,\ u := u/H$$

and

$$\Psi := \Psi/(\Gamma H)\ ,\ t = t\Gamma/(\varepsilon H).$$

This leads to a set of equations as above, but with all constants replaced by unity.

3.3.1 Discretization of the Ψ-Problem

We approximate the stream function by means of the finite-element method on the bounded rectangle

$$\Omega_R = (-R,R)\times(0,1),\ \text{with } R \text{ large enough,}$$

for which we impose the boundary condition $\Psi|_{\partial\Omega_R} = 0$. Let u^n denote the interface at time t^n, u_h^n a piecewise linear approximation and $\mathcal{T}_h$ a triangularization of $\bar{\Omega}_R$. Then the finite-element formulation for piecewise linear functions is given by:
Find $\Psi_h \in V_h$ satisfying

$$\int_{\Omega_R} \text{grad } \Psi_h \cdot \text{ grad } v_h \, dA = \int_{-R}^{R} v_h(x,u_h^n(x))\frac{du_h^n}{dx}(x)dx \text{ for all } v_h \in V_h \qquad (3.33)$$

where

$$V_h = \{v_h \in C(\bar{\Omega}_R)\ :\ \text{ for all } K \in \mathcal{T}_h, v_h \text{ is linear on } K \text{ and } v_h = 0 \text{ on } \partial\Omega_R\}.$$

Let $\{\phi_i\}$, $i = 1, \cdots, \mathcal{N}$, be the piecewise linear basis functions (they take the value one at one node and vanish at all other nodes). Then Problem (**3.33**) is equivalent to the linear system

$$\sum_{i=1}^{\mathcal{N}} \Psi_i \int_{\Omega_R} \text{grad } \phi_i \cdot \text{ grad } \phi_j \, dA = \int_{-R}^{R} \phi_j(x, u_h^n(x)) \frac{du_h^n}{dx}(x) dx, \qquad (3.34)$$

for all $j = 1, \cdots, \mathcal{N}$, where the constants Ψ_i, $i = 1, \cdots, \mathcal{N}$, are defined by

$$\Psi_h(x, z) = \sum_{i=1}^{\mathcal{N}} \Psi_i \phi_i(x, z), \quad (x, z) \in \Omega_R.$$

For reasons for accuracy we construct a time dependent triangularization of Ω_R in such a way that the discretized interface coincides with sides of triangles of $\mathcal{T}_h$ at each time step. In particular, let

$$\Omega_{hf}^n = \{(x, z) \in \Omega_R : \; z > u_h^n(x)\}$$

and

$$\Omega_{hs}^n = \{(x, z) \in \Omega_R : \; z < u_h^n(x)\}.$$

Then we construct triangularizations $\mathcal{T}_{hf}^n$ and $\mathcal{T}_{hs}^n$ of Ω_{hf}^n and Ω_{hs}^n is such a way that the discretized interface coincides with sides of triangles from $\mathcal{T}_{hf}^n$ and $\mathcal{T}_{hs}^n$. In practice the meshes in Ω_{hf}^n and in Ω_{hs}^n are generated by the mesh generator of the SEPRAN package of the Delft University of Technology. An example is shown in Figure 3.2.

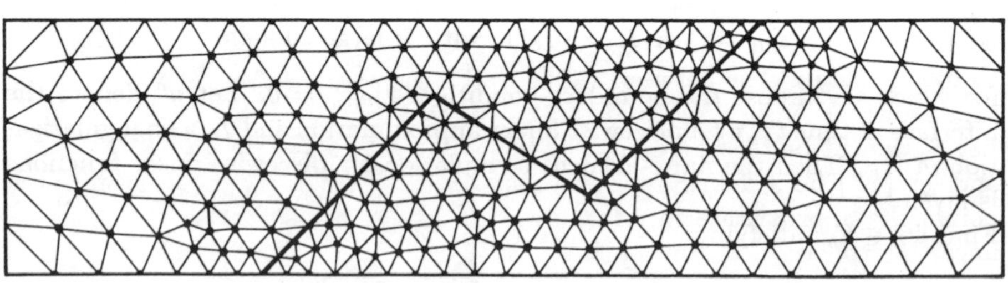

Fig. 3.2: Construction of triangularizations in fresh and salt domain.

We choose a fine mesh in the neighbourhood of the numerical interface

$$S_h^n = \bar{\Omega}_{hf}^n \cap \bar{\Omega}_{hs}^n$$

and in particular near the endpoints of S_h^n and a coarse mesh further away from S_h^n. Setting $S_h^n = \cup_{k=1}^{\mathcal{M}} S_k$, where the S_k are sides of triangles, then another form for the right-hand side of (3.34) is given by

$$\int_{-R}^{R} (u_h^n)'(x)\phi_j(x, u_h^n(x))dx = \sum_{k=1}^{\mathcal{M}} \int_{S_k} \frac{(u_h^n)'}{(1+\{(u_h^n)'\}^2)^{1/2}}\phi_j d\gamma. \qquad (3.35)$$

Here the primes denote differentiation with respect to x.
The linear system (3.34), (3.35) is solved by means of the LU decomposition. To compute an approximation of the discharge field $\boldsymbol{q}$, we set $\boldsymbol{q}_h = \mathbf{curl}\ \Psi_h$. Since Ψ_h is piecewise linear, $\boldsymbol{q}_h$ is constant on each triangle. On a mesh point outside S_h^n, $\boldsymbol{q}_h$ is taken as a weighted average of the discharges in the triangles around that point. However, since $\boldsymbol{q}$ is discontinuous across the interface, we must define two approximated discharge vectors at mesh points on S_h^n corresponding to the discretized fresh water region and to the discretized salt water region. On a mesh point on S_h^n, $\boldsymbol{q}_h^f$ (respectively $\boldsymbol{q}_h^s$) is taken as a weighted average of the discharges on the triangles around that point which lie in Ω_{hf}^n (respectively Ω_{hs}^n).

3.3.2 Discretization of Problem (CP)

Motivated by the fact that, if Ψ is known, the partial differential equation (3.32) is hyperbolic, we use the $S^{\alpha,\beta}$ explicit scheme of Lerat & Peyret (1973) with α, β "optimal" for its discretization, see also Wilders (1983). In addition we make use of the extra equations (3.30), (3.31) for S_f and S_s. This allows us to calculate u only between S_f and S_s. The procedure we propose to use is the following.
Let S_f^n and S_s^n be the approximated values of $S_f(t^n)$ and $S_s(t^n)$ for some $t^n \geq 0$ and let $\{x_i^n\}$ be a distribution of points in the interval (S_f^n, S_s^n). From these points the approximate interface function u_h^u, at time t^n, is constructed by linear interpolation and by setting

$$u_h^n(x) = \begin{cases} 0 & \text{for all } x \leq S_f^n, \\ 1 & \text{for all } x \geq S_s^n. \end{cases} \qquad (3.36)$$

Then we use the $S^{\alpha,\beta}$ scheme to compute an approximation of the interface function, denoted by $\tilde{\tilde{u}}_i^{n+1}$, at time $t^{n+1} > t^n$ and at the "old" points x_i^n. We also use a Runge-Kutta method to compute the new position of the toe and top of the interface, S_f^{n+1} and S_s^{n+1}. This yields a first discretized interface at time t^{n+1}. Then we make a new distribution $\{x_i^{n+1}\}$ of the interval (S_f^{n+1}, S_s^{n+1}), based on a new distribution of coordinates along the interface, and use linear interpolation and (3.36) at time t^{n+1} to compute u_h^{n+1}.

The $S^{\alpha,\beta}$ scheme with α,β optimal

Let

$$u_i^n = u_h^n(x_i)\ ,\ h_i^n = x_{i+1}^n - x_i^n \quad \text{and} \quad \Delta t^n = t^{n+1} - t^n.$$

Then the scheme, which is of second-order and explicit, is given by

$$\tilde{u}_i^n = (1-\beta)u_i^n + \beta u_{i+1}^n + \alpha \frac{\Delta t^n}{h_i^n}\{\Psi_h(x_{i+1}^n, u_{i+1}^n, t^n) - \Psi_h(x_i^n, u_i^n, t^n)\}, \quad (3.37)$$

$$\tilde{\tilde{u}}_i^{n+1} - u_i^n = \frac{\Delta t^n}{\alpha(h_i^n + h_{i-1}^n)}\{(\alpha-\beta)\Psi_h(x_{i+1}^n, u_{i+1}^n, t^n) + (2\beta-1)\Psi_h(x_i^n, u_i^n, t^n)$$

$$+(1-\alpha-\beta)\Psi_h(x_{i-1}^n, u_{i-1}^n, t^n) + \tilde{\Psi}_h(\tilde{x}_i^n, \tilde{u}_i^n, \tilde{t}^n) - \tilde{\Psi}_h(\tilde{x}_{i-1}^n, \tilde{u}_{i-1}^n, \tilde{t}^n)\} \quad (3.38)$$

where $\tilde{x}_i^n = x_i^n + \beta h_i^n$, $\tilde{t}^n = t^n + \alpha(t^{n+1} - t^n)$, the predictor term $\tilde{u}_i^n$ is an approximation of $u(\tilde{x}_i^n, \tilde{t}^n)$ and where $\Psi_h \in V_h$, $\tilde{\Psi}_h \in \tilde{V}_h$ are respectively the solution of

$$\int_\Omega \text{grad}\ \Psi_h \cdot \text{grad}\ v_h\ dA = \int_{-R}^{R} \frac{du_h^n}{dx}(x) v_h(x, u_h^n(x))dx \quad \text{for all } v_h \in V_h \quad (3.39)$$

and

$$\int_\Omega \text{grad}\ \tilde{\Psi}_h \cdot \text{grad}\ \tilde{v}_h\ dA = \int_{-R}^{R} \frac{d\tilde{u}_h^n}{dx}(x) \tilde{v}_h(x, \tilde{u}_h^n(x))dx \quad \text{for all } \tilde{v}_h \in \tilde{V}_h, \quad (3.40)$$

where $\tilde{u}_h^n$ and $\tilde{V}_h$ are the analogs of u_h^n and V_h. The choice $\beta = \frac{1}{2}$ ensures that the scheme (3.37), (3.38) is space centered. Furthermore, when applied to Burgers' equation, the choice $\alpha = 1 + \sqrt{5/2}$ minimizes the dissipative effect of the scheme. Therefore one calls these values of the parameters "optimal".

If follows from (3.39), (3.40) that two new meshes must be generated at each time step: one for the computation of Ψ_h at time t^n, the other for the computation of $\tilde{\Psi}_h$ at time $\tilde{t}^n$.
In order to ensure the stability of the $S^{\alpha,\beta}$ scheme, we choose Δt^n such that it satisfies the Courant-Friedrichs-Lewy (CFL) condition

$$C^n \Delta t^n / h_{\min}^n \leq 1, \quad (3.41)$$

where $h^n_{\min} = \min_i\{h^n_i\}$, and where C^n is an approximation of the maximum of $|\partial\Psi_h/\partial z|$ on both sides on S^n_h. Since $q_{hx} = -\partial\Psi_h/\partial z$, we calculate that

$$C^n = \max_{(x^n_i, u^n_i)\in S^n_h} [\max\{|q^f_{hx}(x^n_i, u^n_i, t^n)|, |q^s_{hx}(x^n_i, u^n_i, t^n)|\}].$$

Next we turn to

The computation of S_f and S_s

We assume that $S^n_f > -R$ and that $S^n_s < R$ and define k^n_f and k^n_s by

$$k^n_f = \min\{i\ |x^n_i > S^n_f\},\ k^n_s = \max\{i\ |x^n_i < S^n_s\}.$$

The second-order Runge-Kutta method applied to equations (3.30) and (3.31) yields

$$\tilde{S}^n_f = S^n_f - \alpha\Delta t^n \frac{\Psi_h(x^n_{k^n_f}, u^n_{k^n_f}, t^n)}{u^n_{k^n_f}},$$

$$\tilde{S}^n_s = S^n_s + \alpha\Delta t^n \frac{\Psi_h(x^n_{k^n_s}, u^n_{k^n_s}, t^n)}{1 - u^n_{k^n_s}},$$

and

$$S^{n+1}_f = S^n_f - \Delta t^n \left[\left(1 - \frac{1}{2\alpha}\right)\frac{\Psi_h(x^n_{k^n_f}, u^n_{k^n_f}, t^n)}{u^n_{k^n_f}} + \frac{1}{2\alpha}\frac{\tilde{\Psi}_h(\tilde{x}^n_{\tilde{k}^n_f}, \tilde{u}^n_{\tilde{k}^n_f}, \tilde{t}^n)}{\tilde{u}^n_{\tilde{k}^n_f}}\right],$$

$$S^{n+1}_s = S^n_s - \Delta t^n \left[\left(1 - \frac{1}{2\alpha}\right)\frac{\Psi_h(x^n_{k^n_s}, u^n_{k^n_s}, t^n)}{u^n_{k^n_s}} + \frac{1}{2\alpha}\frac{\tilde{\Psi}_h(\tilde{x}^n_{\tilde{k}^n_s}, \tilde{u}^n_{\tilde{k}^n_s}, \tilde{t}^n)}{\tilde{u}^n_{\tilde{k}^n_s}}\right],$$

where

$$\tilde{k}^n_f = \min\{i\ |\tilde{x}^n_i > \tilde{S}^n_f\},\ \tilde{k}^n_s = \max\{i\ |\tilde{x}^n_i < \tilde{S}^n_s\}.$$

We impose no linearized stability condition on Δt^n since it would be less constraining than the CFL condition (3.41).

3.3.3 An example

As a test case we computed the evolution of the interface in $\Omega_1 = (-1,1)\times(0,1)$ starting from the initial distribution

$$u_0(x) = \begin{cases} 0, & -1 \le x < -\frac{1}{2}, \\ 2x+1, & -\frac{1}{2} \le x < -\frac{1}{6}, \\ -x+\frac{1}{2}, & -\frac{1}{6} \le x < \frac{1}{6}, \\ 2x, & \frac{1}{6} \le x < \frac{1}{2}, \\ 1, & \frac{1}{2} \le x \le 1. \end{cases} \tag{3.42}$$

The corresponding initial triangularization is shown in Figure (3.2). It was used to determine the approximate stream function $\Psi_h(x,z,t=0)$ and the corresponding approximate discharge field $\boldsymbol{q}_h = \textbf{curl}\ \Psi_h$. Figure 3.3 gives $\Psi_h(x,z,t=0)$ for several values of the height z and Figure 3.4 shows the induced $\boldsymbol{q}_h$. As to be expected from the interface conditions (3.13), $\boldsymbol{q}_h$ is discontinuous along the interface. Consequently $\partial\Psi_h/\partial x$ and $\partial\Psi_h/\partial z$ are also discontinuous. Figure 3.3 clearly shows the jump in $\partial\Psi_h/\partial x$ at the interface points.

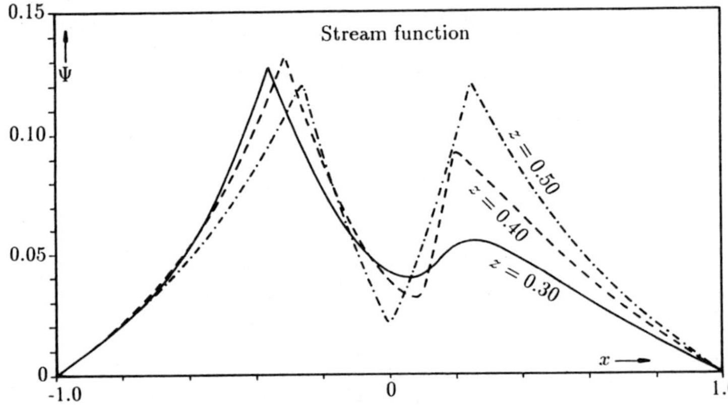

Fig. 3.3: The stream function induced by the initial distribution (3.42).

Due to these discontinuities the space discretization must be fine (at least in a neighbourhood of the interface) in order to obtain an accurate approximation of the stream function along S. With the adaptive mesh, the triangles generated near the interface are smaller than those in the far field, see Figure (3.2).

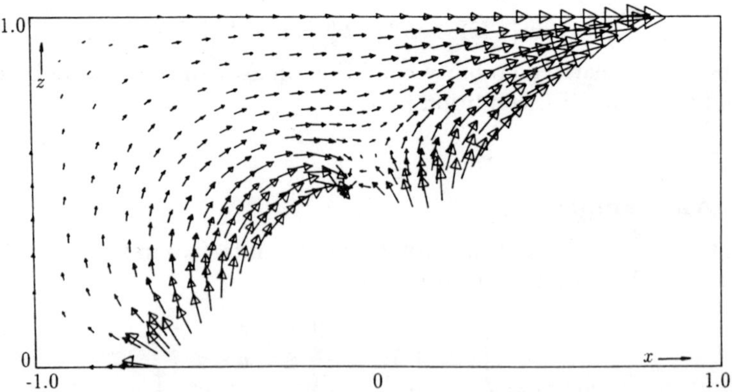

Fig. 3.4: The discharge field in the fresh water induced by the initial distribution (3.42).

In Figures 3.5 and 3.6 the time evolution of the interface u and the corresponding functions S_f and S_s are shown.

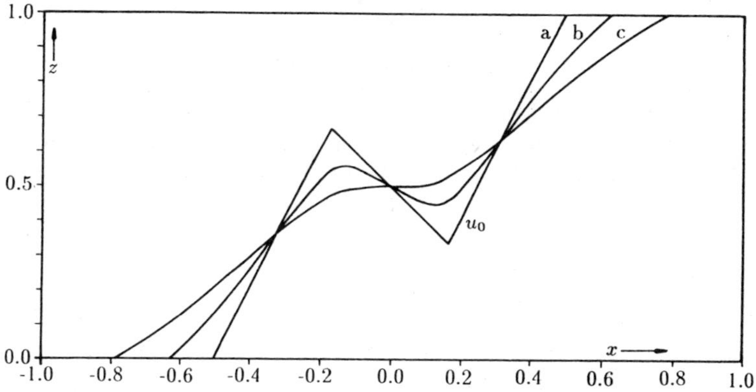

Fig. 3.5: Interfaces computed between S_f and S_s with the moving mesh method. The initial value (a) is given by (3.42). The interfaces are given at the dimensionless times $t = 0 \cdot 14$ (b) and $t = 0 \cdot 38$ (c). We used here $\alpha = 2 \cdot 118$, $\beta = 0 \cdot 50$, and $\Delta t^n = \frac{1}{2}\Delta t^n_{\text{CFL}}$.

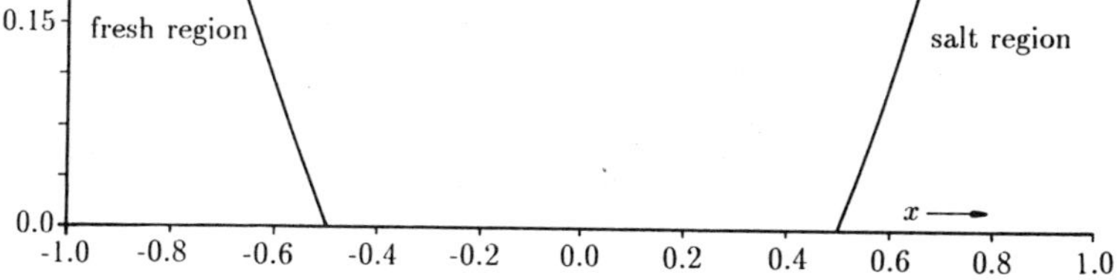

Fig. 3.6: The evolution of the corresponding free boundaries S_f and S_s, see also Figure 3.5

In computing these results, we have taken for the time step

$$\Delta t^n = \frac{1}{2}\Delta t^n_{\text{CFL}}, \tag{3.43}$$

where $\Delta t^n_{\text{CFL}} = h^n_{\min}/C^n$, see inequality (3.41).

Computations were also done with $\Delta t^n = \Delta t^n_{\text{CFL}}$. Then we observed small jumps in the position of the toe and the top, which seem to occur at values of t where the difference $|S_f(t^{n+1}) - S_f(t^n)|$ or $|S_s(t^{n+1}) - S_s(t^n)|$ is equal to the local space step: this difference then has the maximum value allowed by the CFL condition. This has motivated us to perform the numerical computations with time steps satisfying (3.43).

In Figure 3.7 we determined numerically the large time behaviour of the fresh-salt interface, now in the extended domain $\Omega_3 = (-3{,}3) \times (0,1)$. At $t = 0$, we choose again the initial distribution (3.42) which we extended by $u_0 = 0$ to the left and by $u_0 = 1$ to the right. One observes that for large values of t the interface converges towards a rotating line.
In Section 4 we will also encounter a rotating line. There it arises as a self-similar solution of a simplified flow problem (from a Dupuit approximation).

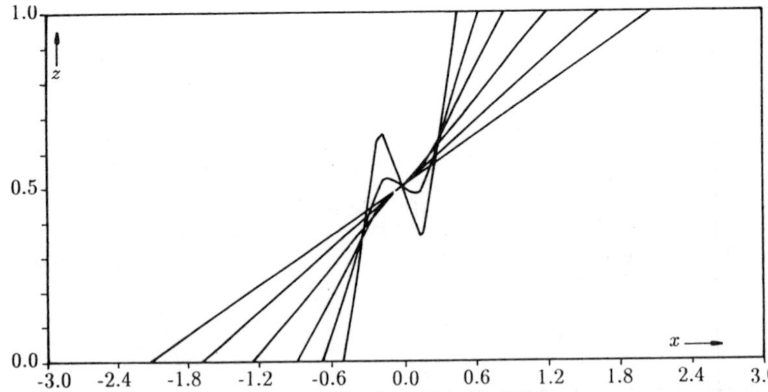

Fig. 3.7: Convergence to the self-similar solution. The interfaces are computed with the moving mesh method. The initial distribution is given by (3.42). The interfaces are shown at dimensionless times $t = 0 \cdot 24$; $t = 0 \cdot 60$; $t = 1 \cdot 43$; $t = 2 \cdot 77$ and $t = 4 \cdot 57$.

3.4 The coning problem

When fresh water is stored in aquifers in coastal areas, as a result of sea water intrusion, one may encounter situations where wells pump back the fresh water from regions above a fresh-salt interface.
As a result of the pumping, and the pressure distribution established in the fluids, the interface below a well will rise, at first slowly and later faster, towards that well. Under certain boundary conditions (e.g. see Bear (1972, p. 553) and Alt & van Duijn (1992)) a new equilibrium may be established before the upconed interface reaches the pumping well. Otherwise, the interface will eventually touch the well in a cusp-like form. The latter situation will be investigated in this section.

In our discussion we consider the reservoir to occupy the full lower half space $\Omega = \{z < 0\}$ with only one well present at the top of the formation, say at the origin. Further, the upper bound $\{z = 0\}$ represents an impermeable wall at which there is a no-flow boundary condition, see Figure 3.8. With respect to the porous medium and the fluids we assume, as in Section 3.1, that all relevant flow parameters are constant.

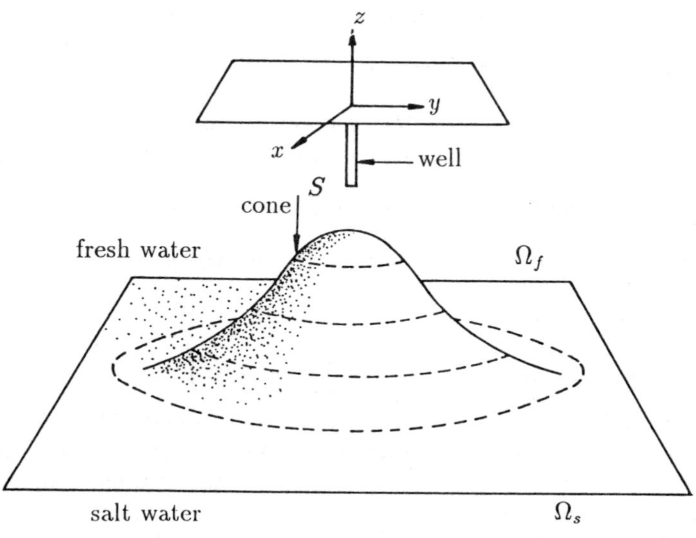

Fig. 3.8: A typical fresh-salt interface system studied in this section. Again Ω_f denotes the fresh water region, Ω_s the salt water region and S the interface.

We study the time evolution of the interface, starting from a flat horizontal initial position. It leads to an interface which is axially symmetric with respect to the z-axis.
We first give a general description of the problem of finding the specific discharge correspondig to a given salt water cone.

3.4.1 General flow problem

For the specific discharge in the reservoir we write

$$\boldsymbol{q} = \boldsymbol{q}_c + \boldsymbol{q}_{\text{well}} \tag{3.44}$$

where $\boldsymbol{q}_c$ represents that part of the flow originating from the salt water cone and where

$$\boldsymbol{q}_{\text{well}} = -\frac{Q}{2\pi r^2}\boldsymbol{e}_r, \tag{3.45}$$

see also ecpressions (2.30). Here the factor 2 arises to ensure that a constant fresh water volume per unit time, Q, is withdrawn from the reservoir. Because the upper boundary is impermeable, we require

$$\boldsymbol{q}_c \cdot \boldsymbol{e}_z = 0 \quad \text{along} \quad \partial\Omega = \{z = 0\}. \tag{3.46}$$

As in Section 3.1 we also decompose

$$\boldsymbol{q} = \begin{cases} \boldsymbol{q}_f & \text{on} \quad \Omega_f \quad (\text{ above the interface } S), \\ \boldsymbol{q}_s & \text{on} \quad \Omega_s \quad (\text{ below the interface } S). \end{cases}$$

Here, again as a result of Darcy's law and fluid imcompressibility,

$$\text{div } \boldsymbol{q}_i = 0 \text{ and } \text{curl } \boldsymbol{q}_i = 0 \text{ on } \Omega_i \ (i = f, s) \tag{3.47}$$

$$\left.\begin{aligned} &(\boldsymbol{q}_s - \boldsymbol{q}_f) \cdot \boldsymbol{n} = 0 && (3.48a) \\ &\left\{\tfrac{\mu}{\kappa}(\boldsymbol{q}_s - \boldsymbol{q}_f) + (\gamma_s - \gamma_f)\boldsymbol{e}_z\right\} \times \boldsymbol{n} = \boldsymbol{0} && (3.48b) \end{aligned}\right\} \text{ along } S$$

where $\boldsymbol{n}$ is the unit normal to S pointing into domain Ω_f. As was noted in Remark 3.1, the shear flow expression (3.48b) follows after substituting $-\text{ grad } p = \frac{\mu}{\kappa}\boldsymbol{q} + \gamma\boldsymbol{e}_z$ into (2.47), working out the details leading to (2.50) and setting the result equal to zero.
The overall equation for the discharge is, compare (2.51),

$$\begin{aligned} \text{curl } \boldsymbol{q} &= -\int_{X_s \in S} \{\boldsymbol{n} \times (\boldsymbol{q}_s - \boldsymbol{q}_f)\}(X_s)\delta_{X_s}\, dA \\ &= \Gamma \int_{X_s \in S} (\boldsymbol{n} \times \boldsymbol{e}_z)(X_s)\delta_{X_s}\, dA \quad \text{on } \Omega, \end{aligned}$$

in sense of distributions. The constant Γ is defined below (3.18). Since curl $\boldsymbol{q}_{\text{well}} = 0$, there results

$$\text{curl } \boldsymbol{q}_c = \Gamma \int_{X_s \in S} (\boldsymbol{n} \times \boldsymbol{e}_z)(X_s)\delta_{X_s}\, dA \quad \text{on } \Omega.$$

To solve for $\boldsymbol{q}_c$, we introduce the vector function (the stream vector) $\boldsymbol{\Psi} : \bar{\Omega} \to \mathbb{R}^3$, with components (Ψ_x, Ψ_y, Ψ_z), which satisfies

$$\boldsymbol{q}_c = \text{ curl } \boldsymbol{\Psi} \text{ and div } \boldsymbol{\Psi} = 0,$$

see also Bruining et al. (1991). Hence

$$\Delta \boldsymbol{\Psi} = -\Gamma \int_{X_s \in S} (\boldsymbol{n} \times \boldsymbol{e}_z)(X_s)\delta_{X_s} dA \quad \text{in } \Omega.$$

By placing an image cone S^{im} in the upper half space, we obtain an equation in the whole three dimensional space.

$$\Delta \boldsymbol{\Psi} = -\Gamma \int_{X_s \in S \cup S^{im}} (\boldsymbol{n} \times \boldsymbol{e}_z)(X_s)\delta_{X_s} dA \quad \text{in } \mathbb{R}^3, \tag{3.49}$$

where the symmetry of the right-hand side implies the no-flow condition (3.46), which in terms of the stream vector means $\Psi_x = \Psi_y = 0$ in $\{z = 0\}$. The solution of (3.49) is given by, see also (2.33) and (2.35)

$$\boldsymbol{\Psi}(X) = \frac{-\Gamma}{4\pi} \int_{X_s \in S \cup S^{im}} (\boldsymbol{n} \times \boldsymbol{e}_z)(X_s) \frac{1}{|X - X_s|} dA \ , \ X \in \mathbb{R}^3. \tag{3.50}$$

The discharge $\boldsymbol{q}$ in the reservoir can now be determined from the expression

$$\boldsymbol{q} = \text{ curl } \boldsymbol{\Psi} + \boldsymbol{q}_{\text{well}} \quad \text{in } \Omega. \tag{3.51}$$

3.4.2 Axial symmetry

Starting from an initially flat interface, located at a distance H below the impermeable upper boundary, we expect that an axially symmetric salt water cone will develop in time. Suppressing the time dependence for the moment, we shall therefore assume that S is of the form

$$S = \big\{(r, \varphi, z) : 0 \le r < \infty, 0 \le \varphi < 2\pi \text{ and } z = h(r) - H\big\} \tag{3.52}$$

where h is a smooth function that satisfies $0 \le h < H$ and $\lim_{r \to \infty} h(r) = 0$.

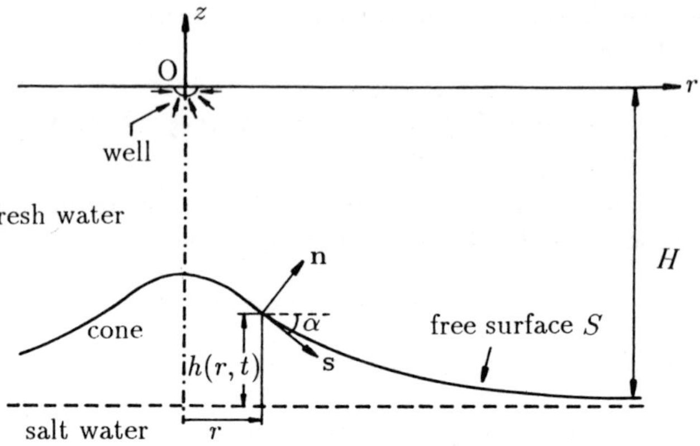

Fig. 3.9: A vertical cross-section of the salt water cone. The height of the interface with repect to the $z = -H$ plane is given as a function of the distance r for a fixed time t.

The symmetry implies that

$$\boldsymbol{n} \times \boldsymbol{e}_z = \sin\alpha(r)\boldsymbol{e}_\varphi \quad \text{along } S. \tag{3.53}$$

Here $\tan\alpha(r) = dh/dr$ and $\boldsymbol{e}_\varphi$ is the unit vector pointing locally in the φ-direction.
Relation (3.53) is used in (3.50), in which we first consider the contribution of the integral along S, denoted by $\boldsymbol{\Psi}_{\text{cone}}$. After some manipulations, for details we refer to Bruining et al. (1991), we obtain that

$$\boldsymbol{\Psi}_{\text{cone}} = \Psi_{\text{cone}}\ \boldsymbol{e}_\varphi,$$

where the function Ψ_{cone} depends only on r and z. It is given by

$$\Psi_{\text{cone}}(r, z) = -\Gamma \int_0^\infty \tan\alpha(r')f(r, z; r')dr'. \tag{3.54a}$$

Here

$$f(r, z; r') = 1/(k\pi)(r'/r)^{1/2}\left[(1 - k/2)K(k) - E(k)\right], \tag{3.54b}$$

where $k = 4r'r/\left[(r' + r)^2 + (z - h(r') + H)^2\right]$, $k \leq 1$, and where K and E are complete elliptic integrals of the first and second kind

$$K(k) = \int_0^{\pi/2} (1 - k\sin^2\theta)^{-1/2}d\theta \text{ and } E(k) = \int_0^{\pi/2} (1 - k\sin^2\theta)^{1/2}d\theta.$$

The function Ψ_{cone} is smooth (infinitely many times differentiable) in the reservoir Ω, away from the interface S. At S it has discontinuous derivatives, which reflects the occurrence of the shear flow. Thus for any point (r, z) not on S, i.e. $z \neq h(r) - H$, we have for the corresponding discharge

$$\boldsymbol{q}_{\text{cone}}(r,z) = -\Gamma \int_0^\infty \tan\alpha(r')\boldsymbol{G}_{\text{cone}}(r,z;r')dr', \qquad (3.55a)$$

in which

$$\boldsymbol{G}_{\text{cone}}(r,z;r') = \text{ curl } \big(f(r,z;r')\mathbf{e}_\varphi\big). \qquad (3.55b)$$

The discharge due to the image cone ($\boldsymbol{q}_{\text{im}}$), placed outside the flow domain, is obtained in a similar way. Together there results for $r > 0$ and $z \neq h(r) - H$

$$(\boldsymbol{q}_{\text{cone}} + \boldsymbol{q}_{\text{im}})(r,z) = -\Gamma \int_0^\infty \tan\alpha(r')\boldsymbol{G}_{\text{tot}}(r,z;r')dr', \qquad (3.56a)$$

where

$$\boldsymbol{G}_{\text{tot}}(r,z;r') = \boldsymbol{G}_{\text{cone}}(r,z;r') - \boldsymbol{G}_{\text{cone}}\big(r, z + 2(h(r') - H); r'\big). \qquad (3.56b)$$

To evaluate the discharge at the interface, necessary to determine its displacement, we follow again Bruining et al. (1991). There it was shown that the fresh water and the salt water discharge at the interface (i.e. the limits $z \downarrow h(r) - H$ and $z \uparrow h(r) - H$) are given by

$$\lim_{z\downarrow(\uparrow)h(r)-H}(\boldsymbol{q}_{\text{cone}} + \boldsymbol{q}_{\text{im}})(r,z) = \boldsymbol{q}_{\text{av}}\big(r, h(r) - H\big) \pm \frac{\Gamma}{2}\sin\alpha(r)\boldsymbol{s}, \qquad (3.57)$$

where the top sign has to be used when we approach the interface from the fresh water region and the lower sign is used when coming from the salt water region. In expression (3.57) we have

$$\boldsymbol{q}_{\text{av}}\big(r, h(r) - H\big) = -\Gamma\ PV \int_0^\infty \tan\alpha(r')\boldsymbol{G}_{\text{tot}}\big(r, h(r) - H; r'\big)dr', \qquad (3.58)$$

in which PV denotes the principle value taken with respect to $r > 0$, and $\boldsymbol{s}$ is the tangential unit vector in the r, z plane, pointing in the direction of increasing r.

The limits in (3.57) account for $\boldsymbol{q}_c$ (from (3.44)) or curl $\boldsymbol{\Psi}$ (from (3.51)) at the interface. To obtain the total discharge at an interface point, the well contribution $\boldsymbol{q}_{\text{well}}$ has to be added.

3.4.3 Time evolution and a numerical result

So far we considered the problem of finding the discharge due to a fresh-salt interface of a given shape: i.e. the function $h(r)$ was supposed to be given. However, our goal is to follow the movement of the interface in time. To emphasize the time dependence, we write now for the height of the interface with respect to the $z = -H$ plane : $h = h(r,t)$, see Figure 3.9. In the computations we start from a situation in which this height is zero, leading to the initial condition

$$h(r,0) = 0 \quad \text{for all } r \geq 0. \tag{3.59}$$

In Section 3.4.1, see the interface condition (3.48a), we observed that the normal component of the specific discharge is continuous across the interface. This we use to determine the time dependent position of the interface. Following de Josselin de Jong (1981) we arrive at the interface motion equation

$$\varepsilon \frac{\partial h}{\partial t}(r,t) = q_n(r,z,t)/\cos\big(\alpha(r,t)\big), \tag{3.60}$$

for $r \geq 0$ and $t > 0$. In this equation we have $z = h(r,t) - H$, ε is the porosity of the medium and q_n is the normal component of the specific discharge at S. For q_n we write

$$q_n = \frac{1}{2}(\boldsymbol{q}_f + \boldsymbol{q}_s)\cdot\boldsymbol{n} \quad \text{at } S.$$

Substituting (3.57) and (3.58) into this expression and adding the well-contribution, results in

$$q_n\big(r, h(r,t) - H, t\big) = \boldsymbol{n}\cdot\boldsymbol{q}_{\text{well}}\big(r, h(r,t) - H\big)$$

$$-\Gamma\ PV \int_0^\infty \tan\alpha(r',t)\ \boldsymbol{n}\cdot\boldsymbol{G}_{\text{tot}}\big(r, h(r,t) - H; r'\big)\ dr'. \tag{3.61}$$

In this equation $\boldsymbol{n} = \boldsymbol{n}(r,t)$.

Again it is convenient to scale the equations. Here we use

$$r := r/H,\ z := z/H,\ h = h/H$$

and

$$q = qH^2/Q,\ t = tQ/(\varepsilon H^3).$$

This procedure introduces a dimensionless number in the scaled equation (3.61), i.e. the gravity factor Γ is replaced by the dimensionless gravity number

$$G = \Gamma H^2/Q, \tag{3.62}$$

denoting the ratio between the gravity forces and the viscous forces. The scaled initial value problem (3.59-3.62) is solved numerically by discretizing the salt water cone into finitely many (N) rings of constant slope. Since the interface is expected to expand in time, this number N may increase in time. The main difficulty in the numerical treatment of (3.61) is to find an accurate approximation of the singular behaviour of $\boldsymbol{n} \cdot \boldsymbol{G}_{\text{tot}}\big(r, h(r,t) - H, r'\big)$ for r' in a small neighbourhood of the observation variable r. We use for this an asymptotic expansion with respect to the relative size of the ring centered at r. This expansion is given in full detail by Bruining et al. (1990).

For the description of the discretization we use two sets of points on the r-axis: nodal points, which are the r-components of the end-points of the rings and central points, halfway between the nodal points. Thus related to the ith ring $(1 \leq i \leq N)$ is the central point r_i and two nodal points $r_{i-1/2} < r_{i+1/2}$. For reasons of accuracy a denser spacing of the rings near the well is required. We choose r_i proportional to i^2

$$r_i = i^2 r_1,$$

where r_1 denotes the central point of the first ring starting from the z-axis. It implies that $r_{i+1/2} - r_{i-1/2} = 2ir_1$ and thus $(r_{i+1/2} - r_{i-1/2})/r_i = 2/i$. The time interval is discretized into points t_n, $n \geq 0$, where the time step $\Delta t = t_{n+1} - t_n$ is not necessarily constant throughout the computations. For $0 \leq i \leq N$ and $n \geq 0$, the following (dimensionless) notation is introduced

$$\alpha_i^n := \text{ slope of the } i^{th} \text{ ring at time } t_n,$$

$$\boldsymbol{n}_i^n := (\sin\alpha_i^n, \cos\alpha_i^n),$$

$$h_i^n := h(r_i, t_n),$$

$$r_i' := i^2 r_1/H, \; r_{i+1/2}' := i(i+1)r_1/H, \; r' := r'/H,$$

$$\boldsymbol{g}_{\text{tot,i}}^n(r') := H\boldsymbol{G}_{\text{tot}}\big(r_i H, (h_i^n - 1)H; r'H\big),$$

$$\boldsymbol{w}_i^n := \boldsymbol{q}_{\text{well}}\big(r_i H, (h_i^n - 1)H\big)H^2/Q.$$

With these expressions, the discharge normal to each ring is approximated by

$$v_i^n = \boldsymbol{n}_i^n \cdot \boldsymbol{w}_i^n - G \sum_{\substack{k=1,\\ k\neq i}}^{N} \tan\alpha_k^n \int_{r'_{k-\frac{1}{2}}}^{r'_{k+\frac{1}{2}}} \boldsymbol{n}_i^n \cdot \boldsymbol{g}_{\text{tot,i}}^n(r')dr' + R_i^n$$

for $1 \leq i \leq N$ and $n \geq 0$. Here the term R_i^n accounts for the contribution of the ith-ring and its image. An expression for this term is given in Bruining et al. (1990).

Having evaluated the v_i^n, we are in a position to determine the displacement of the interface. The positions h_i^{n+1} of the interface at the central points are obtained by discretizing the interface motion equation (3.60) in the following explicit way

$$h_i^{n+1} = h_i^n + \Delta t v_i^n / \cos \alpha_i^n \quad \text{for } 1 \leq i \leq N \quad \text{and} \quad n \geq 0,$$

with $h_i^0 = 0$ for $i \leq i \leq N$. The values of α_i^n, for $2 \leq i \leq N-1$, are obtained by applying a quadratic interpolation between the central points at r_{i-1}, r_i, r_{i+1} and by differentiating the resulting parabola at r_i. For $i = 1$ and $i = N$ the derivative is determined by using the adjacent points only.
We note here that this construction generally leads to vortex rings that are disconnected at the nodal points.

In Figure 3.10 we show the result of a cone computation. The data used here is

Porosity: $\varepsilon = 0.4$;
Hydraulic conductivity: $K = \frac{\gamma_f \kappa}{\mu} = 8.64\ [m/day]$;
Relative density difference: $\frac{\Delta \rho}{\rho_f} = \frac{1}{40}$;
Initial depth: $H = 30\ [m]$;
Initial time step: $\Delta t_{init} = 1\ [day]$;
Minimal time step: $\Delta t_{\min} = 0.02\ [day]$;
Initial number of rings: $N = 30$;
Radial distance of initial last nodal point: $r_{N+\frac{1}{2}} = 150\ [m]$;
Fresh water production rate: $Q = 194.4\ [m^3/day]$.

Using for the gravity number from (3.62) the relation

$$G = \frac{\Delta \rho}{\rho_f} K \frac{H^2}{Q},$$

we obtain $G = 1$ for these values of the parameters. The time conversion factor, necessary to go from real time $[day]$ to dimensionless time, denoted by t_D in Figures 3.10 and 3.11, is

$$T = \frac{Q}{\varepsilon H^3} = 1 \cdot 8\ 10^{-2}\ [day^{-1}].$$

Observe that the interface eventually reaches the well at the top of the formation. When approaching the well, the displacement velocity will increase. This forces us to take smaller time steps down to $\Delta t_{\min} = 0.02[day]$. Also note that the interface becomes steeper after arriving in smaller neighbourhoods of the well.

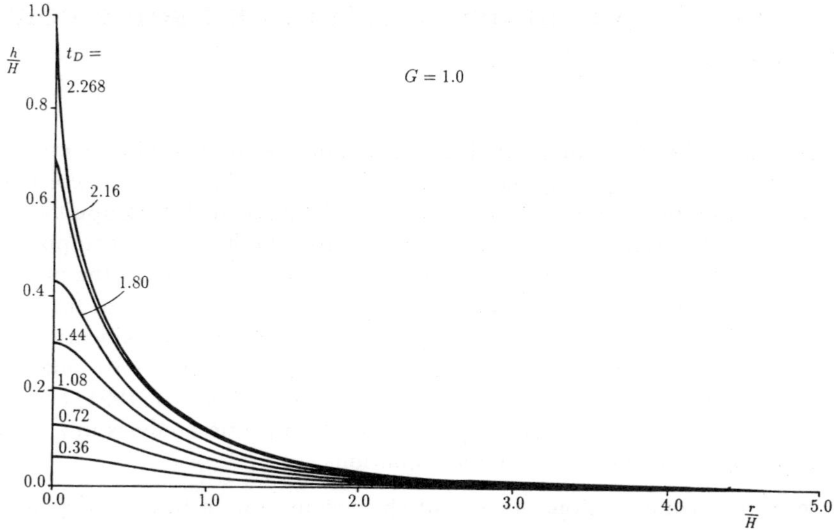

Fig. 3.10: The shape of the salt water cone at different dimensionless times, for gravity number $G = 1$.

In Figure 3.11 we compare the shape of the salt water cone for different values of the gravity number and the same dimensionless time. As to be expected, the figure shows a slower movement of the salt water cone when the influence of gravity increases.

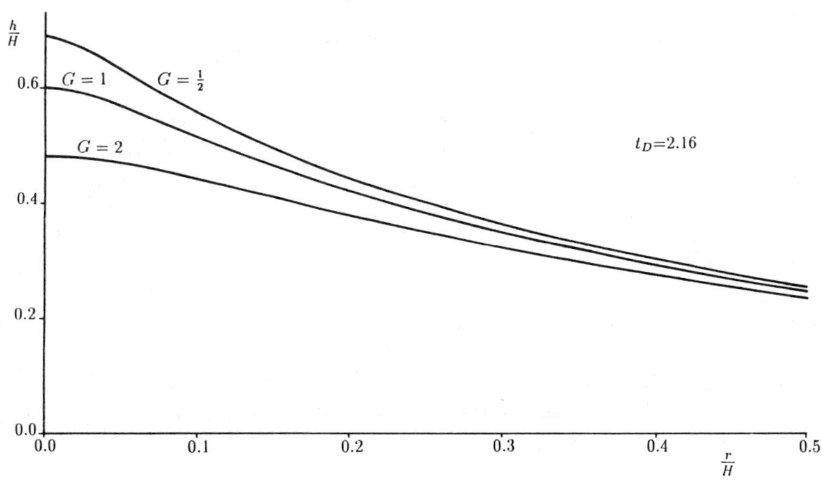

Fig. 3.11: The shape of the salt water cone for different gravity numbers, at the same dimensionless time.

4 The Dupuit approximation and some consequences

We return to the two dimensional interface model as discussed in Sections 3.1 - 3.3. However, here a simplified form of the flow problem will be considered which arises in many practical situations and which is called the Dupuit approximation. In this approximation one assumes that the horizontal component of the specific discharge is constant in each fluid and has a jump at the interface. Since $q_x = -\partial\Psi/\partial z$, an equivalent assumption is that the stream function is linear in z in both fluids. In hydrology one often uses this simplification when the aquifers are long and thin and when the occurring interfaces are sufficiently flat, e.g. see Bear (1972) and Verruijt (1987). In fact, Yortsos (1991) arrives at the Dupuit approximation after applying an asymptotic analysis in $\varepsilon = H/L$ (height/length) to the full system of equations.

However, our computations show that the assumption of linear z-dependence of the stream function, above and below the interface, does not necessarily require very flat interfaces and is in particular valid away from the top and the toe (i.e. for $S_f(t) < x < S_s(t)$, see definitions (3.26)). Figure 4.1 is an example of such computations. It shows the finite element approximation of the scaled stream function in the rectangle $\Omega_1 = (-1,1) \times (0,1)$ for an interface of the form $u(x) = \big(\tanh(x) + 1\big)/2$.

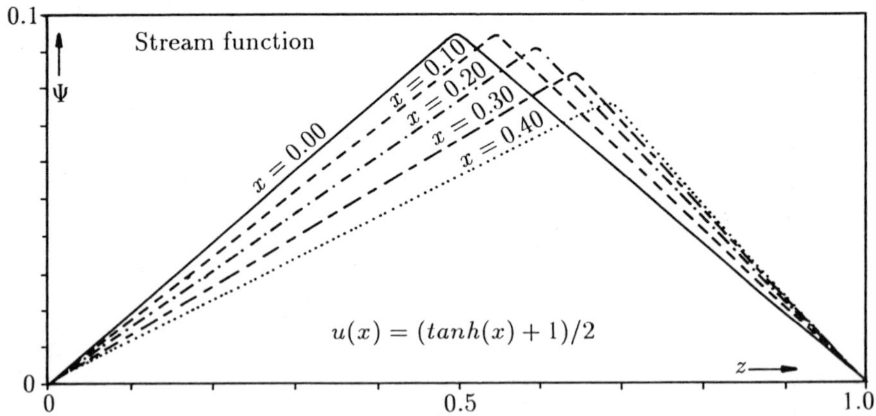

Fig. 4.1: The stream function $\Psi(x,\cdot)$ for various values of x, induced by the interface $u(x)$. The linear behaviour of Ψ in z clearly justifies the Dupuit approximation.

Clearly the Dupuit assumption is valid here, even though $\max_{-1 \leq x \leq 1} u'(x) = 0.5$.

In the analysis below we need for mathematical purposes the interface-slope limitation $|\partial u/\partial x| \leq 1$. This does not cause any significant restriction for practical applications because there one often encounters interfaces with small slopes. This limitation also corresponds to our numerical validation of the Dupuit approximation.
The particular form used here is due to de Josselin de Jong (1981). It leads to a nonlinear diffusion equation for the interface u.

4.1 The nonlinear diffusion equation

Our starting point is the rescaled interface motion equation, see also (3.22),

$$\frac{\partial u}{\partial t} = \frac{\partial}{\partial x}\{\Psi(x, u(x,t), t)\}, \tag{4.1}$$

and the observation that

$$\Psi(x, u, t) = -\int_0^u q_x(x, z, t)dz,$$

or

$$\Psi(x, u, t) = \int_u^1 q_x(x, z, t)dz.$$

Using the simplifying assumption on the flow, as shown in Figure 4.2, the last two equations can be written as

$$\Psi(x, u, t) = -uq_{s_x}(x, u, t) = (1-u)q_{f_x}(x, u, t). \tag{4.2}$$

Further, it follows form the interface conditions (3.13), with $\Gamma = 1$, that

$$-(q_{f_x} - q_{s_x})\sin\alpha + (q_{f_z} - q_{s_z})\cos\alpha = 0 \quad \text{on} \quad S$$

and

$$(q_{f_x} - q_{s_x})\cos\alpha + (q_{f_z} - q_{s_z})\sin\alpha = \sin\alpha \quad \text{on} \quad S.$$

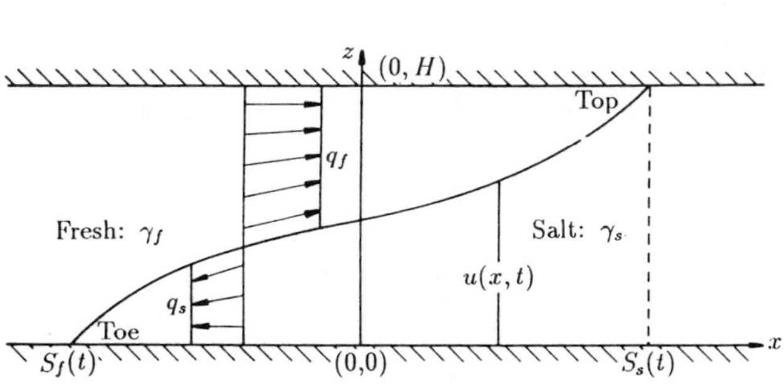

Fig. 4.2: The Dupuit approximation of the flow.

Eliminating $(q_{f_z} - q_{s_z})$ gives

$$q_{f_x} - q_{s_x} = \frac{\tan\alpha}{1+\tan^2\alpha} = \frac{\partial u/\partial x}{1+(\partial u/\partial x)^2}. \tag{4.3}$$

Combining (4.2) and (4.3) gives

$$q_{f_x} = u\frac{\partial u/\partial x}{1+(\partial u/\partial x)^2} \quad \text{and} \quad q_{s_x} = -(1-u)\frac{\partial u/\partial x}{1+(\partial u/\partial x)^2}, \tag{4.4}$$

and finally

$$\Psi(x,u,t) = u(1-u)\frac{\partial u/\partial x}{1+(\partial u/\partial x)^2}. \tag{4.5}$$

We deduce from (4.1) and (4.5), and also from (3.24), that u satisfies the initial value problem

$$\text{(IVP)} \quad \begin{cases} \dfrac{\partial u}{\partial t} = \dfrac{\partial}{\partial x}\left\{u(1-u)\dfrac{\partial u/\partial x}{1+(\partial u/\partial x)^2}\right\} \quad \text{for } (x,t) \in Q = \mathbb{R}\times\mathbb{R}^+ & (4.6) \\ u(\cdot,0) = u_0(\cdot) \qquad \text{with } x \in \mathbb{R}. & (4.7) \end{cases}$$

The partial differential equation (4.6) is of degenerate parabolic type. At points where $u \in (0,1)$ and $\partial u/\partial x \in (-1,1)$ it is parabolic, but at points where $u = 0,1$ or $\partial u/\partial x = -1,1$ it degenerates: that is, the coefficient of the second-order derivative vanishes.
For values of u and $\partial u/\partial x$ outside these intervals this coefficient becomes negative, leading to a negative diffusivity. This we reject and therefore we shall impose the conditions $u \in [0,1]$ and $\partial u/\partial x \in [-1,1]$. The first condition is natural for the problem since u represents the height of the fresh-salt interface in an aquifer of constant thickness 1. The second one has no clear physical interpretation. However it ensures the small slope limitation of the Dupuit approximation.
The degeneracy of equation (4.6) means that in general one cannot expect to find classical solutions. This is shown by the self-similar solution

$$u_s = \bar{Q} \to [0,1],$$

defined by

$$u_s(x,t) = \begin{cases} 1 & x \geq +\frac{1}{2f(t)}; \\ \frac{1}{2} + f(t)\cdot x & |x| < +\frac{1}{2f(t)}; \\ 0 & x \leq -\frac{1}{2f(t)}, \end{cases} \tag{4.8}$$

where $f:[0,\infty)\to\mathbb{R}$ is the solution of the initial value problem

$$\begin{cases} f' = -2\cdot\frac{f^3}{1+f^2} \text{ on } (0,\infty), \\ f(0) = f_0 \in [-1,1]. \end{cases} \tag{4.9}$$

Then $u_s(x,t)$ satisfies equation (4.6) except at the points $x = \pm 1/(2f(t))$. The solution has the form of a line, rotating around the point $x = 0, u = 1/2$, with initial slope f_0. By the monotonicity of f, it rotates clockwise, when $f_0 \in (0,1]$ and counterclockwise when $f_0 \in [-1,0)$. Also without the restriction $f_0 \in [-1,1]$, the function u_s defined by (4.8) satisfies the equation. However imposing it ensures that u_s falls within the mathematical theory for equation (4.6).

Associated with the solution u_s there arise two curves in the domain Q: i.e. the location of the toe of the interface

$$x = S_f^*(t) := -\frac{1}{2f(t)},\ t > 0, \tag{4.10a}$$

separating the region where fresh water is present ($u = 0$) from the region where both fluids are present ($0 < u < 1$), and the location of the top of the interface

$$x = S_s^*(t) := \frac{1}{2f(t)},\ t > 0, \tag{4.10b}$$

separating the region where salt water is present ($u = 1$) from the region where both fresh and salt are present ($0 < u < 1$). Using equation (4.9) for f one finds that $f(t) = O(t^{-1/2})$ as $t \to \infty$. Hence

$$S_i^*(t) = O(t^{1/2}) \text{ as } t\to\infty \text{ for } i = s, f.$$

Further, the set

$$M(u_s) := \{(x,t) \in Q : 0 < u_s(x,t) < 1\},$$

is given by

$$M(u_s) = \{(x,t) \in Q : S_f(t) < x < S_s(t),\ t > 0\}.$$

For the speed of the interface end points one obtains, after differentiating (4.10) and using (4.9) and (4.5),

$$\frac{dS_f^*}{dt} = \frac{f'}{2f^2} = -\frac{f}{1+f^2} = -\lim_{x\downarrow S_f(t)} \frac{\Psi(x,u_s(x,t),t)}{u_s(x,t)} \tag{4.11a}$$

and

$$\frac{dS_s^*}{dt} = -\frac{f'}{2f^2} = \frac{f}{1+f^2} = \lim_{x \uparrow S_s(t)} \frac{\Psi(x, u_s(x,t), t)}{1 - u_s(x,t)}. \tag{4.11b}$$

Note that these expressions are of the same form as the interface equations (3.30), (3.31), derived for the full problem. The curves (4.10) are the *interfaces* or *free boundaries* for Problem (IVP).

The explicit solution u_s serves as a model for the general case in which the initial distribution satisfies

$$u_0(x) = \begin{cases} 0 & -\infty < x \leq a_f \\ \in (0,1) & a_f < x < a_s \\ 1 & a_s \leq x < \infty \end{cases} \tag{4.12}$$

for some numbers $-\infty < a_f < a_s < \infty$. Let us introduce

Definition 4.1 A function $u : \bar{Q} \to \mathbb{R}$ is called a solution of Problem (IVP), with u_0 satisfying (4.12), if there exist functions (for $i = f, s$) $S_i : [0, \infty) \to \mathbb{R}$, with $S_i(0) = a_i$ and $S_f(t) < S_s(t)$ for all $t > 0$, such that

(i) $\dfrac{\partial u}{\partial t} - \dfrac{\partial \Psi}{\partial x} = 0$ in $\{(x,t) : S_f(t) < x < S_s(t) \text{ and } t > 0\}$;

(ii) for all $t > 0$

$u(x,t) \to 0$ and $\Psi(x,t) \to 0$ as $x \downarrow S_f(t)$ and $u(x,t) = 0$ for $x \leq S_f(t)$,
$u(x,t) \to 1$ and $\Psi(x,t) \to 0$ as $x \uparrow S_s(t)$ and $u(x,t) = 1$ for $x \geq S_s(t)$;

(iii) $u(x,0) = u_0(x)$ for $-\infty < x < \infty$.

The function Ψ in the definition is given by (4.5). In terms of the model it represents the horizontal salt water flux, see equation (3.21). The functions $S_i (i = f, s)$ are called the free boundaries for Problem (IVP). The second statement (ii) in Definition 4.1 says that the salt water flux vanishes at the toe and the top of the fresh-salt distribution.

The existence and uniqueness of such solutions was proven in papers by van Duijn & Hilhorst (1987) and van Duijn & Zhang (1988). We discuss some of the mathematical results and techniques in the next section.

4.2 The mathematics of Problem (IVP)

To establish directly the existence of solutions in the sense of Definition 4.1 appears difficult because of the occurrence of the apriori unknown free boundaries S_f and S_s. Therefore one first introduces global solutions which satisfy equation (4.6) in some integrated sense (often called generalized solutions). For such solutions it is easier to prove existence results. Then one has to show, step

by step, that these solutions indeed satisfy Definition 4.1 and have additional relevant physical properties. This is often the hard part of the analysis. Below we outline these steps.
Generalized solutions for Problem (IVP) were first introduced by van Duijn & Hilhorst (1987). The formal definition is

Definition 4.2. A function $u : \bar{Q} \to \mathbb{R}$ is called a (generalized) solution of Problem (IVP), if it satisfies

(i) $u \in L^\infty(\mathbb{R}^+; W^{1,\infty}(\mathbb{R}))$; $\partial u/\partial t$, $\partial \Psi/\partial x \in L^2((-R,R)\times(0,T))$ for all $R, T > 0$;

(ii) $u \in [0,1]$ and $\partial u/\partial x \in [-1,1]$ a.e. in Q;

(iii) $u(\cdot,0) = u_0(\cdot)$ on $\mathbb{R}$;

(iv) $\dfrac{\partial u}{\partial t} - \dfrac{\partial \Psi}{\partial x} = 0$ a.e. in Q.

They showed that if the initial distribution u_0 satisfies (4.12) and is sufficiently flat, i.e. in mathematical terms

$$u_0 \in W^{1,\infty}(\mathbb{R}) \text{ such that } -1 \leq u_0' \leq 1 \text{ on } \mathbb{R}, \qquad (4.13)$$

then Problem (IVP) has a unique generalized solution which is continuous on $\bar{Q}$. Such solutions satisfy a *comparison principle* and a *contraction property.* By this we mean the following. Let u_{01} and u_{02} be two initial functions satisfying (4.12), (4.13) and let u_1 and u_2 denote the corresponding generalized solutions. Then

(i) $u_{01} \leq u_{01}$ on $\mathbb{R}$ implies $u_1 \leq u_2$ on $\bar{Q}$ (comparison principle);

(ii) $\int_{\mathbb{R}} |u_1(x,t) - u_2(x,t)| dx \leq \int_{\mathbb{R}} |u_{01}(x) - u_{02}(x)| dx$ for all $t > 0$ (contraction property).

In particular the last property implies the uniqueness. In deriving qualitative properties of solutions, the comparison principle is an essential tool. As indicated by the self-similar solution u_s, one may expect that solutions become flatter in time. This was shown by van Duijn & Zhang (1988) for arbitrary generalized solutions with u_0 satisfying (4.13). To be precise, they showed.

Lemma 4.3. Let u be a generalized solution of Problem (IVP) with u_0 as in (4.12), (4.13). Then for each $t > 0$,

$$\left|\frac{\partial u}{\partial x}(x,t)\right| \leq f(t) \text{ for almost every } x \in \mathbb{R}, \qquad (4.14)$$

where f is the solution of equation (4.9) satisfying $f(0) = f_0 = ess \sup_{\mathbb{R}} |u_0'|$.

Since f is strictly decreasing for $0 < f_0 \leq 1$, and thus $f(t) < 1$ for all $t > 0$, we conclude from Lemma 4.3 that the degeneracy in the spatial derivative vanishes instantaneously. As a consequence u satisfies equation (4.6) in the classical sense at points where $u \in (0,1)$, i.e. on the set

$$M(u) = \{(x,t) \in Q : 0 < u(x,t) < 1\}.$$

We are interested in the behaviour of the free boundaries induced by the degeneracy at $u = 0$ and at $u = 1$. Using mainly comparison arguments, van Duijn & Zhang (1988) proved the following.
Given any generalized solution u, with u_0 satisfying (4.12) and (4.13), there exist Lipschitz continuous functions (for $i = f, s$) $S_i : [0, \infty) \to \mathbb{R}$ such that $S_i(0) = a_i$ and $S_f(t) < S_s(t)$ for all $t > 0$, S_f is non-increasing and S_s is non-decreasing, and

$$M(u) = \{(x,t) \in Q : S_f(t) < x < S_s(t), t > 0\}.$$

Clearly these properties imply that a generalized solution is also a classical solution in the sense of Definition 4.1. About the monotonicity of the free boundaries S_i, it was also shown that there exist *waiting times* $T_i \geq 0$ such that

$$S_i(t) = a_i \text{ for } \; 0 \leq t \leq T_i$$

and

$$S_i \text{ is strictly monotone on } (T_i, \infty).$$

These waiting times are positive provided $u_0(x)$ is sufficiently flat near $x = a_i$: if

(i) $u_0(x) \leq C_f(x - a_f)^2$ for some $C_f > 0$, then $T_f > 0$;

(ii) $u_0(x) \geq 1 - C_s(x - a_s)^2$ for some $C_s > 0$, then $T_s > 0$.

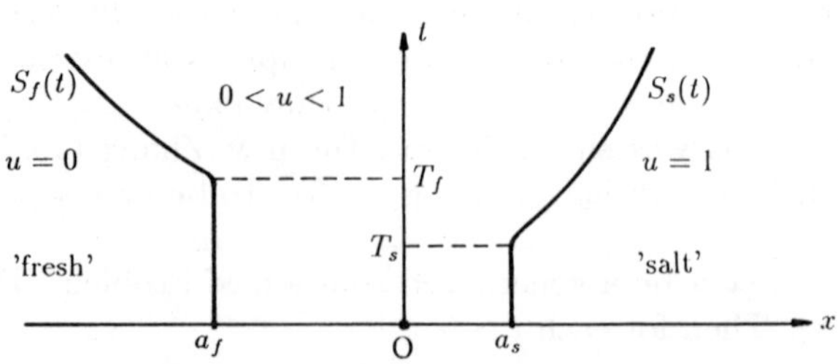

Fig. 4.3: The occurrence of waiting times in the free boundaries.

In a later paper, Bertsch et al. (1989) proved that the free boundaries S_i are smooth after the waiting times, $S_i \in C^1\big((T_i,\infty)\big)$, and satisfy in the classical sense the free boundary equations (3.30), (3.31).

In a second paper Bertsch et al. (1992) discussed the large time behaviour for Problem (IVP). They showed that shifted self-similar solutions are asymptotic profiles in the following sense. Given a u_0 satisfying (4.12), (4.13), find the unique $a \in \mathbb{R}$ such that the shifted self-similar solution $u_{s_a}(x,t) := u_s(x-a,t)$ has initially the equal mass property

$$\int_{\mathbb{R}} \big\{u_0(x) - u_{s_a}(x,0)\big\}\,dx = 0.$$

Then

$$\max_{-\infty<x<\infty} \big|u(x,t) - u_{s_a}(x,t)\big| = o\big(f(t)\big) \text{ as } t \to \infty,$$

where f is the solution of the ordinary differential equation (4.9) for any initial slope $f_0 \in (0,1]$. This type of convergence was also observed numerically in Section 3.3 for the full problem, see the results in Figure 3.7.
In his thesis, Zhang (1988) presents a numerical method for Problem (IVP), based on front-tracking techniques developed by Hoff (1985).

4.3 Other Dupuit problems

As in Section 4.1 we consider a two dimensional flow problem in a vertical cross-section of a horizontally extended aquifer. Here, however, fluid is being injected from one of the sides and through the top, see Figure 4.4.

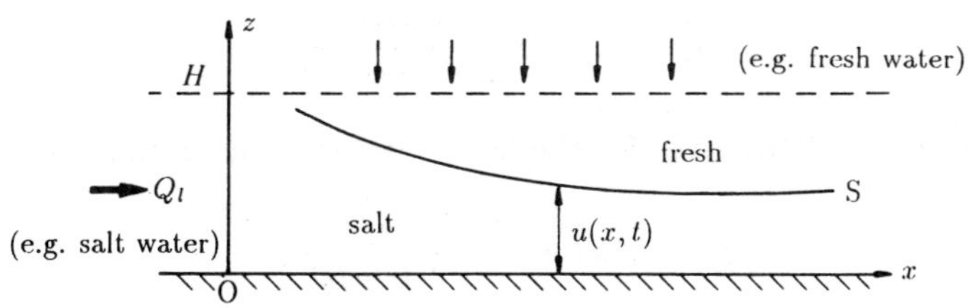

Fig. 4.4: Flow problem with infiltration.

We suppose that fluid is injected from the left, at a rate $Q_l[m^2/s]$, and through the top, where $i(x) \geq 0[m/s]$ denotes the infiltration rate per unit length: i.e. if $-\infty < x_1 < x_2 < \infty$, then

$$\int_{x_1}^{x_2} i(s)ds$$

denotes the amount of fluid that enters the aquifer between the points x_1 and x_2, per unit time. We shall assume that $i(x) \to 0$ as $x \to -\infty$, sufficiently fast, so that $\int_{-\infty}^{x} i(s)ds$ exists for all $x \in \mathbb{R}$. Further we assume that the bottom boundary is impervions. In terms of the stream function, $\boldsymbol{q} = \mathbf{curl}\Psi$, the boundary conditions become

$$\left.\begin{array}{l} q_z = 0 \ \text{ along } \ \{z=0\} \Rightarrow \Psi(x,0,t) = \ \text{constant} \ = Q_l \\ q_z = -i(x) \ \text{ along } \ \{z=H\} \Rightarrow \Psi(x,H,t) = -\int_{-\infty}^{x} i(s)ds \end{array}\right\} \tag{4.15}$$

for $-\infty < x < \infty$ and for $t > 0$. With reference to Figures 4.2 and 4.4, we deduce from $q_x = -\partial\Psi/\partial z$ that

$$-\Psi(x,u,t) + Q_l = uq_{sx} \tag{4.16a}$$

and

$$\int_{-\infty}^{x} i(s)ds + \Psi(x,u,t) = (H-u)q_{fx}. \tag{4.16b}$$

Adding these equations gives

$$Q_\ell + \int_{-\infty}^{x} i(s)ds = uq_{sx} + (H-u)q_{fx}. \tag{4.17}$$

Hence

$$Q(x) := Q_\ell + \int_{-\infty}^{x} i(s)ds \quad \text{for} \ -\infty < x < \infty \tag{4.18}$$

defines the total horizontal discharge through the aquifer. As in Section 4.1, see equation (4.3), if follows from the interface conditions (3.13) that

$$q_{fx} - q_{sx} = \Gamma\frac{\partial u/\partial x}{1+(\partial u/\partial x)^2}. \tag{4.19}$$

Combining (4.17)-(4.19) gives for the horizontal specific discharges

$$q_{fx} = Q(x) + \Gamma u\frac{\partial u/\partial x}{1+(\partial u/\partial x)^2}, \tag{4.20a}$$

$$q_{sx} = Q(x) - \Gamma(H-u)\frac{\partial u/\partial x}{1+(\partial u/\partial x)^2}, \tag{4.20b}$$

and this implies for the stream function along the interface

$$\Psi(x,u,t) = Q_\ell - uQ(x) + \Gamma u(H-u)\frac{\partial u/\partial x}{1+(\partial u/\partial x)^2}. \tag{4.21}$$

It leads to the extended interface motion equation

$$\varepsilon\frac{\partial u}{\partial t}+\frac{\partial}{\partial x}\{uQ(x)\}=\Gamma\frac{\partial}{\partial x}\left\{u(H-u)\frac{\partial u/\partial x}{1+(\partial u/\partial x)^2}\right\} \text{ in } Q. \tag{4.22}$$

This equation was studied in papers by van Duijn & Hilhorst (1987), Blanc & Li (1991), Blanc (1991) and Li (1992). To illustrate the behaviour of solutions we consider here two steady-state situations. They lead to boundary value problems for u : one on the interval $(-\infty, 0)$ with boundary conditions at $-\infty, 0$ and one on the whole real line with boundary conditions at $\pm\infty$.

4.3.1 A problem on $\{x < 0\}$.

Consider a steady flow situation, as in Figure 4.5, where fresh water is injected from the left. It flows through the aquifer in positive x-direction, on top of the stagnant salt water, and it leaves the system at $x = 0$ where the height of the salt water is kept fixed. Both the top and bottom boundary of the aquifer are impermeable, leading to $i(x) = 0$ for all $x < 0$ in expression (4.18).

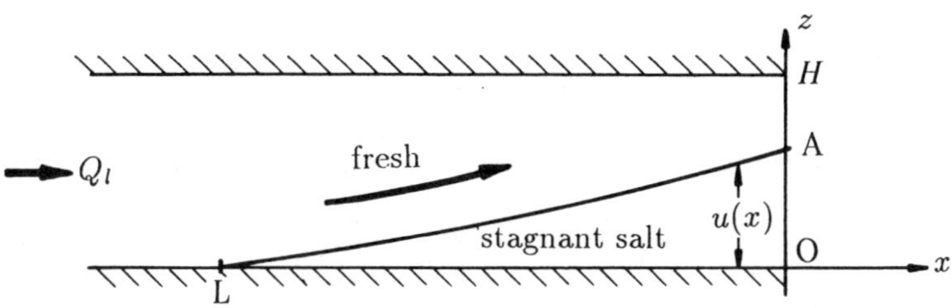

Fig. 4.5: Steady fresh-salt distribution in half-strip.

It gives the problem

$$(\mathrm{BVP}_1)\begin{cases}\left\{Q_\ell u-\Gamma u(H-u)\dfrac{u'}{1+(u')^2}\right\}=0 \text{ for } -\infty<x<0, & (4.23a)\\ u(-\infty)=0,\ u(0)=A\in(0,H), & (4.23b)\end{cases}$$

where the primes denote differentiation with respect to x. As in the solution definition of the time dependent problems, Definition 4.2, we impose the additional condition

$$|u'| \le 1 \text{ on } (-\infty, 0). \tag{4.24}$$

This ensures that the solutions considered here fall into the mathematical theory developed for the time dependent case. It is also useful from a practical point of view, because the Dupuit - solutions only have a physical interpretation for interfaces with relatively small slopes. Obviously we also require

$$0 \leq u \leq H \quad \text{on } (-\infty, 0). \tag{4.25}$$

Equation (4.23a) gives

$$Q_\ell u - \Gamma u(H-u)\frac{u'}{1+(u')^2} = K \quad \text{for } -\infty < x < 0,$$

where K is the constant of integration. By the boundary condition $u(-\infty) = 0$ we obtain $K = 0$. Hence

$$u\left\{Q_\ell - \Gamma(H-u)\frac{u'}{1+(u')^2}\right\} = 0. \tag{4.26}$$

This means that

$$\frac{Q_\ell}{\Gamma} = (H-u)\frac{u'}{1+(u')^2} \quad \text{whenever } u > 0, \tag{4.27}$$

and because $Q_\ell/\Gamma > 0$ it follows that

$$u' > 0 \quad \text{whenever} \quad u > 0.$$

These inequalities imply that there exists a constant $L < 0$, possibly $L = -\infty$, such that

$$\begin{cases} (4.27) \text{ holds on } (L, 0), & (4.28a) \\ u(L) = 0, u(0) = A. & (4.28b) \end{cases}$$

The constant L, denoting the position of the salt water toe, has to be determined from (4.28). Before doing this we note the following. Differentiating (4.27) on $(L, 0)$ and applying condition (4.24) to the result, gives $u'' > 0$ on $(L, 0)$. Hence u attains its maximal derivative at $x = 0$, where is satisfies

$$\frac{Q_\ell}{\Gamma(H-A)} = \frac{u'(0)}{1+\left(u'(0)\right)^2}.$$

The right-hand side is bounded by 1/2. Hence we obtain a *non-existence* result if

$$Q_\ell > \frac{1}{2}\Gamma(H-A).$$

On the other hand, if

$$Q_\ell \leq Q_{\ell_{\max}} := \frac{1}{2}\Gamma(H-A) \tag{4.29}$$

we have

$$0 \le u'(x) < u'(0) \le 1.$$

To solve (4.28) we invert the solution. This is allowed because $u' > 0$ on $(L, 0)$. Thus let $\sigma : [0, A] \to [L, 0]$ be given by $\sigma = u^{-1}$. Since $d\sigma/du = 1/(du/dx)$, we obtain for σ the equation

$$\frac{2Q_\ell}{\Gamma}\frac{d\sigma}{du} = H - u - \sqrt{(H-u)^2 - \frac{4Q_\ell^2}{\Gamma^2}} \quad \text{on} \quad [0, A]. \tag{4.30}$$

Here the minus-sign was chosen because $d^2\sigma/du^2 < 0$, by the convexity of u. With Q_ℓ satisfying (4.29), we have $(H-u)^2 > 4Q_\ell^2/\Gamma^2$ for all $u \in [0, A]$. Therefore expression (4.30) is well-defined. We want to solve it for σ satisfying $\sigma(A) = 0$. Integration gives the implicit expression

$$\sigma(A) - \sigma\big(u(x)\big) = -x = \frac{2Q_\ell}{\Gamma}\int_{\Gamma(H-A)/2Q_\ell}^{\Gamma\left(H-u(x)\right)/2Q_\ell} \{p - \sqrt{p^2-1}\}dp$$

or

$$-x = \frac{Q_\ell}{\Gamma}\left[p^2 + p\sqrt{p^2-1} - \ln\{p + \sqrt{p^2-1}\}\right]_{\Gamma(H-A)/2Q_\ell}^{\Gamma\left(H-u(x)\right)/2Q_\ell}, \tag{4.31}$$

from which we obtain for the position of the salt water toe

$$-L = \frac{Q_\ell}{\Gamma}\left[p^2 + p\sqrt{p^2-1} - \ln\{p + \sqrt{p^2-1}\}\right]_{\Gamma(H-A)/2Q_\ell}^{\Gamma H/2Q_\ell}.$$

Remark 4.4

(i) Two-dimensional, steady flow problems with one fluid being stagnant, as in Figure 4.5, can be solved by the hodograph method via transformations in the complex plane, see Bear (1972). This yields exact solutions to the full problem.

(ii) The steady state solution defined by (4.31) is a stable equilibrium solution of the corresponding time dependent problem. This was shown by van Duijn & Hilhorst (1987).

4.3.2 A problem on $\mathbb{R}$

Again we consider a two-dimensional steady flow situation, now with injection from the top, see Figure 4.6.

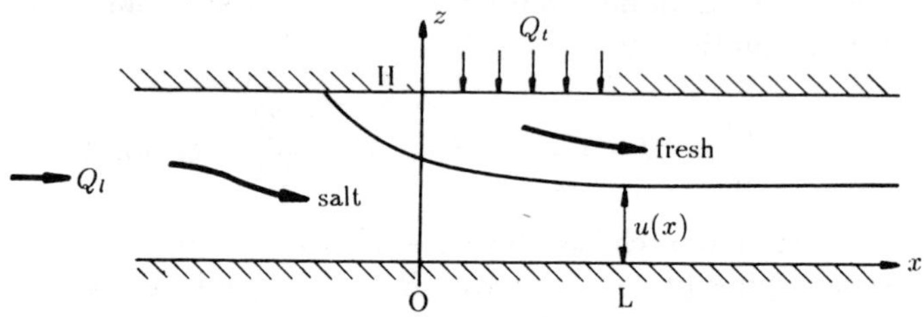

Fig. 4.6: A steady fresh-salt distribution with injection from the top.

As before, salt water is injected from the left at the rate $Q_\ell[m^2/s]$. The top is impermeable, except on the interval $0 < x < L$ where fresh water is being injected, uniformly with respect to x. Let $Q_t[m^2/s]$ denote the total fresh water injection rate. The bottom boundary is again impermeable. For this situation, the total horizontal discharge is given by, see (4.18):

$$Q(x) = \begin{cases} Q_\ell & \text{for} \quad -\infty < x \leq 0 \\ Q_\ell + \frac{Q_t}{L}x & \text{for} \quad 0 < x < L \\ Q_\ell + Q_t & \text{for} \quad L \leq x < \infty. \end{cases} \tag{4.32}$$

We now find for the stationary interface $u = u(x)$, the boundary value problem

$$(\text{BVP}_2)\begin{cases} \left\{Q(x)u - \Gamma u(H-u)\dfrac{u'}{1+(u')^2}\right\}' = 0 \quad \text{on} \quad \mathbb{R} & (4.33a) \\ u(-\infty) = H, \; u'(+\infty) = 0, & (4.33b) \end{cases}$$

where again u satisfies the additional conditions (4.24), (4.25). The boundary condition $u'(\infty) = 0$ is chosen here because the value of $u(\infty)$ is apriori unknown.

Integrating the equation and applying the boundary condition at $x = -\infty$ gives

$$Q(x)u - \Gamma u(H-u)\frac{u'}{1+(u')^2} = Q_\ell H. \tag{4.34}$$

Hence, letting $x \to \infty$, we find

$$u(\infty) = \frac{Q_\ell}{Q_\ell + Q_t}H. \tag{4.35}$$

Equation (4.34) can be written as

$$\Gamma\frac{u'}{1+(u')^2} = \frac{Q(x)-Q_\ell}{H-u} - \frac{Q_\ell}{u} =: F(x;u). \tag{4.36}$$

The right-hand side in (4.36), i.e.the function $F(x;u)$, is strictly increasing with respect to u (for each fixed $x \in \mathbb{R}$). Next define the function $\bar{u} : \mathbb{R} \to [0, H]$ by

$$\bar{u}(x) = \frac{Q_\ell}{Q(x)}H. \tag{4.37}$$

Then $F\big(x;\bar{u}(x)\big) = 0$. We prove

Claim 4.5. $\bar{u}$ is an upper bound for the solution u.

Proof. Suppose not. Then there exists $\hat{x} > 0$ such that $u(\hat{x}) > \bar{u}(\hat{x})$. Since F is strictly increasing in u, we have $F\big(\hat{x}, u(\hat{x})\big) > F\big(\hat{x}, \bar{u}(\hat{x})\big) = 0$. Hence $u'(\hat{x}) > 0$. Because $\bar{u}$ is non-increasing in x this means that $u(x) > \bar{u}(x)$ for all $x > \hat{x}$. This, however, contradicts the boundary condition (4.35). Therefore $u(x) \le \bar{u}(x)$ for all $x \in \mathbb{R}$.

The monotonicity of F also implies that $u' \le 0$ whenever $u \le \bar{u}$. Therefore, by Claim 4.5,

$$u \text{ is non-increasing on } \mathbb{R}. \tag{4.38}$$

As a direct consequence we have that u is constant for $x \ge L$:

$$u(x) = \frac{Q_\ell}{Q_\ell + Q_t}H \quad \text{for} \quad x \ge L. \tag{4.39}$$

This observation can be used for numerical purposes: take as initial conditions

$$u(L) = \frac{Q_\ell}{Q_\ell + Q_t}H \quad \text{and} \quad u'(L) = 0,$$

and continu the solution backwards in space. This may not always be possible. Certain combinations of the parameters lead to interfaces for which the slope becomes too negative. Then the Dupuit condition (4.24) is violated. One can show that

(i) If $\dfrac{Q_\ell}{H} \ge \dfrac{\Gamma}{2} + \dfrac{Q_t}{L}$, then Problem ($\mathrm{BVP}_2$) has a solution such that

$$u(x) = \begin{cases} H & \text{for} \quad -\infty < x \le 0, \\ \text{strictly decreasing} & \text{for} \quad 0 < x < L, \\ \frac{Q_\ell}{Q_\ell + Q_t}H & \text{for} \quad L \le x < \infty. \end{cases} \tag{4.40}$$

(ii) If $\frac{\Gamma}{2} \leq \frac{Q_\ell}{H} < \frac{\Gamma}{2} + \frac{Q_t}{L}$, then no solution exists: i.e. condition (4.24) is violated.
(iii) If $0 < \frac{Q_\ell}{H} < \frac{\Gamma}{2}$ and $\frac{Q_t}{L} \geq \frac{\Gamma}{2}$, then solutions may occur with $u(0) < H$: i.e. the corresponding interfaces intersect the top boundary to the left of the origin.

A stationary fresh-salt distribution as in Figure 4.6, thus with both fluids moving, was studied by Alt & van Duijn (1990) for the full problem (without the Dupuit approximation). They showed that for any combination of parameters there is a solution with an interface that touches tangentially the top boundary to the left of the origin. They also showed that an interface cannot have horizontal segments. Thus interfaces with the structure of (4.40) do not occur and are a direct consequence of the Dupuit approximation.

References

1. Alt, H.W. & van Duijn, C.J. (1990). *A stationary flow of fresh and salt groundwater in a coastal aquifer.* Nonlinear Anal. Theory Math. Appl. **14**, 625-656.

2. Alt, H.W. & van Duijn, C.J. (1992). *A free boundary problem involving a cusp. Part I: Global analysis.* European J. Appl. Math., in print.

3. Bear, J. (1972). *Dynamics of fluids in porous media.* American Elsevier (New York).

4. Bear, J. (1979). *Hydraulics of Groundwater.* McGraw-Hill Series in Water Resources and Environmental Engineering (New York).

5. Bear, J. & Verruijt, A. (1987). *Modeling Groundwater Flow and Pollution.* D. Reidel Publishing Company (Dordrecht).

6. Bertsch, M., van Duijn, C.J., Esteban, J.R. & Zhang, H. (1989). *Regularity of the free boundary in a doubly degenerate parabolic equation.* Comm. Part. Diff. Eqns. **14**, 391-412.

7. Bertsch, M., Esteban, J.R. & Zhang, H. (1992). *On the asymptotic behaviour of the solutions of a degenerate equation in hydrology.* Nonlinear Anal. Theory Math. Appl. (accepted for publication).

8. Blanc, Ph. (1991). *Flow of fresh and salt water in an aquifer with injection from the top.* Report 91-68 of the Faculty of Technical Mathematics and Informatics, Delft University of Technology.

9. Blanc, Ph. & Li, S. (1991). *On a doubly degenerate equation arising in fresh-salt water flow with pumping.* Report 91-30 of the Faculty of Technical Mathematics and Informatics, Delft University of Technology.

10. Bruining, J., van Duijn, C.J., Schotting, R.J. & de Villeneuve, A.H. (1990). *Simulation of coning in bottom water-driven reservoirs: Extended version.* Delft Progress Report **14**, 265-326.

11. Bruining, J., van Duijn, C.J. & Schotting, R.J. (1991). *Simulation of coning in bottom water-driven reservoirs.* Transport in Porous Media **6**, 35-69.

12. Chan Hong, J.R., van Duijn, C.J., Hilhorst, D. & van Kester, J. (1989). *The interface between fresh and salt groundwater: a numerical study.* IMA of Applied Mathematics **42**, 209-240.

13. Van Duijn, C.J. & Hilhorst, D. (1987). *On a doubly nonlinear diffusion equation in hydrology.* Nonlinear Anal. Theory Math. Appl. **11**, 305-333.

14. Van Duijn, C.J. & Zhang. H. (1988). *Regularity properties of a doubly degenerate equation in hydrology.* Comm. Part. Diff. Eqns. **13**, 261-319.

15. Van Duijn, C.J., Peletier, L.A. & Schotting, R.J. (1992). *On the analysis of brine transport in porous media.* European J. Appl. Math., in print.

16. Hoff, D. (1985). *A scheme for computing solutions and interface curves for a doubly degenerate parabolic equation.* SIAM J. Numer. Anal. **12**, 687-712.

17. De Josselin de Jong, G. (1960). *Singularity distributions for the analysis of multiple fluid flow through porous media.* J. Geophys. Res. **65**, 3739-3758.

18. De Josselin de Jong, G. (1981). *The simultaneous flow of fresh and salt water in aquifers of large horizontal extension determined by shear flow and vortex theory.* Proceedings, Euromech. **143** (eds A. Verruijt and F.B.J. Barends). Rotterdam: Balkema.

19. Lerat, A., & Peyret, R. (1973). *Sur le choix de schémas aux différences finies du second ordre fournissant des profils de choc sans oscillation.* C.R. Acad. Sci. Paris **277**, 363-366.

20. Li, S. (1992). *Quasilinear evolution problems.* Ph. D. Thesis, Delft University of Technology.

21. Pinder, G.F. & Gray, W.G. (1977). *Finite element simulation in surface and subsurface hydrology.* Academic Press (New York).

22. Verruijt, A. (1980). *The rotation of a vertical interface in a porous medium.* Water Resoures Research **16**, 239-240.

23. Verruijt, A. (1987). *A finite element model for interface problems in groundwater flow.* In: Microcomputers in Engineering Application. Eds. B.A. Schrefler & R.W. Lewis. John Wiley & Sons, 251-271.

24. Wilders, P. (1983). *Minimization of dispersion in difference methods for hyperbolic conservation laws.* Ph. D. Thesis, University of Amsterdam.

25. Yortsos, Y.C. (1991). *A theoretical analysis of vertical flow equilibrium.* Paper presented at the 66th Annual Technical Conference and Exhibition of the SPE, Dallas (TX), October 6-11.

26. Zhang, H. (1988). *Nonlinear Degenerate Diffusion Problems*. Ph. D. Thesis, Delft University of Technology.